T0139896

Advances in Intelligent Systems and Computing

Volume 1062

The series "Advances in Intelligent Systems and Computing" contains publications on theory, applications, and design methods of Intelligent Systems and Intelligent Computing. Virtually all disciplines such as engineering, natural sciences, computer and information science, ICT, economics, business, e-commerce, environment, healthcare, life science are covered. The list of topics spans all the areas of modern intelligent systems and computing such as: computational intelligence, soft computing including neural networks, fuzzy systems, evolutionary computing and the fusion of these paradigms, social intelligence, ambient intelligence, computational neuroscience, artificial life, virtual worlds and society, cognitive science and systems, Perception and Vision, DNA and immune based systems, self-organizing and adaptive systems, e-Learning and teaching, human-centered and human-centric computing, recommender systems, intelligent control, robotics and mechatronics including human-machine teaming, knowledge-based paradigms, learning paradigms, machine ethics, intelligent data analysis, knowledge management, intelligent agents, intelligent decision making and support, intelligent network security, trust management, interactive entertainment, Web intelligence and multimedia.

The publications within "Advances in Intelligent Systems and Computing" are primarily proceedings of important conferences, symposia and congresses. They cover significant recent developments in the field, both of a foundational and applicable character. An important characteristic feature of the series is the short publication time and world-wide distribution. This permits a rapid and broad dissemination of research results.

** Indexing: The books of this series are submitted to ISI Proceedings, EI-Compendex, DBLP, SCOPUS, Google Scholar and Springerlink **

More information about this series at http://www.springer.com/series/11156

Michał Choraś · Ryszard S. Choraś
Editors

Image Processing and Communications

Techniques, Algorithms and Applications

 Springer

Editors
Michał Choraś
Institute of Telecommunications
and Computer Science
University of Science
and Technology (UTP)
Bydgoszcz, Poland

Ryszard S. Choraś
Department of Telecommunications,
Computer Sciences
and Electrical Engineering
University of Science
and Technology (UTP)
Bydgoszcz, Poland

ISSN 2194-5357   ISSN 2194-5365 (electronic)
Advances in Intelligent Systems and Computing
ISBN 978-3-030-31253-4   ISBN 978-3-030-31254-1 (eBook)
https://doi.org/10.1007/978-3-030-31254-1

This Springer imprint is published by the registered company Springer Nature Switzerland AG
The registered company address is: Gewerbestrasse 11, 6330 Cham, Switzerland

Preface

The monograph contains high-level papers which address all aspects of image processing (from topics concerning low-level to high-level image processing), pattern recognition, novel methods and algorithms as well as modern communications.

We would like to thank all the authors and also the reviewers for the effort they put into their submissions and evaluation.

We are grateful to Agata Gielczyk and Dr Karolina Skowron for their management work, to Dr Adam Marchewka for hard work as Publication Chair, and also to Springer for publishing this book in their Advances in Intelligent Systems and Computing series.

Those papers have also been presented at IP&C 2019 Conference in Bydgoszcz.

Michał Choraś
Conference Chair

Organization

Organization Committee

Conference Chair

Michał Choraś, Poland

Honorary Chairs

Ryszard Tadeusiewicz, Poland
Ryszard S. Choraś, Poland

International Program Committee

Kevin W. Bowyer, USA
Dumitru Dan Burdescu, Romania
Christophe Charrier, France
Leszek Chmielewski, Poland
Michał Choraś, Poland
Andrzej Dobrogowski, Poland
Marek Domański, Poland
Kalman Fazekas, Hungary
Ewa Grabska, Poland
Andrzej Kasiński, Poland
Andrzej Kasprzak, Poland
Marek Kurzyński, Poland
Witold Malina, Poland
Andrzej Materka, Poland
Wojciech Mokrzycki, Poland
Sławomir Nikiel, Poland
Zdzisław Papir, Poland
Jens M. Pedersen, Denmark
Jerzy Pejaś, Poland

Leszek Rutkowski, Poland
Khalid Saeed, Poland
Abdel-Badeeh M. Salem, Egypt

Organizing Committee

Łukasz Apiecionek
Sławomir Bujnowski
Piotr Kiedrowski
Rafał Kozik
Damian Ledziński
Zbigniew Lutowski
Adam Marchewka (Publication Chair)
Beata Marciniak
Tomasz Marciniak
Ireneusz Olszewski
Karolina Skowron (Conference Secretary)
Mścisław Śrutek
Łukasz Zabłudowski

Contents

Image Processing and Communications

Overview of Tensor Methods for Multi-dimensional Signals Change Detection and Compression

Bogusław Cyganek$^{(\boxtimes)}$

AGH University of Science and Technology,
Al. Mickiewicza 30, 30-059 Kraków, Poland
cyganek@agh.edu.pl
http://www.agh.edu.pl

Abstract. An overview of modern tensor based methods for multi-dimensional signal processing is presented. Special focus is laid on recent achievements in signal change detection, as well as on efficient methods of their compression based on various tensor decompositions. Apart from theory, applications as well as implementation issues are presented as well.

Keywords: Tensor change detection · Video shot detection · Orthogonal tensor space · Tensor decomposition · Tensor compression · HOSVD · Tucker decomposition · Artificial intelligence · Deep learning

1 Introduction

Contemporary sensors produce huge amounts of multi-dimensional signals. The most popular are ubiquitous video recordings produced on mobiles, but also signals arising in various branches of industry, such as surveillance cameras, process control, finance, as well as in science, in such domains as particle physics, astronomy, seismology, biology, variety of experimental simulations, to name a few. With no much exaggeration we can say that we live in times of big data. Processing of big data was recently underpinned with artificial intelligence methods, such as deep learning and widely applied convolutional neural networks (CNN). These also entail processing of huge amounts of data for which high computational power computers and GPU are employed [10,11,13,14]. All these can be characterized as high streams of multi-dimensional data [3]. Hence, development of methods for their efficient processing is one of the important research topics. In this context the methods of signal change detection, as well as signal compression seem to be very useful. Tensor based methods offer a natural tool for multi-dimensional signal processing [3–7]. Originally proposed by Tucker [21], then adopted to the signal processing domain [6,12,15,16], tensor decomposition methods play the key role. In this context the special stress is put upon methods of signal clustering based on abrupt change detection of various

© Springer Nature Switzerland AG 2020
M. Choraś and R. S. Choraś (Eds.): IP&C 2019, AISC 1062, pp. 3–5, 2020.
https://doi.org/10.1007/978-3-030-31254-1_1

forms and duration [4, 7, 20]. The second actively investigated are methods for multi-dimensional signal compression. Connection of the two domains offers new possibilities as well [5]. Such hybrid methods can be used for compression of signal chunks. They can find broad applications in compression of video or CNN weights [1, 2, 18, 19], to name a few. However, tensor processing is not free from computational problems such as curse of dimensionality, missing data, storage limitations and computational complexity to name a few. In this keynote, an overview of the above methods will be presented with special focus upon applications and further developments. The basic theory behind tensor analysis, with underpinned few examples, will be presented. Then, an overview of the basic tensor decompositions will be discussed, underlying those especially suited for signal change detection and compression. The talk will conclude with examples as well as ideas for future research in these areas.

Acknowledgments. This work was supported by the National Science Centre, Poland, under the grant NCN no. 2016/21/B/ST6/01461.

References

1. Asghar, M.N., Hussain, F., Manton, R.: Video indexing: a survey. Int. J. Comput. Inf. Technol. **03**(01), 148–169 (2014)
2. de Avila, S.E.F., Lopes, A.P.B., da Luz Jr., A., Araújo, A.A.: VSUMM: a mechanism designed to produce static video summaries and a novel evaluation method. Pattern Recogn. Lett. **32**, 56–68 (2011)
3. Cyganek, B.: Recognition of road signs with mixture of neural networks and arbitration modules. In: Advances in Neural Networks, ISNN 2006. Lecture Notes in Computer Science, vol. 3973, pp. 52–57. Springer (2006)
4. Cyganek, B., Woźniak, M.: Tensor-based shot boundary detection in video streams. New Gener. Comput. **35**(4), 311–340 (2017)
5. Cyganek, B., Woźniak, M.: A tensor framework for data stream clustering and compression. In: International Conference on Image Analysis and Processing, ICIAP 2017, Part I. LNCS, vol. 10484, pp. 1–11 (2017)
6. Cyganek, B., Krawczyk, B., Woźniak, M.: Multidimensional data classification with chordal distance based kernel and support vector machines. J. Eng. Appl. Artif. Intell. **46**, 10–22 (2015). Part A
7. Cyganek, B.: Change detection in multidimensional data streams with efficient tensor subspace model. In: Hybrid Artificial Intelligent Systems: 13th International Conference, HAIS 2018, Lecture Notes in Artificial Intelligence, LNAI, Oviedo, Spain, 20–22 June, vol. 10870, pp. 694–705. Springer (2018)
8. Del Fabro, M., Böszörmenyi, L.: State-of-the-art and future challenges in video scene detection: a survey. Multimedia Syst. **19**(5), 427–454 (2013)
9. Fu, Y., Guo, Y., Zhu, Y., Liu, F., Song, C., Zhou, Z.-H.: Multi-view video summarization. IEEE Trans. Multimedia **12**(7), 717–729 (2010)
10. Gama, J.: Knowledge Discovery from Data Streams. CRC Press, Boca Raton (2010)
11. Gama, J., Žliobaitė, I., Bifet, A., Pechenizkiy, M., Bouchachia, A.: A survey on concept drift adaptation. ACM Comput. Surv. (CSUR) **46**(4), 44:1–44:37 (2014)

12. Kolda, T.G., Bader, B.W.: Tensor decompositions and applications. SIAM Rev. **51**, 455–500 (2008)
13. Krizhevsky, A., Sutskever, I., Hinton, G.E.: ImageNet classification with deep convolutional neural networks. In: Proceedings of the 25th International Conference on Neural Information Processing Systems - Volume 1, NIPS 2012, pp. 1097–1105 (2012)
14. Ksieniewicz, P., Woźniak, M., Cyganek, B., Kasprzak, A., Walkowiak, K.: Data stream classification using active learned neural networks. Neurocomputing **353**, 74–82 (2019)
15. de Lathauwer, L.: Signal processing based on multilinear algebra. Ph.D. dissertation. Katholieke Universiteit Leuven (1997)
16. de Lathauwer, L., de Moor, B., Vandewalle, J.: A multilinear singular value decomposition. SIAM J. Matrix Anal. Appl. **21**(4), 1253–1278 (2000)
17. Lee, H., Yu, J., Im, Y., Gil, J.-M., Park, D.: A unified scheme of shot boundary detection and anchor shot detection in news video story parsing. Multimedia Tools Appl. **51**, 1127–1145 (2011)
18. Mahmoud, K.A., Ismail, M.A., Ghanem, N.M.: VSCAN: an enhanced video summarization using density-based spatial clustering. In: Image Analysis and Processing, ICIAP 2013. LNCS, vol. 1, pp. 733–742. Springer (2013)
19. Medentzidou, P., Kotropoulos, C.: Video summarization based on shot boundary detection with penalized contrasts. In: IEEE 9th International Symposium on Image and Signal Processing and Analysis (ISPA), pp. 199–203 (2015)
20. Sun, J., Tao, D., Faloutsos, C.: Incremental tensor analysis: theory and applications. ACM Trans. Knowl. Discov. Data **2**(3), 11 (2008)
21. Tucker, L.R.: Some mathematical notes on three-mode factor analysis. Psychometrika **31**, 279–311 (1966)

Head Motion – Based Robot's Controlling System Using Virtual Reality Glasses

Tomasz Hachaj$^{(\boxtimes)}$ (iD)

Institute of Computer Science, Pedagogical University of Cracow,
ul. Podchorążych 2, 30-084 Cracow, Poland
tomekhachaj@o2.pl

Abstract. This paper proposes head motion – based robot's controlling system using virtual reality glasses that was implemented using various up-to date software and hardware solutions. System is consisted of
robotic platform with DC motors controlled by micro controller with Wi-
fi network interface. The second micro controller is used as access point
and host of MJPG stereo vision camera stream. Virtual reality glasses
are used to control motor by analyzing user's head motion and to display stereo image from camera mounted in the front of the chassis. This
article also introduces a user - centered head rotation system similar to
yaw – pitch – roll that can be used to intuitively design functions of user
interface. All source codes that were made for system implementation
can be downloaded and tested.

Keywords: Head motion analysis · Signal processing ·
Remote controlling · Virtual reality glasses

1 Introduction

Remote wireless controlling is among basic functionalities of robotic platforms.
Handheld controllers are most popular and reliable type of controlling devices.
However there might be a need to operate a robot without using hands. It might
be necessary if a person controlling it wants to have free hands or if this person
has some hands disabilities. Additionally, if an operator cannot follow the robot
or observe it all the time it might be necessary to receive a broadcast from
cameras installed on the robot. Those two functionalists: hands-free controlling
and camera view displaying can be implemented with the help of virtual reality
(VR) glasses (goggles).

Virtual reality glasses contain optical system that enables displaying stereo-
graphic image and to monitor user's head rotation using gyroscope. Cheap goggles often utilize smartphones which commonly has accelerometers. The stere-
ographic image is obtained by displaying image on smartphone screen in split
mode and contains adjustable lenses which together with binocular disparity
(obtained from stereo vision camera) are used to generate three dimensional
image.

© Springer Nature Switzerland AG 2020
M. Choraś and R. S. Choraś (Eds.): IP&C 2019, AISC 1062, pp. 6–13, 2020.
https://doi.org/10.1007/978-3-030-31254-1_2

In state of the art papers we did not find a solution that has all functionalities we have mentioned, however separate elements are present in number of papers. The virtual reality controlling systems are present in robotic from many years [10,12]. Among possible visualization techniques it is known that virtual reality glasses give sense of immersion and helps with remote navigation of mechanical devices [3,7,8,13] in medical robotic applications [9,16], rehabilitation and physical training [1,2].

This paper proposes head motion – based robot's controlling system using virtual reality glasses that can be implemented using various up-to date software and hardware solutions. This article also describes a user - centered head rotation system similar to yaw – pitch – roll that can be used to intuitively design functions of user interface. All source codes that were made for system implementation: motor control program, Android VR Google Cardboard application and R language implementation of method that recalculates quaternion rotation into proposed rotation coordinates system can be downloaded and tested [6].

2   Materials and Methods

In this section the proposed system implementation and mathematical model of head rotation angles calculation is presented.

2.1   System Implementation

System is consisted of robotic platform on tank chassis with two DC motors controlled by micro controller with Wi-fi network interface. The second micro controller is used as access point and host of MJPG stereo vision camera stream. VR glasses in Google Cardboard technology is used to control motor by analyzing user head motion and to display stereo image from camera mounted in the front of the chassis.

The system prototype has been implemented using several mechanical and electronic components. Robot uses tank chassis T300, which size with camera box and wiring is $28 \times 27 \times 23$ cm. Total weight of the system is about 2020 g. Chassis has two 9 V 150 rpm motors powered by two 3.7 V, 3000 mAh accumulators. Motors allows robot to turn left, right and to move forward or backward. They are controlled by NodeMCU v3 with Esp8266 and Motor Shield (compatible with Arduino sketch). Robot has USB 2.0 MJPEG dual lens ELP-960P2CAM-V90 camera connected to Raspberry Pi 3 B micro controller. Raspberry Pi is also a Wi-fi access point and MJPEG server for camera that operates in resolution $2 \times 320 \times 240$. The streaming application is MJPG Streamer installed on Raspbian operating system. This second controller is powered by 12 000 mAh power bank. Virtual reality glasses uses Google Cardboard technology and requires smartphone with Android v. > 4.1. Head rotation in Google Cardboard API is returned in quaternions [5]. Also a cell phone magnetometer inside VR glasses has to be calibrated (I have used Samsung Galaxy A3 207). The overall schema of this system is presented in Fig. 1.

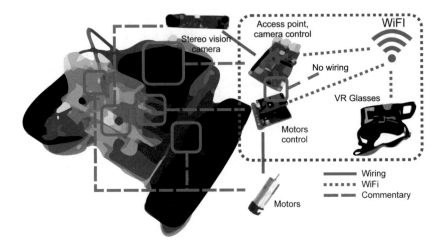

Fig. 1. This figure presents overall schema of system.

2.2   Head Rotation Calculation

For the purpose of creating head motion controlled system it might be more convenient to recalculate head rotation coordinates from quaternions into three-parameter angle-based system. To make this coordinates system user-centered we use coordinates similar to yaw – pitch – roll triplet, however, with different definition of axis and rotations directions which are more 'intuitive' for a user. This type of description is more straightforward while defining behavior of the human - computer interface. There are three rotation axes, which initial position is dependent to user initial position (see Fig. 2). The vertical y axis which governs horizontal rotation is defined as the normal vector of the top of the user's head (in fact it is normal vector of the side part of the smartphone). The horizontal rotation is defined with left-handed screw order. When a person is looking straight, the rotation horizontal rotation angle equals 0, when he or she is looking left, rotation angle is negative, when he or she is looking right rotation angle is positive. x axis which governs vertical rotation is perpendicular to y axis and is defined by a vector which *more or less* links ears of a user (in fact it is normal vector of the top part of the smartphone). The vertical rotation is defined with right-handed screw order. When a person is looking straight, the rotation vertical rotation angle equals 0, when he or she is looking up, rotation angle is positive, when he or she is looking down rotation angle is negative. z axis is a cross product of unit vectors of x and y. z governs sideways head bending. The bending rotation is defined with left-handed screw order. When a person is looking straight, the rotation vertical rotation angle equals 0, when he or she is bending head right, rotation angle is positive, when he or she is bending head left rotation angle is negative. The coordinates system is left-handed. To recalculate quaternion-based rotation returned by a VR goggles the following calculations have to be done. Let us assume that the output rotation angles are restricted to

$[-\pi, \pi]$ which is enough to define head rotation of stationary (for example sitting or standing in one place) person.

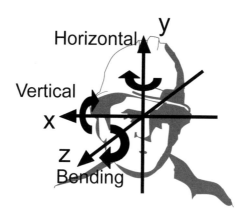

Fig. 2. This figure presents definition of user-centered coordinate system with 'intuitive' rotations definition.

The quaternion multiplication equation is defined as follows:

$$Q_1 \cdot Q_2 = \begin{bmatrix} Q_1.W * Q_2.X + Q_1.X * Q_2.W + Q_1.Y * Q_2.Z - Q_1.Z * Q_2.Y \\ Q_1.W * Q_2.Y + Q_1.Y * Q_2.W + Q_1.Z * Q_2.X - Q_1.X * Q_2.Z \\ Q_1.W * Q_2.Z + Q_1.Z * Q_2.W + Q_1.X * Q_2.Y - Q_1.Y * Q_2.X \\ Q_1.W * Q_2.W - Q_1.X * Q_2.X - Q_1.Y * Q_2.Y - Q_1.Z * Q_2.Z \end{bmatrix} \quad (1)$$

Quaternion conjugate equals:

$$\overline{Q} = [-Q.X, -Q.Y, -Q.Z, Q.W]. \quad (2)$$

In order to rotate vector $V = [x, y, z]$ by quaternion Q we apply the following calculation $T(Q, V)$:

$$T(Q, V) \Leftarrow \begin{cases} Q_V \leftarrow [x, y, z, 0]; \\ Q_V' \leftarrow Q \cdot (Q_V \cdot \overline{Q}); \\ V' \leftarrow [Q_V'.X, Q_V'.Y, Q_V'.Z]. \end{cases} \quad (3)$$

Let us assume that the initial head rotation is Q_H. In order to recalculate quaternion rotation Q to Vertical – Horizontal – Bending relatively to Q_H we apply the following calculation:

$$Vertical \leftarrow acos(T(Q_H, [1, 0, 0]) \circ T(Q, [0, 0, 1])) - \frac{\pi}{2}$$

$$Horizontal \leftarrow \pi - acos(T(Q_H, [0, 0, 1]) \circ T(Q, [0, 1, 0])) - \frac{\pi}{2} \quad (4)$$

$$Bending \leftarrow \pi - acos(T(Q_H, [1, 0, 0]) \circ T(Q, [0, 1, 0])) - \frac{\pi}{2}$$

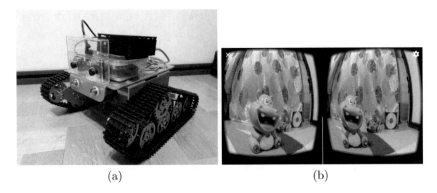

(a)   (b)

Fig. 3. This figure presents final implementation of the robot (a) and screenshot from VR glasses (b).

After those operations we can recalculate the quaternion-based head rotation Q to 'intuitive' three-dimensional description in domain $[-\pi, \pi]$, taking into account the initial head rotation Q_H.

2.3 Interpretation of Head Motions and Remote Navigation

The communication between robot and VR glasses is done via Wi-fi using TCP/IP protocol. Both smartphone with Android OS that is used by Google Cardboard technology and NodeMCU micro controller can communicate via HTTP however this transmission protocol is too slow for real-time motors control with this hardware. Because of it TCP/IP socket-level communication protocol has been applied. The Android application running on smartphone is monitoring head motions. The applications determines initial rotation of the user's head (Q_H) and for each following data acquisition Vertical – Horizontal – Bending head rotation is calculated. A user can reinitialize Q_H by touching smartphone screen and the current head rotation Q becomes Q_H. The Horizontal rotation angle governs platform's turning and Vertical rotation is responsible for moving forward or backward. If the head position changes above a threshold values (those values are defined in Android application) application sends a TCP/IP package to NodeMCU with appropriate command. Firmware on NodeMCU processes those messages and changes voltages on output pins of motor shield. The robot continues doing particular motion until a message to stop arrives from Android application. The stop message is send when a head position of the user is within the 'neutral' threshold value. Messages from smartphone are sent no more often than a certain time span, which is also defined in application. This prevents robot platform to be overflown by incoming commands. A user in VR glasses see the view from stereo vision camera that is mounted on the front of the robot.

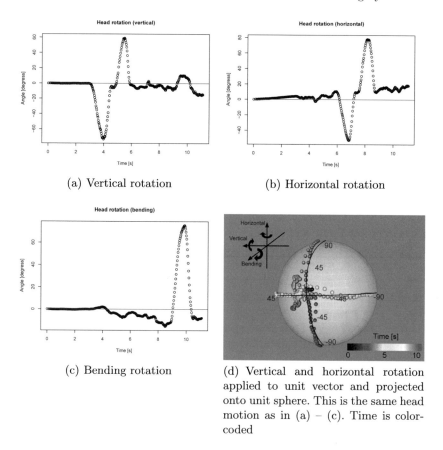

(a) Vertical rotation

(b) Horizontal rotation

(c) Bending rotation

(d) Vertical and horizontal rotation applied to unit vector and projected onto unit sphere. This is the same head motion as in (a) – (c). Time is color-coded

Fig. 4. This figure presents plots of example head rotations of user in VR glasses calculated using (4).

3   Results

After implementing the system (see Fig. 3(a)), tests have been performed on its remote head-motion based controlling module. As can be seen in Fig. 3(b) view in VR glasses is additional distorted by Google Cardboard API in order to strengthen effect of stereo vision and immersive experience of a user. The remote controlling systems worked smoothly allowing remote control of robot with nearly close to real-time preview from stereo vision camera. Beside navigation commands mathematical model (4) was tested in order to check if it can be used to produce intuitive motions descriptions and visualizations. B. In order to do so a user operating VR glasses was asked to move head down and up, then turn it left and right, and finally to bend it clockwise and return to initial head position. Plots of obtained angles descriptions are presented in Fig. 4(a)–(c). Additionally Fig. 4(d) presents vertical and horizontal rotation applied to unit vector and its projection onto unit sphere. That last plot visualizes in 3D what the head motion trajectory was.

4   Discussion and Conclusions

As could be seen in previous section the proposed head motion – based robot's controlling system satisfies needs of remote controlling of robotic platform. The mathematical model of head rotation calculation enables to generate intuitive user – centered rotation description. In this description positive directions of axis are right, up and towards front user's head. The positive direction of rotations is right, up, and clockwise in case of head side banding, opposite rotations have negative angle coordinates. With the help of this description it is straightforward to define how head motion should be interpreted and translated to remote commands to robot. The communication protocol takes into account system latency and prevents robot from being overflown by incoming commands.

Due to applying a stereo vision camera in proposed systems can be utilized as the prototype in many scientific researches. After stereo calibration this system can be used for measuring distances between robot and elements of environment and three-dimensional points cloud generation. The system can be also applied for developing and testing vision – based robotic odometry [14,15] or simultaneous localization and mapping (SLAM) algorithms [4,11]. There is also a large potential in developing and testing methods for head motion analysis and recognition. Many commands to system might be coded in head gestures that could be classified by appropriate algorithm.

References

1. Borowska-Terka, A., Strumiłło, P.: Algorithms for head movements' recognition in an electronic human computer interface. Przegląd Elektrotechniczny **93**(8), 131–134 (2017)
2. Brütsch, K., Koenig, A., Zimmerli, L., Mérillat-Koeneke, S., Riener, R., Jäncke, L., van Hedel, H.J., Meyer-Heim, A.: Virtual reality for enhancement of robot-assisted gait training in children with neurological gait disorders. J. Rehabil. Med. **43**(6), 493–499 (2011)
3. Dornberger, R., Korkut, S., Lutz, J., Berga, J., Jäger, J.: Prototype-based research on immersive virtual reality and on self-replicating robots. In: Business Information Systems and Technology 4.0 Studies in Systems, Decision and Control, pp. 257–274 (2018). https://doi.org/10.1007/978-3-319-74322-6_17
4. Engel, J., Stückler, J., Cremers, D.: Large-scale direct slam with stereo cameras. In: 2015 IEEE/RSJ International Conference on Intelligent Robots and Systems (IROS), pp. 1935–1942, September 2015. https://doi.org/10.1109/IROS.2015. 7353631
5. Grygiel, R., Bieda, R., Wojciechowski, K.: Angles from gyroscope to complementary filter in IMU. Przegląd Elektrotechniczny **90**(9), 217–224 (2014)
6. Hachaj, T.: GitHub repository of the project (2019). https://github.com/browarsoftware/rpm_rotation_calculation. Accessed 22 Mar 2019

7. Kato, Y.: A remote navigation system for a simple tele-presence robot with virtual reality. In: 2015 IEEE/RSJ International Conference on Intelligent Robots and Systems (IROS), pp. 4524–4529, September 2015. https://doi.org/10.1109/IROS. 2015.7354020

8. Kurup, P., Liu, K.: Telepresence robot with autonomous navigation and virtual reality: demo abstract. In: SenSys (2016)

9. Lin, L., Shi, Y., Tan, A., Bogari, M., Zhu, M., Xin, Y., Xu, H., Zhang, Y., Xie, L., Chai, G.: Mandibular angle split osteotomy based on a novel augmented reality navigation using specialized robot-assisted arms - a feasibility study. J. Cranio-Maxillofac. Surg. **44**(2), 215–223 (2016). https://doi.org/10.1016/j.jcms.2015.10. 024. http://www.sciencedirect.com/science/article/pii/S1010518215003674

10. Monferrer, A., Bonyuet, D.: Cooperative robot teleoperation through virtual reality interfaces. In: Proceedings Sixth International Conference on Information Visualisation, pp. 243–248, July 2002. https://doi.org/10.1109/IV.2002.1028783

11. Mur-Artal, R., Tardüs, J.D.: ORB-SLAM2: an open-source SLAM system for monocular, stereo, and RGB-D cameras. IEEE Trans. Robot. **33**(5), 1255–1262 (2017). https://doi.org/10.1109/TRO.2017.2705103

12. Nguyen, L., Bualat, M., Edwards, L., Flueckiger, L., Neveu, C., Schwehr, K., Wagner, M., Zbinden, E.: Virtual reality interfaces for visualization and control of remote vehicles. Auton. Robots **11**(1), 59–68 (2001). https://doi.org/10.1023/a: 1011208212722

13. Regenbrecht, J., Tavakkoli, A., Loffredo, D.: A robust and intuitive 3D interface for teleoperation of autonomous robotic agents through immersive virtual reality environments. In: 2017 IEEE Symposium on 3D User Interfaces (3DUI), pp. 199–200, March 2017. https://doi.org/10.1109/3DUI.2017.7893340

14. Usenko, V., Engel, J., Stückler, J., Cremers, D.: Direct visual-inertial odometry with stereo cameras. In: 2016 IEEE International Conference on Robotics and Automation (ICRA), p. 1885 (2016). https://doi.org/10.1109/ICRA.2016.7487335

15. Wang, R., Schworer, M., Cremers, D.: Stereo DSO: large-scale direct sparse visual odometry with stereo cameras. In: IEEE International Conference on Computer Vision (ICCV), October 2017

16. Zinchenko, K., Komarov, O., Song, K.: Virtual reality control of a robotic camera holder for minimally invasive surgery. In: 2017 11th Asian Control Conference (ASCC), pp. 970–975, December 2017. https://doi.org/10.1109/ASCC.2017. 8287302

Robustness of Haar Feature-Based Cascade Classifier for Face Detection Under Presence of Image Distortions

Patryk Mazurek$^{(\boxtimes)}$ and Tomasz Hachaj

Institute of Computer Science, Pedagogical University of Cracow,
Podchorazych 2, 30-084 Cracow, Poland
{patryk.mazurek,tomasz.hachaj}@up.krakow.pl

Abstract. In this paper examines effectiveness of HAAR feature-based cascade classifier for face detection in the presence of various image distortions. In the article we have focused on picture distortions that are likely to be met in everyday life, namely blurring, salt and pepper noise, contrast and brightness shifts and "fisheye" type distortion typical for wide-angle lens. In the paper present the mathematical model of the classifier and distortions, the training procedure and finally results of segmentation under various level of distortion. The test dataset is a large publicly available "Labelled Faces in the Wild" (LFW). Results show that Cascade Classifier finds it most difficult to recognize images that contain 70% noise type salt and pepper. The least impact on the effectiveness of the method use of blurred images even though the high parameter of blurring. From the obtained results it appears that the effectiveness of face detection is also affected by the adequate parameters of contrast and brightness.

Keywords: Cascade classifier · Face detection · Haar features · Low-quality

1 Introduction

Face detection is a highly developed computer technology being used in variety of applications and multiple sciences. Face detection can be used in the process of identification of human faces, access control on mobile devices or for the purpose of entertainment. Each face detection system, the same as any other object-class detection method, tackles with a problem with the quality of image upon which the system is operating. In the majority of detected images, quality is a factor that considerably affects the effectiveness of system operation. In order to achieve higher effectiveness of image detection, systems ought to be resistant to low image quality [5].

This paper focuses on one of the most popular methods of face detection, that is Cascade Classifier [1]. When the method was presented, the basic classifier was

© Springer Nature Switzerland AG 2020
M. Choraś and R. S. Choraś (Eds.): IP&C 2019, AISC 1062, pp. 14–21, 2020.
https://doi.org/10.1007/978-3-030-31254-1_3

Haar, subsequently Ojala proposed a new LBP classifier [2], which was added to Cascade Classifier. Herein, we intend only to concentrate on HAAR classifier which operates by selection of squares containing dark and bright fragments. In case of face detection, division of a square into dark and fair regions serves the purpose of calculation of the sum of pixels' value from under the selected square. A dark square may represent fragments of eyes, mouth, and the fair square may represent a nose, forehead or cheeks.

Each classifier consists of a few stages where multiple functions are performed. In order for the classifier to consider an examined area to be the face under search, the sum obtained from the tested areas must exceed a certain threshold defined in a file containing Haar cascade model. Sometimes, even this type of testing is not capable for errors elimination and the classifier is bound to accept the rectangles that does not contains faces (false positive error).

Regardless the foregoing errors, scientists use Cascade Classifier in various tasks and come up with brand new solutions based on this method. With a view to improve results of face detection, another classifier was proposed with features that better reflect the face structure [3,16,17]. Another extension on the Cascade Classifier for the purpose of more accurate results is to apply a few weaker models and to verify the results obtained after each stage [4]. Next issue where Haar classifier has been tested is to detect the area around eyes and to make use of the driver's fatigue [6,7]. Cascade Classifier not only assists the process of face or body composition detection but may also be applied in medicine, e.g. for the purpose of recognition of leukocytes on ultrasound scans [11] or detection of objects from the bottom of the sea [12].

Development of new technologies as well as neural networks accounts for the fact that recognition of various objects or faces becomes more and more accurate and faster, each of those technologies, however, depending significantly upon the quality of images supplied for the purpose of detection [9,10].

In this paper have been prepared 4 test groups in which it has been checked how the Cascade Classifier work for face detection under presence of image distortions. Each of test contains a specially prepared images with the appropriate level of distortion. In the first test it was prepared a images with salt and pepper noise. This noise can be find in images who taken in low light. In next test, it was use low-quality images. To reduce the quality images was used the Gauss filter and Median filter. This type of distortion can be obtained when taking low-resolution images. The third test involves the modifying images for changes in brightness and contrast values. This type of images can be obtained by manually configuration of camera or when the lighting changes dynamically and the camera has to adjust the settings. In the last test it was use images with "fisheye" effect, To get this effect can be used a specific lens or application witch will added this effect to images. The "fisheye" effect can be found in art photography and cameras with a wide of view.

The research presented in this article can help to creating a new solutions to face detection or it will be helpful when creating a images database for learning models in cascade classifier method.

2   Materials and Methods

2.1   Databases and Research Methods

For this research a Haar feature-based cascade classifier available together with OpenCV 3.4 library was used. A model was trained based on 9832 positive images containing faces (including mirrored, vertical) and 1000 negative images without faces. Model was trained based on 24 × 24 pixel images and Adaboost learning algorithm [13].

In order to check effectiveness of the tested model we have used LFW (Labelled Faces in the Wild) [8] that contains 13233 images of 5479 various people. LFW has pictures of public personalities in their everyday routine which enables to more thoroughly verify the model effectiveness.

The effectiveness of Cascade Classifier is affected by the following factors: model accuracy, in the test we use pre-training model, so we have no influence on the model's learning effectiveness. The model who was used achieves efficiency at the level 90%. And the quality of images supplied. The common types on image distortions can be divided into following groups:

(a) **Salt and pepper** - during "salt and pepper" noise simulation 11 tests have been carried out with noise being inserted into the image within the range of 0–100, and a constant pace of 10. The noise level was the percentage of pixels against all pixels in the image. Noise pixels have been inserted randomly into the tested images.

(b) **Blurring** - blurring simulation has been conducted with the use of Gaussian blur filter and median filter.

$$G(x, y) = \frac{1}{2\pi\sigma^2} e^{-\frac{x^2+y^2}{2\sigma^2}} \tag{1}$$

For each filter a value between 1–17 was applied, adjusting input data every 2, thanks to which 9 tests for each filter have been completed. The use of Gaussian and median filters of high calibre causes image blurring but not loss of edges or important data in the image.

(c) **Contrast and brightness** - in brightness and contrast tests 11 tests have been prepared with the value of pixels changed accordingly.

$$f(i, j) = a * x(i, j) + b \tag{2}$$

The value of $x(i, j)$ corresponds to the pixels of the basic image, the value of b is responsible for brightness, which change was carried out by adding an appropriate value from the range between $[-127:127]$ to pixel. In case of contrast (variable a), the value of each pixel was multiplied by an adequate value ranging between $[0.0:2.0]$ with the constant pace of 0.2. In each test, after the pixel value change, it was necessary to adjust it to appropriate range $[0:255]$.

(d) **Lenses** - the "fisheye" effect can be obtained by applying special lenses or adequate algorithms which help to modify and transform the traditional image into the one with "fisheye" effect. During the test, an image was modified by means of setting up a new Q coordinate system with its centre being located halfway through the height and width of the modified image. Subsequently, Cartesian coordinates are replaced with polar coordinates using.

$$r = \sqrt{(x^2 + y^2)} \qquad (3)$$

$$\theta = atan2(x, y) \qquad (4)$$

Next, image is mapped onto a spherical or elliptical object. Value of distortion of an image is between a vector and a polar coordinate from a "virtual camera" to the beginning of a coordinate system. Mathematically, the above phenomenon can be presented in the following way:

$$x = sin(\theta) * cos(\theta) \qquad (5)$$

$$y = sin(\theta) * sin(\theta) \qquad (6)$$

$$z = sin(k) \qquad (7)$$

Where k parameter specifies an angle of distortion which it proved to be ranging between 0.002–0.02, with a constant pace of 0.002 in the test. The ultimate stage requires to transfer an adequately distorted image onto the normal 2-D image.

3   Results

The present section is devoted to presentation of results obtained in the course of carried out tests. Each test checked the impact of deterioration of image quality on effectiveness of operation of the tested model.

Table 1. This table presents results of "salt and paper" test.

Noise level	0	10	20	30	40	50	60	70	80	90	100
Positive results	95.6%	89.1%	76.9%	60.5%	41.7%	24.8%	13.0%	5.8%	2.1%	0.6%	0.2%

During test 1 measurements have been made to check how various noise levels in an image affect effectiveness of detection rate. Based on the presented table (Table 1) it can be concluded that images with noise level up to 20% are likely to get high effectiveness of detection. An increase of the noise level accounts for a decrease in effectiveness of image detection. In case of Haar classifier that was used in the test, images containing noise above 60% are extremely difficult to be detected or simply impossible. As it appears from the test, Haar classifier is ultra-sensitive to noise, because salt and pepper adds to image black and white pixel and this disrupts the effect of Haar wavelet. The Haar wavelet's work involves calculating the differences between square. High content of white and

Table 2. This table compares results of the tests with Median filter and Gaussian filter.

Filter size	0	1	3	5	7	9	11	13	15	17
Median filter	95.6%	96.8%	96.0%	96.1%	95.9%	95.7%	95.3%	94.9%	94.4%	93.9%
Gaussian filter	95.6%	95.8%	95.8%	95.8%	95.0%	93.6%	92.9%	90.9%	88.7%	86.1%

black pixels in images results in incorrect calculation results. The most frequent circumstances in which noise is generated is taking photos in poor illumination.

Another task was to apply the blurring effect in images and to check effectiveness of the tested model. The effect in question can be seen in the photos or films of low resolution. For that purpose median filter and Gaussian filter was used. On the presented table (Table 2) it can be observed that the use of filters of low values does not cause deterioration of effectiveness or even, as in case of Gaussian filter, is responsible for an increase of detection rate (Table 2). Use of filters of high value is bound to result in a decrease of image detection rate but every filter responds to an increase of value in a different way. As far as median filter is concerned (Table 2), application of value of 17 caused detection effectiveness to drop to 86% whereas in case of Gaussian filter effectiveness decreased to 94%. Each filter removes certain minor information from images causing them to blur but leaving, however, edges and data vital for HAAR classifier which enables face detection.

Table 3. This table presents results of the contrast test for values from 0.0 to 2.0.

Filter size	0.0	0.2	0.4	0.6	0.8	1.0	1.2	1.4	1.6	1.8	2.0
Positive results	0%	44.5%	94.7%	94.8%	94.9%	95.6%	96.0%	95.7%	94.5%	92.0%	87.7%

Table 4. This table presents results of the brightness test for values from −127 to 127.

Value	−127	−104	−81	−58	−35	0	35	58	81	104	127
Positive results	48.9%	77.6%	87.3%	94.1%	95.7%	95.6%	96.0%	95.6%	94.8%	89.9%	70.8%

Another task required to check effectiveness of detection with images being modified in terms of contrast and brightness. As it is shown in the table (Tables 3 and 4), a model achieves low effectiveness with images with low level of contrast and brightness. An increase of contrast (Table 3) up to the value of 0.4 significantly affects detection rate. The highest effectiveness of 96% was achieved applying 1.2 value, further increase of contrast, however, was responsible for a decrease of detection rate. Next Table (Table 4) seems to demonstrate a similar tendency as in case of contrast, namely an increase of brightness ranging between −127–35 results in increased detection rate. Adding values ranging between −35–58 and increasing brightness has not proved to significantly affect model with effectiveness maintaining between 95%–96%. Further increases in brightness pixels cause the face detection rate of a model to decline. From the discussed test it appears that HAAR classifier is most reliable in detection of images that are slightly brightened or with the contrast increased. Similar images can be obtained while bad configuration of settings doing taking photo or film making.

Table 5. This table presents results of the "Fisheye" effect test.

Value	0.002	0.004	0.006	0.008	0.01	0.012	0.014	0.016	0.018	0.02
Positive results	69.0%	57.4%	51.2%	47.3%	44.1%	41.7%	40.3%	38.7%	37.0%	36.1%

During the last test it has been checked how HAAR classifier copes with "fisheye" effect. This effect is produced by special lens, applications or image editor (e.g. GIMP) that are capable of adequately transforming a traditional image into a fisheye effect image. As shown table (Table 5), a model has reached the highest effectiveness at the level close to 70% with distortions of 0.002 being applied. With an increase of the value of distortions, model effectiveness in image detection is on the decline. A changed distortion does not cause problems for human eye as far as image detection is concerned but it becomes a considerable impediment for a model. Application of fisheye effect accounts for significant face distortions and change of proportions (size, shape) which turned out to be highly problematic for HAAR classifier.

4 Discussion

In the article it was presented four group of studies whose task was to check effectiveness of the tested model. The obtained results confirm that the increase in noisiness/deformation of the image cause a drop in the number of correctly detected faces. Even the low level of noise causes a significant decrease in effectiveness of the Cascade Classifier method. The solution to the problem could be the use of filters or the preparation of learning databases in witch images with little noise are located. They are often used trained classifiers whose efficiency is high. Therefore, it is more convenient to use image filtering methods, that will allow to remove or reduce the impact of interference. In the case of salt and pepper, can be use a median filter. If we are dealing with a change in contrast (noisy using contrast), can be use dynamic image corrections [14]. From the results obtained the test you can see that the use of slightly enhanced brightness and contrast on input images results in improvement of detection efficiency (on the level around 0.5%). Last type of distortion witch was used to check to effectiveness was "fisheye" effect. The main problem in this type of distortion is the pixel displacement based on a circle or ellipse. In the case of images that are disturbed due to "fisheye" effect can be use method proposed in [15], witch involves calculating the distortion value and then adjusting each pixel to the grid.

5   Conclusions

The test described above was devoted to face detection using Cascade Classifier method. Apart from face detection, this very method can have other applications as well and can also be used for object-oriented detection provided that an adequate model had been prepared. The exemplary situations were supposed to verify if HAAR classifier was capable of detection of faces in images of varying quality. As results demonstrate, face detection rate was significantly influenced by noise, low value of contrast and brightness along with image distorting effects. In order to improve effectiveness of HAAR classifier, it is recommended to use methods that correct image quality or remove effects of distortions. Another solution that could contribute in the future towards enhanced object detection is an application of images with minor distortions in the image learning process so that a model is trained to deal with images of lower quality.

References

1. Voila, P., Jones, M.J.: Rapid object detection using a boosted cascade of simple features. In: Proceedings of the 2001 IEEE Computer Society Conference on Computer Vision and Pattern Recognition (ICCVPR 2001), 8–14 December 2001, Kauai, USA, vol. 1, pp. I-511–I-518 (2001)
2. Ojala, T., Pietikainen, M., Maenpaa, T.: Multiresolution gray-scale and rotation invariant texture classification with local binary pattern. IEEE Trans. Pattern Anal. Mach. Intell. **24**(7), 971–987 (2002)
3. Ma, S., Bai, L.: A face detection algorithm based on Adaboost and new Haar-Like feature. In: IEEE International Conference on Software Engineering and Service Science, 23 March 2017, pp. 651–654 (2017)
4. Li, C., Qi, Z., Jia, N., Wu, J.: Human face detection algorithm via Haar cascade classifier combined with three additional classifiers. In: IEEE International Conference on Electronic Measurement & Instruments (ICEMI), pp. 483–487 (2017)
5. Joshi, P., Prakash, S.: Image quality assessment based on noise detection. In: International Conference on Signal Processing and Integrated Networks (SPIN), 24 March 2014, pp. 755–759 (2014)
6. Fitriyani, N.L., Yang, C.-K., Syafrudin, M.: Real-time eye state detection system using Haar cascade classifier and Circular Hough Transform. In: IEEE 5th Global Conference on Consumer Electronics, 29 December 2016, pp. 1–3 (2016)
7. Xiang, B., Cheng, X.: Eye detection based on improved AD AdaBoost algorithm. In: International Conference on Signal Processing Systems, 23 August 2010, pp. V2-617–V2-620 (2010)
8. Huang, G.B., Ramesh, M., Berg, T., Learned-Miller, E.: Labeled faces in the wild: a database for studying face recognition in unconstrained environments. Technical report 07-49, University of Massachusetts, Amherst (2007)
9. Zhou, Y., Liu, D., Huang, T.: Survey of face detection on low-quality images. In: IEEE International Conference on Automatic Face & Gesture Recognition (FG 2018), 07 June 2018, pp. 769–773 (2018)
10. Grm, K., Štruc, V., Artiges, A., Caron, M., Ekenel, H.K.: Strengths and weaknesses of deep learning models for face recognition against image degradations. IET Biometrics **7**, 81–89 (2018)

11. Budiman, R.A.M., Achmad, B., Faridah, Arif, A., Nopriadi, Zharif, L.: Localization of white blood cell images using Haar cascade classifiers. In: 1st International Conference on Biomedical Engineering (IBIOMED), 20 March 2017
12. Sawas, J., Petillot, Y., Pailhas, Y.: Cascade of boosted classifiers for rapid detection of underwater objects. In: ECUA 2010, Istanbul (2010)
13. Viola, P., Jones, M.J.: Rapid object detection using a boosted cascade of simple features. In: IEEE CVPR (2001)
14. Gu, K., Zhai, G., Liu, M., Min, X., Yang, X., Zhang, W.: Brightness preserving video contrast enhancement using S-shaped transfer function. In: 2013 Visual Communications and Image Processing (VCIP), 09 January 2014 (2014)
15. Dong, X., Zhang, Y., Liu, J., Hu, G.: A fisheye image barrel distortion correction method of the straight slope constraint. In: 2015 8th International Congress on Image and Signal Processing (CISP), 18 February 2016 (2016)
16. Miziolek, W., Sawicki, D.: Face recognition: PCA or ICA. Przeglad Elektrotechniczny 88(7a), 286–288 (2012)
17. Siwek, K., Osowski, S.: Comparison of methods of feature generation for face recognition. Przeglad Elektrotechniczny 90(4), 206–209 (2014)

Eyes State Detection in Thermal Imaging

Paweł Forczmański$^{(\boxtimes)}$ ⓘ and Anton Smoliński ⓘ

Faculty of Computer Science and Information Technology,
West Pomeranian University of Technology, Szczecin,
Żołnierska Str. 49, 71-210 Szczecin, Poland
{pforczmanski,ansmolinski}@wi.zut.edu.pl

Abstract. The paper presents a problem of analysing facial area captured in thermal spectrum in order to estimate the drowsiness of the observed person. In the developed approach the state of eyes is taken into consideration. During the experiments we trained and applied a contemporary general-purpose object detector based on Haar-like features, known to be accurate when working in visible spectrum. Based on a position of a face, we localize eyes and apply a Gabor filtering-based approach to classify the eyes state. In contrast to the traditional, visible light-based methods, by using thermal image we are able to capture eyes state in very adverse lighting conditions. The experiments performed on video sequences taken in simulated cabin environment have shown that the proposed approach gives a very prospective results.

Keywords: Thermal imaging · Face detection · Eyes detection · Haar-like features · Gabor filtering · Drowsiness estimation

1 Introduction

Road safety studies indicate that most traffic accidents are an effect of drivers' behaviour [1] associated with fatigue and drowsiness [2]. Such loss of concentration hinders drivers in terms of quick and correct decisions [3], which leads to approximately 20% of accidents [4]. Existing techniques of driver fatigue assessment are based largely on wearable external sensors and devices. On the other hand, modern machine vision techniques enable continuous observation of the driver without his/her cooperation. Many characteristic behaviours are observable as the movement of the head and eyelids or the way of looking at the road in front of a vehicle [3]. An interesting review of selected indicators can be found in [5].

The paper presents a part of a larger project devoted to the development of a methodology for acquiring, integrating and analyzing selected visual multi-modal visual data in the context of assessing the psychophysical state and the degree of fatigue of drivers or operators of motor vehicles. The proposed methodology and the algorithms will make it possible to assess the fatigue level in respect of uncontrolled environmental conditions (e.g. in the complex lighting conditions or in total darkness) [6].

© Springer Nature Switzerland AG 2020
M. Choraś and R. S. Choraś (Eds.): IP&C 2019, AISC 1062, pp. 22–29, 2020.
https://doi.org/10.1007/978-3-030-31254-1_4

Machine vision techniques allow for a continuous observation of a driver. The first stage of processing is often a detection of face and facial features [7]. Regions-of-interest are then tracked to collect the relevant time characteristics which form the base for the inference mechanism about the driver fatigue. When a fatigue is detected, the Advanced Driver Assistance System can stimulate the driver e.g. with acoustic signals. In a critical situation, after detecting a sleep, such system can even automatically stop the vehicle. Face and face features detection are problems that are considered solved under controlled conditions (e.g. in visible light spectrum) [8,9], yet there are many issues still to be researched, especially in terms of uncontrolled environment. Hence, we propose to solve the problem of uncontrolled lighting conditions by introducing thermal image acquisition and processing. While the detection and tracking of landmarks on the face are fully mature for sequences recorded in the visible band, their analysis for other modes, e.g. thermal image, is currently not explored by many researchers. The analysis of the scientific domain shows that it is possible to find certain facial areas in the thermal images using specially trained detectors based on Haar-like features [10], Histogram of Oriented Gradients [10] and Local Binary Patterns [11], with the help of AdaBoost learning approach [12,13]. While there are many algorithms that are able to detect driver fatigue and drowsiness based on eyes state and blink analysis (e.g. [3,14,15]), they work in visible light, or in infrared band, only. It should be noted, that during travelling, drivers encounter dynamic lighting conditions (blinding sun, night driving) and various environments affecting the conditions of the observation (e.g. passing through a dark tunnel). The above problems show the weaknesses of traditional imaging techniques.

As it was mentioned, the analysis of the visible band image of a face recorded in the unfavourable illumination leads to the high error rate or can even be totally impossible. On the other hand, under the same conditions, using thermal image, we are able to detect eyes blinking. Hence, we propose a method to detect eyes state in video sequences captured in thermal spectrum.

2   Method Description

2.1   Assumptions

Traditional imaging technique, namely capturing image in the visible lighting is the most straightforward method of visual data acquisition. Required hardware is not expensive and its operational parameters can be very high, in terms of spatial resolution, dynamic range and sensitivity. On the other hand, it should be remembered, that such devices can work only in good lighting conditions. It would be impossible to illuminate driver's face during driving with any sort of additional light source, since it could disturb his/her functioning. Therefore, it is reasonable to equip the system with other capturing devices, working in different lighting spectra, e.g. infrared or thermal. Unfortunately, IR band requires an illumination with infrared illuminator, which during a prolonged period may be hazardous to the driver (a risk of retinal and corneal burns, cataracts [16]).

Hence, the thermal spectrum has been selected, since it is free of the above mentioned flaws, allowing face detection algorithms to disregard the necessity of visible/IR light illuminating the subject [17]. Thermal imaging does not depend on the lighting conditions, which is important when dealing with an observation of a driver in uncontrolled environment. In such case, car's interior is often illuminated by directional light, coming from the windows (sun, street and lamps of other cars).

3 Proposed Solution

3.1 General Overview

As it was mentioned previously, the algorithm consists of three main modules: face detection and tracking, eyes detection and tracking, and eyes state analysis. It works in a loop iterated over the frames from the video stream (see Fig. 1).

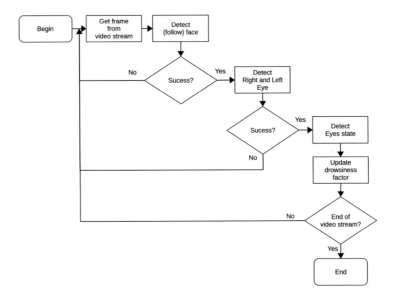

Fig. 1. Algorithm of drowsiness estimation by eyes state analysis.

3.2 Face and Eyes Detection

The processing begins with the detection of a face in the input frame - a pre-trained Viola-Jones detector [9] was used. After the successful detection, the face is tracked using a simple detector, that assumes only small movement of the head in consecutive frames. The initial research conducted by the authors, oriented at the algorithms of face detection in images taken in non-visible bands

[9, 18, 19] showed, that it is possible and the accuracy is sufficient for the further stages of processing. The experiments on thermal images taking into consideration Precision and Recall rates showed, that despite the fact that the traditional descriptors used by Viola-Jones detector are slightly weaker than recent deep-learning approaches [9], they are much faster and do not require any special hardware-based accelerators to work in real-time.

Detected face's bounding box is taken as a starting point of eyes detector. There are two separate detectors, one for each eye, both based on Viola-Jones approach, trained on a set of over 1700 eyes samples extracted from benchmark video streams. It should be noted, that the detection may not always return a precise position of face and/or eyes, because of occlusion or pose change. Therefore, the introduced tracker is based on position approximation [19], assuming that, under regular driving conditions, face and eyes positions should not change significantly across a small number of consecutive frames. Therefore, the resulting coordinates of the face's and eyes' bounding boxes are calculated based on averaged 10 past detections. This is used to calculate a confidence level normalized to the range $\langle 0, 1 \rangle$. Exemplary results are presented in Fig. 2.

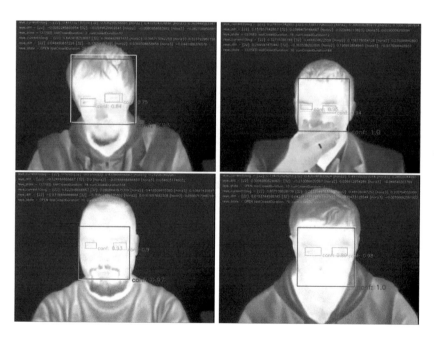

Fig. 2. Exemplary results of face tracking and eyes detection.

3.3   Eyes State Estimation

Extracted eyes are then classified as open or closed. The classification is performed in the following steps (for each eye separately):

1. Extract eye region;
2. Resample the image to 100×50 pixels;
3. Filter the image using Gabor filtering, using 30 sets of parameters, resulting 30 output images;
4. Calculate energy and standard deviation for each resulting image;
5. Concatenate above coefficients into a singe feature vector (of 60 values);
6. Perform the classification.

The filtering uses a standard Gabor filter [20] with kernel described by the following fixed parameters: window size equal to 20×10, standard deviation of the Gaussian envelope equal to 2.5, spatial aspect ratio equal to 5.0 and phase offset equal to $3\pi/2$. We got 30 filtering results by altering orientation angle of the normal to the parallel stripes of a Gabor function in the range $(0, 5\pi/6\rangle$ with a $\pi/6$ step and wavelength of the sinusoidal factor equal to $\{1, 2, 4, 8, 16\}$. Exemplary results of filtering are presented in Fig. 3. The left part of the image is devoted to the closed eye, while the right one to the open eye, respectively.

For each filtering result we calculate the energy (normalized by the image's dimensions) and standard deviation. The feature vectors are then taken as input for the binary classifier.

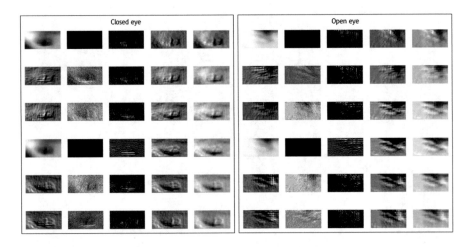

Fig. 3. Selected images after Gabor filtering.

4   Experiments

4.1   Dataset

The video materials have been recorded using an IR camera FLIR SC325 working in infrared band, equipped with 16-bit sensor of 320×240 pixels. It offers 25×18.8 degree FOV, works at 60 Hz and is interfaced by Ethernet. The acquisition protocol of the benchmark data was presented in [21].

4.2   Evaluation Protocol

All algorithms have been implemented in the Python environment using machine vision and machine learning libraries (OpenCV, SciKit-learn). Manually marked ground truth contains over 12900 frames taken from benchmark data [21].

We investigated several classifiers employed at the eyes state detection, and based on the experiments we selected kNN ($k = 1$) as the appropriate one. The experimental protocol involved 10-fold cross validation with the following number of samples: left eye – 5558 (opened), 2414 (closed) and right eye – 8192 (opened), 4013 (closed). The results of the experiments are presented in Table 1.

Table 1. The results of the experiments on the classification of eyes state.

Classifier	True positive rate			False positive rate		
	Opened	Closed	Average	Opened	Closed	Average
1NN	0.973	0.936	0.962	0.064	0.027	0.053
Kstar	0.974	0.935	0.962	0.065	0.026	0.053
Random forest	0.976	0.883	0.947	0.117	0.024	0.089
Bagging	0.958	0.796	0.909	0.204	0.042	0.155
MultiLayerPerceptron	0.936	0.781	0.889	0.219	0.064	0.172
j48 (C4.5 decision tree)	0.919	0.815	0.887	0.185	0.081	0.153
Random tree	0.921	0.806	0.886	0.194	0.079	0.159
Classification via regression	0.929	0.754	0.876	0.246	0.071	0.193
REPTree	0.908	0.732	0.855	0.268	0.092	0.215
SimpleLogistic	0.888	0.425	0.748	0.575	0.112	0.435
SVM with SMO	0.968	0.174	0.728	0.826	0.032	0.586
NaiveBayes	0.575	0.779	0.637	0.221	0.425	0.283

The main experiment was aimed at the verification the ability to discern between images containing open and closed eyes regions taking into consideration face and eyes detection stages. We tested the influence of the confidence level (responsible for the detection accuracy) on the eyes state detection. If we increase the confidence level (resulting in more eyes' candidates being rejected), the classifier performs quite well. On the other hand, if the confidence level is lower, the overall accuracy decreases.

4.3   Results

Compared to the manually marked video, the algorithm achieves an accuracy equal to 89%. Most of the blinking situations are detected, some of them, unfortunately, with a small delay. It is caused by the applied buffer analysis. The accuracy changes in accordance to the detectors's confidence. As it was anticipated, the confidence influences the detector performance. If the confidence is lower, then the total number of detected eyes is higher. The lower the detector confidence, the eyes state classifier's accuracy is also lower (see Fig. 4), since detected eyes are of poorer quality. Hence, if the eyes are detected with a higher confidence, the eyes state classifier performs also with a higher accuracy.

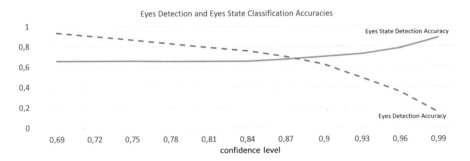

Fig. 4. Accuracy of both modules as a function of confidence level.

5 Summary

In the paper we proposed an algorithm of eyes state estimation that may lead to the driver's drowsiness detection based on thermal image analysis. It consists of two modules: face/eyes detection and eyes state classification. The detection uses Haar-like features and Viola-Jones detector, while eyes state is classified using Gabor-filtered images. The experiments showed that such a solution is capable of accurate eyes state detection in adverse lighting conditions, and has a high practical potential.

Acknowledgement. The authors would like to thank Kacper Kutelski, student of Computer Science at West Pomeranian University of Technology, for his implementation of eyes and face detection algorithms.

References

1. Weller, G., Schlag, B.: Road user behavior model. Deliverable D8 project RIPCORD-ISERET, 6 Framework Programme of the European Union (2007). http://ripcord.bast.de
2. Smolensky, M.H., Di Milia, L., Ohayon, M.M., Philip, P.: Sleep disorders, medical conditions, and road accident risk. Accid. Anal. Prev. **43**(2), 533–548 (2011)
3. Krishnasree, V., Balaji, N., Rao, P.S.: A real time improved driver fatigue monitoring system. WSEAS Trans. Sig. Process. **10**, 146–155 (2014)
4. Virginia Tech Transportation Institute: Day or night, driving while tired a leading cause of accidents (2013). http://www.vtnews.vt.edu/articles/2013/04/041513-vtti-fatigue.html
5. Daza, I.G., Bergasa, L.M., Bronte, S., Yebes, J.J., AlmazÃąn, J., Arroyo, R.: Fusion of optimized indicators from advanced driver assistance systems (ADAS) for driver drowsiness detection. Sensors **14**, 1106–1131 (2014)

6. Małecki, K., Forczmański, P., Nowosielski, A., Smoliński A., Ozga, D.: A new benchmark collection for driver fatigue research based on thermal, depth map and visible light imagery. In: Progress in Computer Recognition Systems, CORES 2019. Advances in Intelligent Systems and Computing, vol. 977. Springer, Cham (2019)
7. Mitas, A., Czapla, Z., Bugdol, M., Rygula, A.: Registration and evaluation of biometric parameters of the driver to improve road safety, pp. 71–79. Scientific Papers of Transport, Silesian University of Technology (2010)
8. Viola, P., Jones, M.J.: Robust real-time face detection. Int. J. Comput. Vis. **57**(2), 137–154 (2004)
9. Forczmański, P.: Performance evaluation of selected thermal imaging-based human face detectors. In: 10th International Conference on Computer Recognition Systems, CORES 2017. Advances in Intelligent Systems and Computing, vol. 578 (2018)
10. Hussien, M.N., Lye, M., Fauzi, M.F.A., Seong, T.C., Mansor, S.: Comparative analysis of eyes detection on face thermal images. In: 2017 IEEE International Conference on Signal and Image Processing Applications (ICSIPA), Kuching, pp. 385–389 (2017). https://doi.org/10.1109/ICSIPA.2017.8120641
11. Zhang, Y., Hua, C.: Driver fatigue recognition based on facial expression analysis using local binary patterns. Optik Int. J. Light Electron Opt. **126**(23), 4501–4505 (2015)
12. Burduk, R.: The AdaBoost algorithm with the imprecision determine the weights of the observations. In: Intelligent Information and Database Systems, Part II. LNCS, vol. 8398, pp. 110–116 (2014)
13. Jo, J., Lee, S.J., Park, K.R., Kim, I.J., Kim, J.: Detecting driver drowsiness using feature-level fusion and user-specific classification. Exp. Syst. Appl. **41**(4), 1139–1152 (2014)
14. Cyganek, B., Gruszczynski, S.: Hybrid computer vision system for drivers' eye recognition and fatigue monitoring. Neurocomputing **126**, 78–94 (2014)
15. Nowosielski, A.: Vision-based solutions for driver assistance. J. Theor. Appl. Comput. Sci. **8**(4), 35–44 (2014)
16. Berglund, B., Rossi, G.B., Townsend, J.T., Pendrill, L.R.: Measurement With Persons: Theory, Methods, and Implementation Areas, p. 422. Psychology Press, New York (2013)
17. Jasiński, P., Forczmański, P.: Combined imaging system for taking facial portraits in visible and thermal spectra. In: Proceedings of the International Conference on Image Processing and Communications 2015. Advances in Intelligent Systems and Computing, vol. 389, pp. 63–71 (2016)
18. Nowosielski, A., Forczmański, P.: Touchless typing with head movements captured in thermal spectrum. Pattern Anal. Appl. **22**(3), 841–855 (2018). https://doi.org/10.1007/s10044-018-0741-0
19. Forczmański, P., Kutelski, K.: Driver drowsiness estimation by means of face depth map analysis. In: Pejas, J., El Fray, I., Hyla, T., Kacprzyk, J. (eds.) Advances in Soft and Hard Computing, ACS 2018. Advances in Intelligent Systems and Computing, vol. 889. Springer, Cham (2019)
20. Andrysiak, T., Choras, M.: Image retrieval based on hierarchical Gabor filters. Int. J. Appl. Math. Comput. Sci. **15**(4), 471–480 (2005)
21. Małecki, K., Nowosielski, A., Forczmański, P.: Multispectral data acquisition in the assessment of driver's fatigue. In: 17th International Conference on Transport Systems Telematics (TST). Communications in Computer and Information Science, vol. 715, pp. 320–332 (2017)

Presentation Attack Detection for Mobile Device-Based Iris Recognition

Ewelina Bartuzi◉ and Mateusz Trokielewicz$^{(\boxtimes)}$◉

Research and Academic Computer Network (NASK),
Kolska 12, 01-045 Warsaw, Poland
{ewelina.bartuzi,mateusz.trokielewicz}@nask.pl
https://eng.nask.pl

Abstract. Apart from ensuring high recognition accuracy, one of the main challenges associated with mobile iris recognition is reliable Presentation Attack Detection (PAD). This paper proposes a method of detecting presentation attacks when the iris image is collected in visible light using mobile devices. We extended the existing database of 909 bonafide iris images acquired with a mobile phone by collecting additional 900 images of irises presented on a color screen. We explore different image channels in both RGB and HSV color spaces, deep learning-based and geometric model-based image segmentation, and use Local Binary Patterns (LBP) along with the selected statistical images features classified by the Support Vector Machine to propose an iris PAD algorithm suitable for mobile iris recognition setups. We found that the red channel in the RGB color space offers the best-quality input samples from the PAD point of view. In subject-disjoint experiments, this method was able to detect 99.78% of screen presentations, and did not reject any live sample.

Keywords: Biometrics · Iris recognition ·
Presentation Attack Detection · Mobile devices

1   Introduction

Iris biometrics is popular in high-security applications, since it provides a decent level of recognition accuracy compared to other biometric traits. Due to the ubiquity of mobile devices, with a 10 year increase in the average number of devices per 100 people from 50.6 to 103.5 [2], currently one of the fastest-growing branches of biometrics is the mobile-based authentication. In addition to high recognition accuracy, these solutions must also be resilient to attacks. This paper proposes an open-source Presentation Attack Detection for biometric systems using iris images collected in a mobile environment with non-specialized sensors, which broadens the possible applications beyond those to which the device manufacturer restricts a given biometric implementation. The method, utilizing local binary patterns and statistical image features selected with PCA and classified with an SVM, is able to reach close-to-perfect performance on a subject-disjoint testing subset consisting of live and screen iris presentations.

© Springer Nature Switzerland AG 2020
M. Choraś and R. S. Choraś (Eds.): IP&C 2019, AISC 1062, pp. 30–40, 2020.
https://doi.org/10.1007/978-3-030-31254-1_5

The study advances the research in iris and periocular biometrics by offering the **following contributions to the state of the art:**

- an open-source code for two PAD methods suitable for detecting color irises presented on a screen: one based on LBP texture descriptor, which detects areas with specific frequencies, the second employing statistical features of the image; this can serve as a useful benchmark method for visible light iris PAD,[1]
- extension of the existing database, consisting of 900 attack iris presentations obtained by imaging an iris sample displayed on a color screen,[2]
- analysis of the optimal color representation of samples collected in visible light employing different color spaces and their individual channels, with respect to both the PAD component as well as the overall recognition accuracy.

This article is organized as follows. Section 2 briefly summarizes existing scientific literature related to the topic of mobile iris recognition and PAD. Section 3 describes a subset of the existing database *MobiBits* used for the purpose of this study and details a complementary set of created fake iris representations. Description of the proposed method is included in Sect. 4, whereas the experimental scenarios and results are discussed in Sect. 5. Finally, Sect. 6 provides relevant conclusions.

2   Related Work Review

The importance of equipping a biometric system with a reliable Presentation Attack Detection component is already well recognized throughout the biometrics community. Researchers have proposed numerous methods for detection of attack irises, including paper printouts, textured contact lenses, or prosthetic eyes. These techniques include employing image texture descriptors, such as Local Binary Patterns (LBP) [14], Binarized Statistical Image Features (BSIF) [15], or Local Phase Quantization [21], keypoint detectors and descriptors such as Scale Invariant Feature Transform (SIFT) [17], as well as calculating image quality metrics [8]. Deep-learning-based PAD techniques have also recently emerged, *e.g.*, [19]. Thavalengal *et al.* proposed a multi-spectral analysis of the iris [22]. Czajka, on the other hand, exploited biological properties of the eye's reaction to light stimuli, introducing a method based on pupil dynamics [12]. Recently, Trokielewicz *et al.* proposed a deep-learning-based PAD component for detecting cadaver iris presentations [4]. In the mobile domain, methods such as exploiting the properties of a light field camera have been proposed [18], or employing magnified phase information [20]. Recently published review paper by Czajka and Bowyer presents a systematic summary of PAD for iris recognition [13].

---

[1] Available for download at http://zbum.ia.pw.edu.pl/EN/node/46.

[2] This dataset of attack iris samples is available to researchers at http://zbum.ia.pw. edu.pl/EN/node/46.

3 Experimental Data

A part of the existing multimodal biometric database, created by these authors in the past, containing eye and periocular images was used in this paper [10]. Photographs have been collected in visible light using Huawei Mate S (13 Mpx, f/2.0). Data acquisition from 53 people (20 women and 33 men), aged from 14 to 71 years, has been divided into two measurement sessions. Example photographs from this database are shown in Fig. 1. This dataset contains both the high quality (referred to as HQ) and low quality (LQ) images, with HQ images being taken with flash, and LQ images being taken without flash illumination.

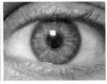

Fig. 1. From left to right: example same-eye images from the smartphone camera acquired with flash (HQ image), without flash (LQ image), attack screen presentations for the same samples (with and without flash), taken with the same smartphone camera as original images.

Part of this study was to extend the live iris dataset with a complementary set of attack iris samples. For this purpose, a popular (in visible light iris recognition) way of creating artificial data was used: displaying photos on the screen of a phone and then taking photographs of the screen. The device used for taking pictures of artifacts was the same device as the one used to collect real samples in the original study. A total of 900 attack representations were created.

The iris images were cropped in accordance with the ISO/EIC 19794-6:2011, up to a resolution of 640 × 480 pixels. Color preprocessing was applied to come up with five different representations, the first two including **RGB**, unmodified images straight from the sensor (Fig. 2, top row), **R** images created by extracting the red channel of the RGB color space (Fig. 2, second row), **S** images created by extracting the saturation component from the HSV representation (Fig. 2, third row), and **GRAY** images (Fig. 2, bottom row), grayscale created from the RGB image.

4 Methodology

4.1 Presentation Attack Detection Methods

For the purpose of mitigating screen presentation attacks, we have implemented two PAD methods, both requiring only a single, static iris image, and both not requiring any exhaustive computations. The first method relies on the statistical

BONA FIDE **ATTACK**

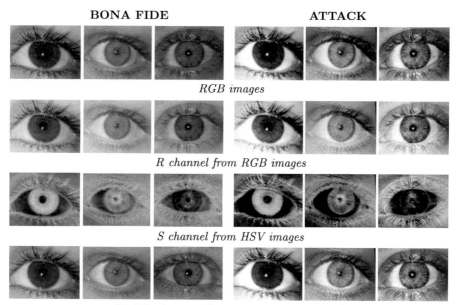

RGB images

R channel from RGB images

S channel from HSV images

GRAY – images after conversion to grayscale

Fig. 2. Examples of real (left) and fake (right) irises of brown/hazel, green and blue/gray eyes.

features of an image, whereas the second one utilizes an LBP-based texture descriptor.

Statistical Image Features: In this method, seven statistical characteristics of the image were calculated: average image intensity (μ), variance of the pixels value (σ), skewness ($\frac{\sum_{i=1}^{M}\sum_{j=1}^{N}(I(i,j)-\mu)^3}{M\cdot N\cdot\sigma^{3/2}}$), kurtosis ($\frac{\sum_{i=1}^{M}\sum_{j=1}^{N}(I(i,j)-\mu)^4}{M\cdot N\cdot\sigma^2}$), the 10th, 50th, and 90th percentile of the pixels value.

Local Binary Patterns: LBP is one of the most popular texture descriptors, which involves the analysis of a pixel in relation to its surroundings [3]. The value of each pixel is compared to the value of the central pixel, and the binary code created in this way is converted into a number in the decimal system: $LBP_{N,R}(I_C) = \sum_{n=1}^{N} s(I_n - I_C)2^{n-1}$ where N, R are the number of surrounding neighbors and the radius, respectively, I_C denotes the central pixel, I_N denotes the n-th pixel from neighborhood of central pixel, and

$$s(I_n - I_C) = \begin{cases} 1 & \text{if } I_n - I_C \geq 0 \\ 0 & \text{otherwise} \end{cases} \quad (1)$$

The obtained values are represented as histograms: one for the entire image, and individual histograms created for each piece of the image divided into 100 parts. In addition, we implemented version of the LBP algorithm resistant to rotation (uniform LBP code) [3]. The best results were obtained for the surroundings of eight neighbors analyzing the whole picture. In this way, the feature vectors counted 59 elements.

Features Selection and Classification: Features obtained from each method were then sorted by relevance using principal component analysis (PCA), the influence of subsequent features on attack detection accuracy was examined, and an optimal number of features is determined, Fig. 3. Table 1 summarizes the optimal number of LBP features for each type of image. The binary classification was carried out using a support vector machine (SVM) classifier with a radial basis function kernel. The data were divided into subject-disjoint training and testing subsets in a ratio of 80:20.

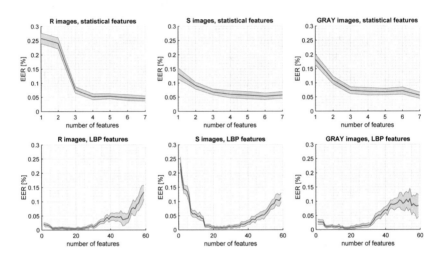

Fig. 3. EER as a function of the number of successive statistical (top row) and LBP (bottom row) features ordered according to the PCA. Average (solid dark blue lines) and standard deviation obtained from 20 training/testing data splits (light blue shades) are shown.

Table 1. APCER, BPCER, EER$_{PAD}$ for two PAD methods for the optimal no. of parameters.

Method	Optimal no. of features	Image type	APCER [%]	BPCER [%]	EER$_{PAD}$ [%]
Statistical features	7	R	1.89	4.22	5.00
	6	S	5.94	4.84	4.95
	7	GRAY	3.92	4.53	4.97
LBP	20	R	0.22	0.00	0.10
	23	S	0.11	0.44	0.28
	17	GRAY	0.44	0.00	0.22

4.2   Iris Recognition: OSIRIS

For the purpose of iris recognition, the Open Source for IRIS (OSIRIS) is employed [5]. The academic-based software was developed as a part of the BioSecure project and implements the original Daugman concept, including segmentation of the iris and its normalization by transformation from cartesian to polar coordinates using the *rubber sheet model*. The encoding of the iris features is performed using phase quantization of a response of Gabor filtering outcomes, and then comparing the binary codes using the XOR operation to obtain the normalized Hamming distance. Values close to zero should indicate data from the same iris, whereas typical results for different irises comparisons oscillate around 0.5 (usually they are concentrated in the range of 0.4–0.45 because of shifting of the iris codes).

OSIRIS Segmentation: The first stage of iris image processing is location and segmentation, with the exception of image noise, among others in the form of eyelids, eyelashes, reflections, shadows. The result is a binary mask, which determines which pixels belong to the iris. In the original OSIRIS, this is performed by a circular Hough transform used to roughly approximate the circles representing the edges of the iris, and then employing active contours algorithm to exclude noisy regions. Examples of segmentation results can be seen in Fig. 4, top row.

DCNN-Based Segmentation: Due to the type of photos that is different from NIR images for which OSIRIS was originally built, a second segmentation method is also used, namely a model based on the SegNet architecture retrained with iris images taken in infrared light, as well as images representing only the R channel of the RGB color scheme, using the implementation from [6].

5   Experiments and Results

Stage 1: Presentation Attack Detection

The LBP descriptor and seven statistical image characteristics were used as a PAD component put in front of the iris recognition pipeline. In both cases, the features were sorted from the most important, and then combined in the order given and treated as a feature vector, which was then classified using a SVM (Fig. 3). The data were divided with 20 subject-disjoint, 80/20 training and test sets. In this part of the analysis, we used iris representations in form of R, S and GRAY images. Table 1 presents the optimal number of parameters for each image type and the results of the binary classification in the form of the error metrics as recommended by the ISO/IEC standard on presentation attack detection.

Both the methods based on texture analysis and statistical image features obtained good results, with LBP allowing for almost perfect discrimination between bona-fide and attack samples. In the case of the R channel image representation, the EER_{PAD} was as low as 0.11%. Since the EER_{PAD} was considered as a minimization target for parameter choice, the APCER here is larger than

BPCER, however, this can be tuned by moving the acceptance threshold in the desired direction. Analysis of the statistical features of the image gave higher errors, but a downward trend can be seen here, therefore adding other image features could also reduce these errors. In each case the most discriminative information seems to be represented in the red channel R. **The APCER and BPCER errors concern LQ images, for the HQ both errors are 0%, and both the detection rate of attack and bona fide samples, is perfect.**

Stage 2: Iris Recognition
In this Section we evaluate the impact that attack samples can have on recognition accuracy. The first stage of the OSIRIS recognition pipeline consists of image segmentation. The original OSIRIS segmentation repeatedly encountered problems with correctly segmenting the images, Fig. 4.

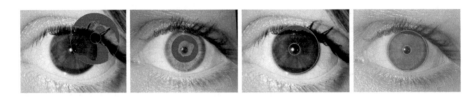

Fig. 4. Examples of incorrect segmentation from the OSIRIS algorithm (left) and corresponding results for DCNN-based approach (right) for the same samples. The iris should ideally be located within the two green circles, and the red regions should denote non-iris portions that lay within these circles.

Since most of these errors will likely lead to erratic iris verification, the segmentation phase has to be altered or replaced to be able to test the effectiveness of the iris feature representation, encoding, and matching of OSIRIS for iris images taken in visible light. The original segmentation was therefore replaced with a solution based on convolutional neural network and the Hough transform, cf. Sect. 4.2, which allowed for close-to-perfect segmentation of most samples, cf. Fig. 4, bottom row.

After the segmentation stage, the OSIRIS method was employed to calculate all possible comparison scores between within-class image pairs (*genuine*), between-class image pairs (*impostors*), and between real samples and their corresponding attack representations (*real vs fake*). This was done for all three image types (R, S, and GRAY), as well as for high quality images only, low quality images only, and between images of different quality (denoted as HQ, LQ, and HQ:LQ, respectively), leading to a total of 9 experiments without, and 9 experiments with the PAD component included. Comparison scores distributions are presented in Fig. 5, whereas the obtained average Equal Error Rates for all 18 experiments are summarized in Table 2. In the experiments run without the

Table 2. Equal Error Rates and their standard deviations obtained using OSIRIS for three different image representation: R, S, and GRAY, for higher- and lower quality images and between different qualities, both without and with the PAD component.

Image Type	R	S	GRAY
Without PAD component			
HQ	29.15 (±0.66)	27.55 (±0.54)	6.86 (±0.29)
LQ	38.71 (±0.48)	46.28 (±0.30)	37.14 (±0.58)
HQ:LQ	40.40 (±0.48)	47.52 (±0.24)	53.45 (±0.53)
Including PAD component			
HQ	7.68 (±0.58)	26.02 (±0.47)	6.57 (±0.29)
LQ	18.09 (±0.41)	43.77 (±0.33)	21.36 (±0.57)
HQ:LQ	20.24 (±0.42)	47.48 (±0.23)	47.83 (±0.53)

PAD method, all comparisons with *attack* samples are considered as impostor comparisons.

Conclusions drawn from comparison score distributions plotted in Fig. 5:

(1) the best separation between genuine and impostor comparisons can be found when matching high quality (HQ) images representing the R channel and the grayscale (GRAY) conversion of the RGB image,

(2) both the low quality (LQ) and mixed quality (HQ:LQ) comparisons do not offer distribution separation that would enable reasonably accurate iris recognition,

(3) using the R channel, **bonafide-vs-attack scores overlap with those obtained from genuine comparisons, thus making the PAD component crucial** (including it improves the EER from almost 30% to 7.68%),

(4) but, surprisingly, for grayscale (GRAY) images the same scores overlap with the *impostor* scores distribution (and including PAD in this case improves the EER from 6.86% to 6.57%).

The best results EER-wise were obtained for high-quality images, in the representations of the R channel and grayscale conversion of the RGB color space, which gave EER = 7.68% and EER = 6.57%, respectively. For lower-quality images, R images yielded better results, giving (still unacceptable) EER of 18.13%, compared against 21.36% obtained for the GRAY images and EER = 43.77% for S images. For mixed quality database, only the R channel allows for a non-random recognition accuracy with EER = 20.24%, whereas S and GRAY images yield EERs close to 50%.

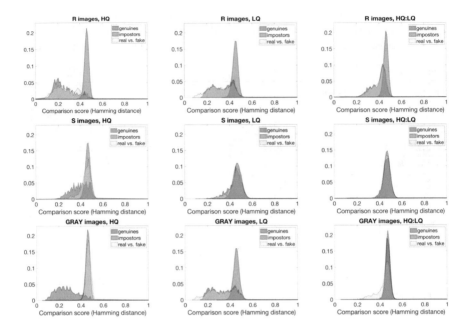

Fig. 5. Score distributions obtained for HQ images only (left), for LQ images only (middle), and between HQ and LQ images (right). Genuine scores (blue), impostor scores (red), and scores between real samples and their fake representations (yellow) are shown.

6   Conclusions

This paper offers an open-source presentation attack detection method designed to detect attack representations of iris samples collected with a mobile phone in visible light. LBP-derived features allow for a nearly perfect attack detection accuracy with APCER = 0.11% and BPCER = 0% with a dataset of attack iris representations consisting of irises displayed on a screen, which was created for this study.

By testing the proposed PAD method coupled with the OSIRIS recognition pipeline with DCNN-based image segmentation stage, we show that by employing the R channel of the RGB color space, recognition accuracy of 7.68% EER can be achieved, compared to EER = 30% obtained prior to the inclusion of a PAD component. Surprisingly, we have found the grayscale representation of the RGB image color space to offer some kind of resilience to this particular attack, as comparison scores obtained from matching real iris samples and their fake counterparts were similar to the scores obtained from matching impostor image pairs. Here, the proposed PAD allowed for a moderate reduction of EER from 6.86% to 6.57%.

This paper follows the guidelines for research reproducibility by making the dataset of attack iris representations, as well as source codes for the PAD methods proposed, open-sourced to serve as a benchmark for visible light iris presentation attack detection, especially with respect to mobile applications.

Acknowledgments. Project CYBERSECIDENT/382354/II/NCBR/2018 financed by the National Centre for Research and Development in the framework of Cyber-SecIdent programme.

References

1. Quinn, G.W., Matey, J.R., Grother, P.: IREX IX Part One: Performance of Iris Recognition Algorithms. In: NIST Interagency Report 8207 (2018). https://doi.org/10.6028/NIST.IR.8207
2. International Telecommunication Union (ITU): Global and regional ICT data: mobile-cellular subscriptions. https://www.itu.int/en/itud/statistics/pages/stat/default.aspx. Accessed June 2018
3. Ojala, T., Pietikainen, M., Maenpaa, T.: Multiresolution gray scale and rotation invariant texture classification with local binary patterns. IEEE Trans. Pattern Anal. Mach. Intell. **24**(7), 971–987 (2002). https://doi.org/10.1109/TPAMI.2002.1017623
4. Trokielewicz, M., Czajka, A., Maciejewicz, P.: Presentation attack detection for cadaver iris. In: 9th IEEE International Conference on Biometrics: Theory, Applications and Systems (BTAS 2018), Los Angeles, California, 22–25 October 2018 (2018)
5. Othman, N., Dorizzi, B., Garcia-Salicetti, S.: OSIRIS: an open source iris recognition software. https://doi.org/10.1016/j.patrec.2015.09.002
6. Trokielewicz, M., Czajka, A., Maciejewicz, P.: Post-mortem iris recognition with deep-learning-based image segmentation (2019). http://arxiv.org/abs/1901.01708
7. ISO/IEC 30107-1:2016: Information technology – biometric presentation attack detection – part 1: framework (2016)
8. Galbally, J., Marcel, S., Fierrez, J.: Image quality assessment for fake biometric detection: application to iris, fingerprint, and face recognition. IEEE Trans. Image Process. **23**(2), 710–724 (2014)
9. Ren, J., Jiang, X., Yuan, J.: Noise-resistant local binary pattern with an embedded error-correction mechanism. IEEE Trans. Image Process. **22**(10) (2013). https://doi.org/10.1109/TIP.2013.2268976
10. Bartuzi, E., Roszczewska, K., Trokielewicz, M., Białobrzeski, R.: MobiBits: multimodal mobile biometric database. In: 17th International Conference of the Biometrics Special Interest Group, BIOSIG 2018, Darmstadt, Germany, 26–29 September 2018 (2018). https://doi.org/10.23919/BIOSIG.2018.8553108
11. ISO/IEC JTC 1/SC 37: Information technology "vocabulary" part 37: biometrics (FDIS), October 2016
12. Czajka, A.: Pupil dynamics for iris liveness detection. IEEE Trans. Inf. Forensics Secur. **10**(4), 726–735 (2015)
13. Czajka, A., Bowyer, K.: Presentation attack detection for iris recognition: an assessment of the state of the art. ACM Comput. Surv. (in Rev.) (2018). https://arxiv.org/abs/1804.00194
14. Doyle, J.S., Flynn, P.J., Bowyer, K.W.: Automated classification of contact lens type in iris images. In: IEEE International Conference on Biometrics (ICB), pp. 1–6, June 2013

15. Komulainen, J., Hadid, A., Pietikäinen, M.: Generalized textured contact lens detection by extracting BSIF description from Cartesian iris images. In: IEEE International Joint Conference on Biometrics (IJCB), pp. 1–7, September 2014

16. Menotti, D., Chiachia, G., Pinto, A., Schwartz, W.R., Pedrini, H., Falcão, A.X., Rocha, A.: Deep representations for iris, face, and fingerprint spoofing detection. IEEE Trans. Inf. Forensics Secur. **10**(4), 864–879 (2015)

17. Pala, F., Bhanu, B.: Iris liveness detection by relative distance comparisons. In: IEEE Conference on Computer Vision and Pattern Recognition (CVPR) Workshops, July 2017

18. Raghavendra, R., Busch, C.: Presentation attack detection on visible spectrum iris recognition by exploring inherent characteristics of light field camera. In: IEEE International Joint Conference on Biometrics, pp. 1–8, September 2014

19. Raghavendra, R., Raja, K.B., Busch, C.: ContlensNet: robust iris contact lens detection using deep convolutional neural networks. In: IEEE Winter Conference on Applications of Computer Vision (WACV), pp. 1160–1167, March 2017

20. Raja, K.B., Raghavendra, R., Busch, C.: Video presentation attack detection in visible spectrum iris recognition using magnified phase information. IEEE Trans. Inf. Forensics Secur. **10**(10), 2048–2056 (2015)

21. Sequeira, A.F., Thavalengal, S., Ferryman, J., Corcoran, P., Cardoso, J.S.: A realistic evaluation of iris presentation attack detection. In: International Conference on Telecommunications and Signal Processing (TSP), pp. 660–664, June 2016

22. Thavalengal, S., Nedelcu, T., Bigioi, P., Corcoran, P.: Iris liveness detection for next generation smartphones. IEEE Trans. Consum. Electron. **62**(2), 95–102 (2016)

Gaze-Based Interaction
for VR Environments

Patryk Piotrowski and Adam Nowosielski$^{(\boxtimes)}$ 🅳

Faculty of Computer Science and Information Technology,
West Pomeranian University of Technology, Szczecin,
Żołnierska 52, 71-210 Szczecin, Poland
patryk.piotrowski19@gmail.com, anowosielski@wi.zut.edu.pl

Abstract. In this paper we propose a steering mechanism for VR headset utilizing eye tracking. Based on the fovea region traced by the eyetracker assembled into VR headset the visible 3D ray is generated towards the focal point of sight. The user can freely look around the virtual scene and is able to interact with objects indicated by the eyes. The paper gives an overview of the proposed interaction system and addresses the effectiveness and precision issues of such interaction modality.

Keywords: Gaze-based interaction · Virtual reality · Eye tracking · Gaze-operated games

1 Introduction

Virtual reality systems are computer-generated environments where an user experiences sensations perceived by the human senses. These systems are based primarily on providing video and audio signals, and offer the opportunity to interact directly with the created scene with the help of touch or other form of manipulation using hands. Vision systems for virtual reality environments consist most frequently of head-mounted goggles, which are equipped with two liquid crystal displays placed opposite the eyes in a way that enables stereoscopic vision. The image displayed in the helmet is rendered independently for the left and right eye, and then combined into the stereopair. More and more solutions appear on the market, and the most popular include: HTC Vive, Oculus Rift CV1, Playstation VR (dedicated for Sony Playstation 4), Google Cardboard (dedicated for Android mobile devices).

The virtual reality solutions are delivered with controllers which aim is to increase the level of user's immersion with elements of the virtual environment. Interestingly, many novel interfaces offer hands-free control of electronic devices. The touchless interaction there is based on recognition of user actions performed by the whole body [1] or specific parts of the body (e.g. hands [2], head [3]). A completely new solution, not widely used and known, is the control through the sight, i.e. gaze-based interaction. The operation of such systems is based on eyetracking, a technique of gathering real-time data concerning gaze direction of

© Springer Nature Switzerland AG 2020
M. Choraś and R. S. Choraś (Eds.): IP&C 2019, AISC 1062, pp. 41–48, 2020.
https://doi.org/10.1007/978-3-030-31254-1_6

human eyes [4]. The technology is based on tracking and analysing the movement of the eyes using cameras or sensors that register the eye area [4]. The latest solutions on the market introduce eye tracking capabilities to virtual reality environments [5].

In the paper we propose a novel steering mechanism based on ray-casting for human-computer interfaces. Based on the fovea region traced by the eye tracker, assembled into VR headset, the visible 3D ray is generated towards the focal point of sight. Thanks to head movements the user can freely look around the virtual scene and is able to interact with objects indicated by eyes.

The paper is structured as follows. In Sect. 2 the related works are addressed. Then in Sect. 3 the concept of gaze-based interaction in virtual reality environment is proposed. An example application is presented in Sect. 4. The proposed system is evaluated in Sect. 5. Final conclusions and a summary are provided in Sect. 6.

2   Related Works

Most of the eye tracking solutions have been utilized for analysis of the eye movement for advertising industry, cognitive research and analysis of individual patterns of behaviours [6–8]. The eye tracking systems are recognized tools for analysis of the layout and operation efficiency of human-computer interaction systems [9]. They are now regarded also as input modality for people with disabilities [10]. For some users who suffer from certain motor impairments, the gaze interaction may be the only option available. Typical way of interaction in such systems assume fixations and dwell times. User is expected to look at specific element of the interface for a predefined time period called the dwell-time and after that the system assumes the selection (equivalent to mouse clicking). Such solution is used for navigating through graphical user interface or for eye-typing using the on-screen keyboard. Some innovations to this technique has been proposed. In [11] a cascading dwell gaze typing technique dynamically adjust the dwell time of keys in an on-screen keyboard. Depending on the probability during typing some keys are easier to select by decreasing their dwell times other are harder to choose (increased dwell times). A completely different approach was presented in [10]. Here, a dwell-free eye-typing technique has been proposed. Users are expected to look at or near the desired letters without stopping to dwell.

New solution replacing the traditional technique of fixations and dwell time has been proposed in [12]. Authors has proposed gaze gestures. In contrast to the classical way of interaction using the eye sight, a gaze gesture uses eye motion and are insensitive to accuracy problems and immune against calibration shift [12]. This is possible because gaze is not used directly for pointing and only the information of relative eye movements are required. The gaze gestures have also been reported to be successful in interacting with games [13].

Apart from using gaze to control computer systems other interesting applications can be found in the scientific literature. To accelerate the raytracing in [14] fewer primary rays are generated in the peripheral regions of vision. For the fovea region traced by the eye tracker the sampling frequency is maximised [14].

The above examples show the multitude of applications of eye tracking systems. The novelty is now installing these systems in virtual reality headsets which offers new possibilities of application.

3   Gaze-Driven Ray-Cast Interface Concept

The concept of using an eye tracker for steering in the virtual reality environment assumes the usage of the eye focus point to interact with objects of the virtual scene. An overview of the system built upon this concept is presented in Fig. 1 and the process of interaction consists of the following steps:

- mapping the direction of the user's eye focus on the screen coordinates,
- generating a primary ray (raycasting using the sphere) from the coordinates of the user's eye focus direction,
- intersection analysis with scene objects,
- indication of the object pointed by the sight,
- handling the event associated with the object,
- rendering a stereopair for virtual reality goggles.

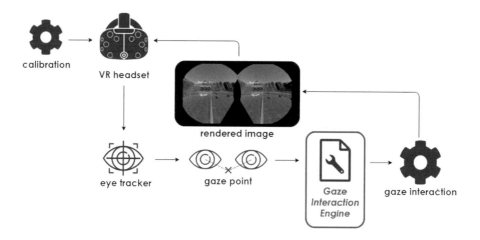

Fig. 1. Scheme of the VR system with gaze-based interaction.

The main idea, then, is to generate a ray which takes into account the position, rotation and direction of the eye's focus on the virtual scene. During the initial mapping, coordinates are taken for the left and right eyes independently, and the final value of the focal point is the result of their averaging. In case of intersection detection with the scene object the appropriate procedures are executed.

Figure 2 presents diagram of the interaction process with a scene object using eye focus direction. Four states can be distinguished for the object: no interaction, beginning, continuing (during), and ending interaction. Start of the interaction is crucial since it might be triggered with the eyesight solely (after a predefined dwell time) or with the use of hand operated controller.

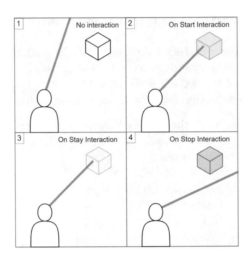

Fig. 2. Diagram showing the interaction process with a scene object using eye focus direction.

4   Implementation and Application

Based on the concept presented in Sect. 3 a sight-operated interaction system, named *Gaze Interaction Engine* (marked with a red border in Fig. 1), for the virtual reality environment has been developed. This solution has the form of a *UnityAsset* module for the Unity environment. It is hereby made available to the public and can be accessed through the web page [15]. The developed gaze-based interaction system is designed for the virtual reality HTC Vive hardware and eye tracker from Pupil Labs [5]. In our research the eye tracker has been set to receive 640 × 480 pixel infrared eye image with 120 frames per second.

Our interface can be employed to create computer games and multimedia applications. A good example of using the tool was presented during the event devoted to games creation *GryfJam* in Szczecin (Poland) on 17th and 18th of May 2019. One of the authors of the paper, Patryk Piotrowski, with the help of Michał Chwesiuk developed a simple game for virtual reality glasses with manipulation only with the gaze. The game, named *KurzoCITY*, belongs to the genre of arcade games. The player's goal is to collect as many grains as possible on the farm under the pressure of competition from virtual poultry (see Fig. 3 for a game preview).

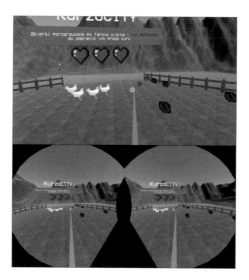

Fig. 3. The use of Gaze Interaction Engine in the *KurzoCITY* game: screen view (top) and stereo pair for virtual reality googles (bottom).

The eye tracker used in the helmet analyzes the movement of the eyeballs. For each frame a ray is generated from the player's eyes to the focal point. A look at the grain allows it to be collected. Over time, the level of difficulty of the game increases by adding new opponents and raising the number of grains to collected by the player. The game ends when the poultry collect a total of three seeds.

5   Evaluations

The game described in the preceding section, as already mentioned, had been developed during the *GryfJam* event. Using the developed game and event's participants, tests of the effectiveness of the *Gaze Interaction Engine* were conducted. We observed high playability which indicates that proposed gaze-steering mechanism is successful. Nevertheless, some problems and imperfections of the eye tracking system have been noticed. Among over 30 participants of all our experiments, we found 2 who were not able to pass the calibration process entirely. The greatest setup difficulty was fitting the helmet and adjusting the distance between the lenses which ensure correct detection of the pupil. With an unmatched arrangement, the position of the pupil can not be determined correctly and the examples are presented in Fig. 4. The top left sample, for comparison purposes, contains a correct case. The eye is in the center, corneal reflections are visible, the center of pupil is annotated with the red dot and the pupil border is surrounded by a red border.

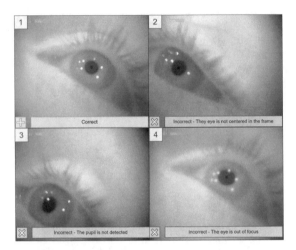

Fig. 4. Calibration problems: the appearance of the eye seen by the eye tracking system mounted in the virtual reality helmet.

The second problem encountered was decalibration of the eye tracker at the time of use. Expressions which appear on the face may cause slight shifts of the entire headset and in the effect render the eye tracker erroneous.

Encountered problems, described above, can be classified as hardware related. To evaluate the accuracy of the eye-based interaction an additional experiment has been conducted. We prepared a grid with 26 separate interactable buttons (divided into three rows, occupying approximately half field of view vertically and 100% field of view horizontally). The goal of each participant was to press the highlighted button by focusing the eyes on it with the dwell-time equal 600 ms and visual progress indicator provided. We measured the time of pressing randomly highlighted buttons and the accuracy of the process itself. There were 17 participants (volunteers from students and employees of our university) who performed between 2 and 6 sessions. There have been 70 sessions in total and each sessions consisted of pressing 23.6 buttons on average. Results are presented in the graphical form in Fig. 5.

The averaged time of pressing a random button equal 1.79 s. It includes the 600 ms dwell-time, required for the interaction to take place. The precision seems to be more problematic here. We registered the averaged (over all participants and sessions) error rate of 5.51%. The error have been calculated as the ratio of pressing the improper button (most often the adjacent one) for the total number of presses. These results indicate that interfaces composed of many components arranged close to each other may be problematic to operate using current eye tracking solutions for the virtual reality helmets. However, when the number of interactive elements in the scene decreases and the size of these elements increase the interaction is quite convenient. The proposed game is a good proof here. With the relatively small dwell time (set to 200 ms compared to 600 ms in the pressing buttons experiment) very high level of interaction among participants have been observed.

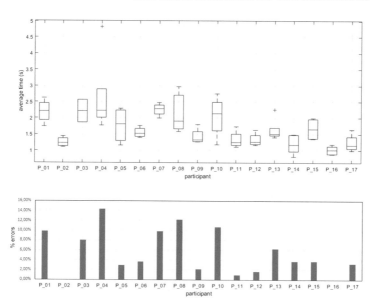

Fig. 5. Evaluation results: performance (top) and accuracy (bottom) of 17 participants.

6 Conclusion

The proposed interaction system for virtual reality environments enables the effective implementation of multimedia applications and games, operated using the eyesight. The visible 3D ray is generated towards the focal point of sight to facilitate user with the interaction process where free head movements are present. The eye control is faster compared to, for example, additional hand-operated controllers. In order for the motor reaction to take place, a stimulus and a nerve impulse are required after visual observation. These stages are eliminated. Eye trackers mounted in the VR headsets can significantly help people with disabilities offering unusual possibilities and for a wide range of recipients can offer new opportunities for interaction in human-computer interfaces and games.

References

1. Giorio, C., Fascinari, M.: Kinect in Motion - Audio and Visual Tracking by Example. Packt Publishing, Birmingham (2013)
2. Nowosielski, A.: Evaluation of touchless typing techniques with hand movement. In: Burduk, R., et al. (eds.) Proceedings of the 9th International Conference on Computer Recognition Systems, CORES 2015. AISC, vol. 403, pp. 441–449. Springer, Cham (2016)
3. Nowosielski, A.: 3-steps keyboard: reduced interaction interface for touchless typing with head movements. In: Kurzynski, M., Wozniak, M., Burduk, R (eds.) Proceedings of the 10th International Conference on Computer Recognition Systems, CORES 2017. AISC, vol. 578, pp. 229–237. Springer, Cham (2018)

4. Mantiuk, R., Kowalik, M., Nowosielski, A., Bazyluk, B.: Do-it-yourself eye tracker: low-cost pupil-based eye tracker for computer graphics applications. In: LNCS, vol. 7131, pp. 115–125 (2012)
5. Pupil Labs GmbH: Eye tracking for virtual and augmented reality. https://pupil-labs.com/vr-ar/. Accessed 15 June 2019
6. Wedel, M., Pieters, R.: A review of eye-tracking research in marketing. In: Malhotra, N.K. (ed.) Review of Marketing Research, vol. 4, pp. 123–147. Emerald Group Publishing Limited (2008)
7. Berkovsky, S., Taib, R., Koprinska, I., Wang, E., Zeng, Y., Li, J., Kleitman, S.: Detecting personality traits using eye-tracking data. In: Proceedings of the 2019 CHI Conference on Human Factors in Computing Systems, CHI 2019, pp. 221:1–221:12. ACM, New York (2019)
8. Jankowski, J., Ziemba, P., Watróbski, J., Kazienko, P.: Towards the tradeoff between online marketing resources exploitation and the user experience with the use of eye tracking. In: Nguyen, N.T., Trawiński, B., Fujita, H., Hong, T.P. (eds.) Intelligent Information and Database Systems, ACIIDS 2016. LNCS, vol. 9621, pp. 330–343. Springer, Berlin (2016)
9. Jacob, R.J.K., Karn, K.S.: Commentary on section 4. Eye tracking in human-computer interaction and usability research: ready to deliver the promises. In: Hyönä, J., Radach, R., Deubel, H. (eds.) The Mind's Eye, pp. 573–605. North-Holland (2003)
10. Kristensson, P.O., Vertanen, K.: The potential of dwell-free eye-typing for fast assistive gaze communication. In: Spencer, S.N. (ed.) Proceedings of the Symposium on Eye Tracking Research and Applications (ETRA 2012), pp. 241–244. ACM, New York (2012)
11. Mott, M.E., Williams, S., Wobbrock, J.O., Morris, M.R.: Improving dwell-based gaze typing with dynamic, cascading dwell times. In: Proceedings of the 2017 CHI Conference on Human Factors in Computing Systems (CHI 2017), pp. 2558–2570. ACM, New York (2017)
12. Drewes, H., Schmidt, A.: Interacting with the computer using gaze gestures. In: Baranauskas, C., Palanque, P., Abascal, J., Barbosa, S.D.J. (eds.) Human-Computer Interaction – INTERACT 2007, INTERACT 2007. LNCS, vol. 4663, pp. 475–488. Springer, Berlin (2007)
13. Istance, H., Hyrskykari, A., Immonen, L., Mansikkamaa, S., Vickers, S.: Designing gaze gestures for gaming: an investigation of performance. In: Proceedings of the 2010 Symposium on Eye-Tracking Research & Applications (ETRA 2010), pp. 323–330. ACM, New York (2010)
14. Siekawa, A., Chwesiuk, M., Mantiuk, R., Piórkowski, R.: Foveated ray tracing for VR headsets. In: MultiMedia Modeling. LNCS, vol. 11295, pp. 106–117 (2019)
15. Piotrowski, P., Nowosielski, A.: Gaze interaction engine (project page) (2019). https://github.com/patryk191129/GazeInteractionEngine

Modified Score Function and Linear Weak Classifiers in LogitBoost Algorithm

Robert Burduk[(✉)] and Wojciech Bozejko[(✉)]

Faculty of Electronic, Wroclaw University of Science and Technology,
Wybrzeze Wyspianskiego 27, 50-370 Wroclaw, Poland
{robert.burduk,wojciech.bozejko}@pwr.wroc.pl

Abstract. This paper presents a new extension of LogitBoost algorithm based on the distance of the object to the decision boundary, which is defined by the weak classifier used in boosting. In the proposed approach this distance is transformed by Gaussian function and defines the value of a score function. The assumed form of transforming functions means that the objects closest or farthest located from the decision boundary of the basic classifier have the lowest value of the scoring function. The described algorithm was tested on four data sets from UCI repository and compared with LogitBoost algorithm.

Keywords: LogitBoost algorithm ·
Distance to the decision boundary · Score function

1 Introduction

Boosting is a machine learning effective method of producing a very accurate classification rule by combining a weak classifier [7]. The weak classifier is defined to be a classifier which is only slightly correlated with the true classification i.e. it can classify the object better than random classifiers. In boosting the weak classifier is learned on various training examples sampled from the original learning set. The sampling procedure is based on the weight of each example. In each iteration, the weights of examples are changing. The final decision of the boosting algorithm is determined on the ensemble of classifiers derived from each iteration of the algorithm. One of the fundamental problems of the development of different boosting algorithms is choosing the weights and defining rules for an ensemble of classifiers. In recent years, many authors presented various concepts based on the boosting idea [1,5,9,11,14] as well as many examples in which such algorithms have been used in practical application [3,8].

In this paper we consider a modification of the LogitBoost algorithm in which linear base classifiers were used. The proposed modification concerns the calculation of the scoring function. The value of the scoring function depends on the modified distance of the object from the decision boundary defined by the linear base classifier.

© Springer Nature Switzerland AG 2020
M. Choraś and R. S. Choraś (Eds.): IP&C 2019, AISC 1062, pp. 49–56, 2020.
https://doi.org/10.1007/978-3-030-31254-1_7

This paper is organized as follows: Sect. 2 introduces the necessary terms of the boosting algorithm. In the next section there is our modification of Logit-Boost algorithm. Section 3 presents the experiment results comparing LogitBoost with our algorithm. Finally, some conclusions are presented.

2   Boosting Algorithms

In the work [4] weak and strong learning algorithms were discussed. The weak classifier is defined as slightly better than random. Schapire formulated the first algorithm to boost a weak classifier, where a verb boost means "to improve". Therefore, the main idea in boosting is to improve the prediction of a weak classifier by creating a set of weak classifiers which is a single strong classifier. The main steps of the well-known AdaBoost algorithm [2] are presented in Table 1.

Table 1. AdaBoost algorithm.

1.		Let $w_{1,1} = ... = w_{1,n} = 1/n$
2.		For $t = 1, 2, ...T$ do:
	a.	Fit f_t using weights $w_{t,1}, ..., w_{t,n}$, and compute the error e_t
	b.	Compute $c_t = \ln((1 - e_t)/e_t)$
	c.	Update the observations weights:
		$w_{t+1,i} = w_{t,i} \exp(c_t, I_{t,i}) / \sum_{j=1}^{n}(w_{t,i} \exp(c_t, I_{t,i})), \quad i = 1, ..n$
3.		Output the final classifier:
		$\hat{y}_i = F(x_i) = sign(\sum_{t=1}^{T} c_t f_t(x_i)).$

Table 2. Notation of the AdaBoost algorithm.

i	Observation number, $i = 1, ..., n$.
t	Stage number, $t = 1, ..., T$.
x_i	A p-dimensional vector containing the quantitative variables of the ith observation.
y_i	A scalar quantity representing the class membership of the ith observation, $y_i = -1, 1$.
f_t	The weak classifier at the tth stage.
$f_t(x_i)$	The class estimate of the ith observation at the tth stage.
$w_{t,i}$	The weight of the ith observation at the tth stage, $\sum_i w_{t,i} = 1$.
$I_{t,i}$	Indicator function, $I(f_t(x_i) \neq y_i)$.
e_t	The classification error at the tth stage, $\sum_i w_{t,i} I_{t,i}$.
c_t	The weight of f_t.
$sign(x)$	$= 1$ if $x \geq 0$ and $= -1$ otherwise.

One of the main steps in the algorithm is to maintain a distribution of the training set using the weights. Initially, all weights of the training set observations are set equally. If an observation is incorrectly classified (at the current stage) the weight of this observation is increased. Similarly, the correctly classified observation receives less weight in the next step. In each step of AdaBoost algorithm the best weak classifier according to the current distribution of the observation weights is found. The goodness of a weak classifier is measured by its error. Based on the value of this error the ratio is calculated. The final prediction of AdaBoost algorithm is a weighted majority vote of all weak classifiers.

The LogitBoost (adaptive logistic regression) is a popular modification of the AdaBoost algorithm. LogitBoost works similarly to AdaBoostM1 and can give better average accuracy than AdaBoostM1 for data with poorly separable classes.

3 Modified Score Function

In this section, we proposed the modification of LogitBoost algorithm. In particular, the method of calculating the weights of objects as well as the weights of the basic classifiers is not changed. We propose that each base classifier has a scoring function, which depends on the distance to the decision boundary defined by this base classifier. In order to calculate the value of the scoring function, the distance of the object from the decision boundary is transformed by a function

$$SC(x_i) = \frac{1}{\sigma\sqrt{2\pi}} \exp\left(-\frac{(d(x_i) - \mu)^2}{2\sigma^2}\right), \tag{1}$$

where $d(x_i)$ is the distance from the object x_i to the decision boundary. Our goal is not to provide a probabilistic interpretation of the score function, therefore, it is farther not transformed. Before the learning process feature scaling is used to bring all values into the range $[0, 1]$. The proposed algorithm is labeled LogitBoostM and it is as follows:

Table 3. LogitBoostM algorithm.

1.	Scale all features into the range $[0, 1]$
2.	Set the initial weights $w_{1,1} = ... = w_{1,n} = 1/N$
3.	For $t = 1, 2, ...T$ do:
a.	Train a base classifier f_t using weights $w_{t,1}, ..., w_{t,n}$
b.	Compute for each object x_i distance to the decision boundary defined by f_t - $d(x_i)_t$
c.	Transform the distance to score $SC(x_i)_t = \frac{1}{\sigma\sqrt{2\pi}} \exp\left(-\frac{(d(x_i)_t - \mu)^2}{2\sigma^2}\right)$
d.	Update the observations weights: $w_{t+1,i} = [\exp(F_t(x_i)) + \exp(-F_t(x_i))]^{-2}$ where $(F_t(x_i)) = (1/2)\sum_{j=1}^{t} SC(x_i)_t$
3.	Output the final classifier: $\hat{y}(x_i) = sign\left(\sum_{t=1}^{T} F_t(x_i)\right).$

The proposed transformation function is based on the normal distribution. We assume that the objects closest or farthest located from the decision boundary of the basic classifier have the lowest value of the scoring function. In addition, we assume that the base (weak) classifier forms a linear decision boundary.

An example of such a weak classifier is recursive partitioning with one split (especially ID3 algorithm with one split). This type of a base classifier was used in experimental research.

4   Experiments

The main aim of the experiments was to compare the proposed modification LogitBoostM algorithm with original one. In the experiments we use the implementation of LogitBoost algorithm described in [2]. The results are obtained via 10-fold-cross-validation method. In the experiments, the value of the parameter μ was tested. The σ parameter was always equal to 5.

In the experimental research we use 4 publicly available binary data sets from UCI and KEEL repository. Table 4 presents the properties of the data sets which we used in experiments. For all data sets the feature selection process [6, 10, 12] was performed to indicate four most informative features.

Table 4. Properties of the data sets used in the experiments.

Data sets	Features	Classes	Observations
Cancer	8	2	699
Pima	8	2	768
Sonar	60	2	208
Wdbc	30	2	569

Results for the fifty iterations of LogitBoost and the proposed modification LogitBoostM algorithms are shown in Figs. 1, 2, 3, 4, 5, 6, 7 and 8. The results were presented for two measures of classification quality – classification error and kappa statistic. For LogitBoostM algorithm the value of the parameter μ has been changed $\mu = 0.25, \mu = 0.5, \mu = 0.75$. Other parameters have not changed.

The obtained results indicate that the proposed algorithm can improve the quality of the classification in comparison to LogitBoost algorithm. In particular, this applies to two data sets (Pima and Sonar). The improvement for Sonar data set concerns two measures of classification quality, and one for Pima data set. The improvement of the quality of the classification is noticeable in the later iterations of the algorithm, about 15 for Pima data set and 35 for Sonar data set. It should also be noted that the proposed algorithm in the last iterations achieves results similar to the LogitBoost algorithm. Additionally, in the range of μ parameter values tested, no significant differences in the obtained results were noticed.

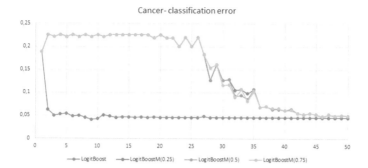

Fig. 1. The error for the Cancer data set.

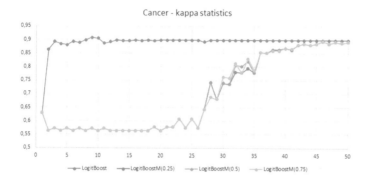

Fig. 2. The kappa statistics for the Cancer data set.

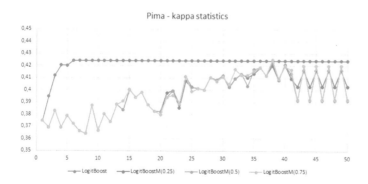

Fig. 3. The error for the Pima data set.

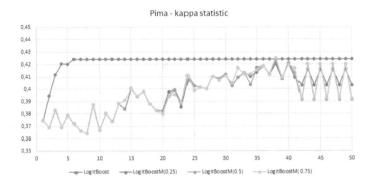

Fig. 4. The kappa statistics for the Pima data set.

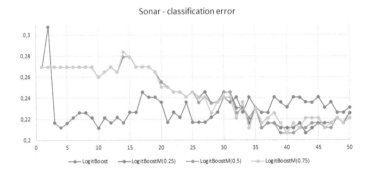

Fig. 5. The error for the Sonar data set.

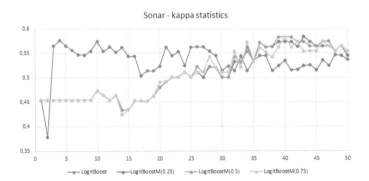

Fig. 6. The kappa statistics for the Sonar data set.

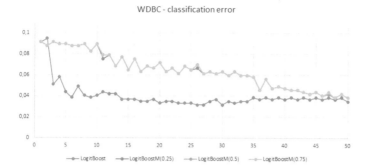

Fig. 7. The error for the Parkinson data set.

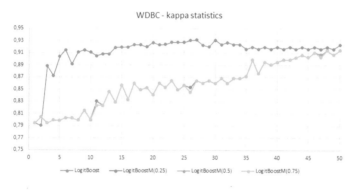

Fig. 8. The kappa statistics for the Parkinson data set.

5   Conclusions

In this paper we presented the modification of LogitBoost algorithm. In the proposed modification the score function depends on the distance from the decision boundary defined by each base (weak) classifier. The proposed score function has no probabilistic interpretation. However, it determines the assurance of the object belonging to a class label that depends on the distance of the object from the decision boundary.

In this paper we focus on one parameter μ of the proposed method. In our future work, we want to extend our analyses to other functions which transform the distance of the object from the decision boundary to the score function and apply for diagnostic of multiple myeloma [13]. In addition, we will adopt the proposed score function to more kinds of boosting algorithms such as Real AdaBoost.

Acknowledgments. This work was supported in part by the National Science Centre, Poland under the grant no. 2017/25/B/ST6/01750.

References

1. Burduk, R.: The AdaBoost algorithm with the imprecision determine the weights of the observations. In: Asian Conference on Intelligent Information and Database Systems, pp. 110–116. Springer (2014)
2. Dmitrienko, A., Chuang-Stein, C., D'Agostino, R.B.: Pharmaceutical Statistics Using SAS: A Practical Guide. SAS Institute (2007)
3. Frejlichowski, D., Gościewska, K., Forczmański, P., Nowosielski, A., Hofman, R.: Applying image features and AdaBoost classification for vehicle detection in the SM4Public system. In: Image Processing and Communications Challenges 7, pp. 81–88. Springer (2016)
4. Freund, Y., Schapire, R.E.: A decision-theoretic generalization of on-line learning and an application to boosting. J. Comput. Syst. Sci. **55**(1), 119–139 (1997)
5. Freund, Y., Schapire, R.E., et al.: Experiments with a new boosting algorithm. In: ICML, vol. 96, pp. 148–156. Citeseer (1996)
6. Guyon, I., Elisseeff, A.: An introduction to variable and feature selection. J. Mach. Learn. Res. **3**(Mar), 1157–1182 (2003)
7. Kearns, M., Valiant, L.: Cryptographic limitations on learning boolean formulae and finite automata. J. ACM (JACM) **41**(1), 67–95 (1994)
8. Kozik, R., Choraś, M.: The http content segmentation method combined with AdaBoost classifier for web-layer anomaly detection system. In: International Joint Conference SOCO 2016-CISIS 2016-ICEUTE 2016, pp. 555–563. Springer (2016)
9. Oza, N.C.: Boosting with averaged weight vectors. In: International Workshop on Multiple Classifier Systems, pp. 15–24. Springer (2003)
10. Rejer, I.: Genetic algorithms for feature selection for brain-computer interface. Int. J. Pattern Recognit Artif Intell. **29**(05), 1559008 (2015)
11. Shen, C., Li, H.: On the dual formulation of boosting algorithms. IEEE Transact. Pattern Anal. Mach. Intell. **32**(12), 2216–2231 (2010)
12. Szenkovits, A., Meszlényi, R., Buza, K., Gaskó, N., Lung, R.I., Suciu, M.: Feature selection with a genetic algorithm for classification of brain imaging data. In: Advances in Feature Selection for Data and Pattern Recognition, pp. 185–202. Springer (2018)
13. Topolski, M.: Algorithm of multidimensional analysis of main features of PCA with blurry observation of facility features detection of carcinoma cells multiple myeloma. In: International Conference on Computer Recognition Systems, pp. 286–294. Springer (2019)
14. Wozniak, M.: Proposition of boosting algorithm for probabilistic decision support system. In: International Conference on Computational Science, pp. 675–678. Springer (2004)

Color Normalization-Based Nuclei Detection in Images of Hematoxylin and Eosin-Stained Multi Organ Tissues

Adam Piórkowski[✉] [ORCID]

Department of Biocybernetics and Biomedical Engineering, AGH University of Science and Technology, A. Mickiewicza 30 Av., 30–059 Cracow, Poland
pioro@agh.edu.pl

Abstract. This article presents an adaptation of a color transfer method for binarization of cell nuclei in Hematoxylin- and Eosin-stained microscopy images. The aim is to check the ability and accuracy of nuclei detection for multi-organ cases using a public dataset. The results are obtained using the Monte Carlo method and then compared to the ground truth segmentations. Some cases are presented in detail and discussed. This method seems to be promising for the further development of classic and deterministic algorithms for H&E nuclei detection.

Keywords: Nuclei detection · Nuclei segmentation · Cell counting · Color transfer · Multi-organ tissue · Hematoxylin and Eosin staining

1 Introduction

Staining with hematoxylin and eosin (H&E) is the most popular method in histopathology. Although this is a very common task, it has still not been fully automated. One of the occurring subproblems is the correct segmentation of cell nuclei.

Classic image processing methods do not always give satisfactory and reproducible results [22]. Many different solutions have been created, although they are limited to the tissues of a given organ [4,7,9,14,17,25].

There are no standardized protocols for H&E staining. This results in significant differences in coloration of specimens, which causes problems with the reproducibility of segmentation. Some methods put more emphasis on color normalization in images based on deconvolution of hematoxylin and eosin [5]. Adaptation of different conditions into a unified space enables robust quantitative tissue analysis [6]. Reinhard's [19], Macenko's [13], and Li's [12] color normalization methods are treated as the standard approaches [1,2,11]. In [21] a deconvolution for hematoxylin channel was presented. State-of-the-art color normalization methods can be found in [16].

This article focuses on segmentation and detection of cell nuclei based on the original method and is a continuation of previous works [15,18]. The focus

© Springer Nature Switzerland AG 2020
M. Choraś and R. S. Choraś (Eds.): IP&C 2019, AISC 1062, pp. 57–64, 2020.
https://doi.org/10.1007/978-3-030-31254-1_8

was on the evaluation of a method used for multi-organ tissues, presented in a standardized, public dataset [10]. Further considerations, such as separation of clustered objects [3,8] or the issue of segmentation accuracy [17,20], will be taken into account in further works. Also other color spaces can be considered [23,24].

2   Modified Color Transfer Transformation

The standard color transfer procedure, presented in [19], assumes transformations from RGB space to Lab space using logarithmic operations. The author of the modified method presented in this article decided not to use logarithmic operations because such a simplification increases the sensitivity and precision of the transformations presented below; it is also simpler to implement and much more efficient. In this case, the $LinearLab$ (LL, La, Lb) space can be calculated using combined transformation with matrix multiplication (1).

$$\begin{bmatrix} LL \\ La \\ Lb \end{bmatrix} = \begin{bmatrix} 0.3475 & 0.8265 & 0.5559 \\ 0.2162 & 0.4267 & -0.6411 \\ 0.1304 & -0.1033 & -0.0269 \end{bmatrix} \begin{bmatrix} R \\ G \\ B \end{bmatrix} \quad (1)$$

Subsequently, the standard color normalization in Lab space is conducted (2) as described in [19], using mean values $(\overline{LL}, \overline{La}, \overline{Lb})$ and standard deviations of channels (σ_s - source and σ_t - target).

$$LL\prime = \frac{\sigma_t^{LL}}{\sigma_s^{LL}}(LL - \overline{LL})$$

$$La\prime = \frac{\sigma_t^{La}}{\sigma_s^{La}}(La - \overline{La}) \quad (2)$$

$$Lb\prime = \frac{\sigma_t^{Lb}}{\sigma_s^{Lb}}(Lb - \overline{Lb})$$

Finally, the space transformation back to the new RGB is performed using the combined matrix for multiplication (3).

$$\begin{bmatrix} R\prime \\ G\prime \\ B\prime \end{bmatrix} = \begin{bmatrix} 0.5773 & 0.2621 & 5.6959 \\ 0.5773 & 0.6071 & -2.5452 \\ 0.5833 & -1.0628 & 0.2076 \end{bmatrix} \begin{bmatrix} LL\prime \\ La\prime \\ Lb\prime \end{bmatrix} \quad (3)$$

To carry out a valid $LLab$ to RGB transformation, the RGB values should be trimmed to the 0–255 range (4). This indirect binarization causes backgrounds or structures to disappear; an example is shown in Fig. 1, where the nuclei area is unified in blue, and the background or other structures are shown in yellow or pink.

$$R = \begin{cases} 0, & R < 0 \\ R, & R \in <0,255> \\ 255, & R > 255 \end{cases}, G = \begin{cases} 0, & G < 0 \\ G, & G \in <0,255> \\ 255, & G > 255 \end{cases}, B = \begin{cases} 0, & B < 0 \\ B, & B \in <0,255> \\ 255, & B > 255 \end{cases}.$$

$$(4)$$

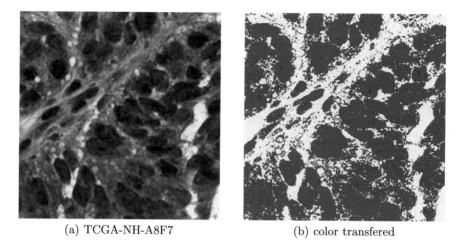

(a) TCGA-NH-A8F7   (b) color transfered

Fig. 1. Indirect binarization - removal of selected structures.

3   Nuclei Detection

To detect cell nuclei, the binarization is performed on the converted color image using the formula (5) [18]. The color space parameters are chosen based on the determination of a set of conversion parameters which for a binarized image give the most accurate mask coverage, as described in the formula (6), taking into account pixel-by-pixel comparison. The design of such a formula, based on the standard concept of $ACCURACY$, is dependent of the density of the detected cell nuclei within the entire image, therefore another approach (Jaccard formula, 7) can be also tested.

$$V_3(x,y) = \begin{cases} 0, R(x,y) > 0 \\ 1, R(x,y) = 0 \end{cases} \tag{5}$$

$$ACC = \frac{TP + TN}{TP + TN + FN + FP} \tag{6}$$

$$JACC = \frac{TP}{TP + FN + FP} \tag{7}$$

In order to determine the best possible matches, for each image of the dataset 150,000 random target value sets $(\overline{R}, \sigma_t^{LL}, \overline{G}, \sigma_t^{La}, \overline{B}, \sigma_t^{Lb})$ of normalization parameters $(\overline{LL}, \sigma_t^{LL}, \overline{La}, \sigma_t^{La}, \overline{Lb}, \sigma_t^{Lb})$ were generated and tested. The ranges for parameters are presented below.

- $\overline{LL}$ - LL calculated for target $(\overline{R}, \overline{G}, \overline{B})$ with range of 0–255,
- σ_t^{LL} - LL standard deviation - 1–1000,
- $\overline{La}$ - La calculated for target $(\overline{R}, \overline{G}, \overline{B})$ with range of 0–255,
- σ_t^{La} - La standard deviation - 1–1000,

- $\overline{Lb}$ - Lb calculated for target $(\overline{R}, \overline{G}, \overline{B})$ with range of 0–255,
- σ_t^{Lb} - Lb standard deviation - 1–1000.

The results are presented in Table 1.

Table 1. The best JACC values and their color transfer parameters for all pictures of the dataset. The best ACC values are also presented.

Label	localization	type	$\overline{R}$	σ_t^{LL}	$\overline{G}$	σ_t^{La}	$\overline{B}$	σ_t^{Lb}	ACC	JACC
TCGA-KB-A93J-01A-01-TS1	Stomach	Stomach adenocarcinoma	195	474	165	3	132	2	0.9170	0.7752
TCGA-RD-A8N9-01A-01-TS1	Stomach	Stomach adenocarcinoma	218	457	77	6	219	4	0.9214	0.7612
TCGA-DK-A2I6-01A-01-TS1	Bladder	Bladder Urothelial Carc.	247	530	76	36	28	11	0.9314	0.7460
TCGA-18-5592-01Z-00-DX1	Liver	Lung squamous cell carcinoma	64	886	161	139	118	117	0.7202	0.7310
TCGA-G2-A2EK-01A-02-TSB	Bladder	Bladder Urothelial Carc.	224	270	72	35	55	16	0.9510	0.6452
TCGA-AR-A1AS-01Z-00-DX1	Breast	Breast invasive carcinoma	201	492	29	19	106	2	0.8942	0.6436
TCGA-B0-5710-01Z-00-DX1	Kidney	Kidney renal clear cell carc.	186	160	92	201	97	12	0.9566	0.6318
TCGA-NH-A8F7-01A-01-TS1	Colon	Colon adenocarcinoma	195	456	61	45	194	20	0.8705	0.6192
TCGA-AY-A8YK-01A-01-TS1	Colon	Colon adenocarcinoma	251	584	218	51	92	16	0.8801	0.6149
TCGA-A7-A13F-01Z-00-DX1	Breast	Breast invasive carcinoma	236	404	86	0	113	0	0.9206	0.5924
TCGA-HE-7130-01Z-00-DX1	Kidney	Kidney renal papillary cell carc.	222	255	29	670	57	45	0.8569	0.5831
TCGA-49-4488-01Z-00-DX1	Liver	Lung adenocarcinoma	196	205	5	464	125	48	0.8264	0.5745
TCGA-A7-A13E-01Z-00-DX1	Breast	Breast invasive carcinoma	247	413	83	1	101	0	0.9177	0.5659
TCGA-21-5786-01Z-00-DX1	Liver	Lung squamous cell carcinoma	123	803	69	14	50	23	0.8347	0.5605
TCGA-B0-5711-01Z-00-DX1	Kidney	Kidney renal clear cell carc.	211	199	109	182	100	5	0.9425	0.5571
TCGA-B0-5698-01Z-00-DX1	Kidney	Kidney renal clear cell carc.	187	182	191	122	88	15	0.9424	0.5489
TCGA-G9-6356-01Z-00-DX1	Prostate	Prostate adenocarcinoma	249	284	74	612	0	18	0.8993	0.5435
TCGA-G9-6336-01Z-00-DX1	Prostate	Prostate adenocarcinoma	67	140	149	712	2	34	0.8051	0.5174
TCGA-AR-A1AK-01Z-00-DX1	Breast	Breast invasive carcinoma	223	365	127	72	103	36	0.8679	0.5123
TCGA-50-5931-01Z-00-DX1	Liver	Lung adenocarcinoma	4	670	170	166	163	47	0.7541	0.5050
TCGA-E2-A1B5-01Z-00-DX1	Breast	Breast invasive carcinoma	39	40	0	53	124	1	0.9235	0.4848
TCGA-21-5784-01Z-00-DX1	Liver	Lung squamouscell carcinoma	110	205	217	5	209	21	0.9172	0.4824
TCGA-E2-A14V-01Z-00-DX1	Breast	Breast invasive carcinoma	224	371	55	526	104	11	0.8752	0.4778
TCGA-38-6178-01Z-00-DX1	Liver	Lung adenocarcinoma	238	403	24	388	45	38	0.8571	0.4664
TCGA-G9-6362-01Z-00-DX1	Prostate	Prostate adenocarcinoma	168	436	34	763	187	54	0.8170	0.4627
TCGA-HE-7128-01Z-00-DX1	Kidney	Kidney renal papillary cell carc.	255	370	0	137	68	15	0.9262	0.4516
TCGA-HE-7129-01Z-00-DX1	Kidney	Kidney renal papillary cell carc.	217	383	75	71	47	43	0.8642	0.4374
TCGA-G9-6348-01Z-00-DX1	Prostate	Prostate adenocarcinoma	78	56	1	557	0	14	0.8499	0.4280
TCGA-CH-5767-01Z-00-DX1	Prostate	Prostate adenocarcinoma	205	331	75	856	221	48	0.8528	0.4154
TCGA-G9-6363-01Z-00-DX1	Prostate	Prostate adenocarcinoma	93	246	4	787	84	27	0.8299	0.4027

4 Conclusions

The presented approach produced rather good results for H&E images in the case of non-damaged cell nuclei that were not involved in the neoplastic process (Fig. 2). The accuracy of the proposed method is very high for a full pixel-by-pixel comparison of objects and backgrounds, as well as for comparison of objects themselves. Unfortunately, for the open chromatin nucleus, this method is unable to generate a full closed cell nucleus mask, which results in poorer outcomes when using the proposed comparison methods (Fig. 3). As a separate observation from the presented research, it can be pointed out that the use of a standard assessment approach based on the accuracy formula (6) fails, and there is a need to reduce the influence of the background area size (7) in order to precisely detect nuclei.

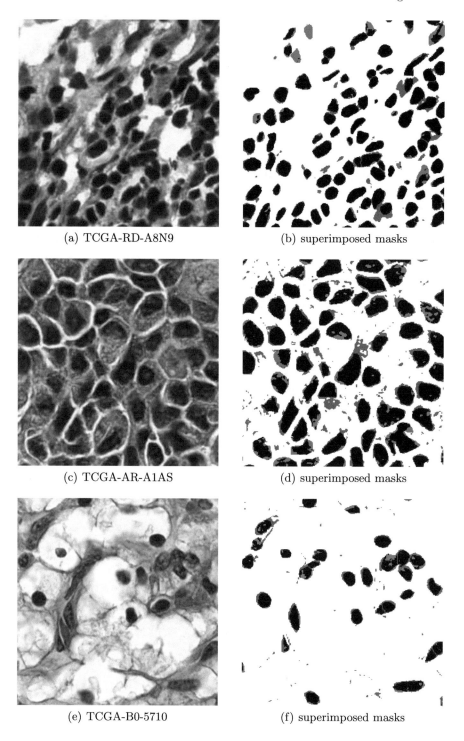

(a) TCGA-RD-A8N9

(b) superimposed masks

(c) TCGA-AR-A1AS

(d) superimposed masks

(e) TCGA-B0-5710

(f) superimposed masks

Fig. 2. The best matches for images with fully filled cell nuclei.

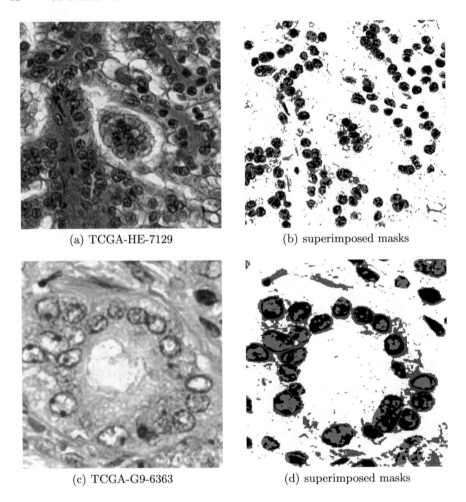

(a) TCGA-HE-7129

(b) superimposed masks

(c) TCGA-G9-6363

(d) superimposed masks

Fig. 3. The worst matches for images with open chromatin nuclei.

The proposed method will be further developed and also applied in other issues of color image segmentation. The preliminary results of the research confirm its benefits. A sample algorithm demo is available on the site: http://home.agh.edu.pl/pioro/hecell/.

Acknowledgments. This research was funded by AGH University of Science and Technology as a research project No. 16.16.120.773.

References

1. Bautista, P., Hashimoto, N., Yagi, Y.: Color standardization in whole slide imaging using a color calibration slide. J. Pathol. Inform. **5**(1), 4 (2014). https://doi.org/10.4103/2153-3539.126153

2. Ing, N., Salman, S., Ma, Z., Walts, A., Knudsen, B., Gertych, A.: Machine learning can reliably distinguish histological patterns of micropapillary and solid lung adenocarcinomas. In: Conference of Information Technologies in Biomedicine. pp. 193–206. Springer (2016)

3. Iwaszenko, S., Nurzynska, K.: Rock grains segmentation using curvilinear structures based features. In: Real-Time Image Processing and Deep Learning 2019, vol. 10996 (2019). https://doi.org/10.1117/12.2519580

4. Kłeczek, P., Dyduch, G., Jaworek-Korjakowska, J., Tadeusiewicz, R.: Automated epidermis segmentation in histopathological images of human skin stained with hematoxylin and eosin. In: Medical Imaging 2017: Digital Pathology, vol. 10140, p. 101400M. International Society for Optics and Photonics (2017)

5. Kłeczek, P., Mól, S., Jaworek-Korjakowska, J.: The accuracy of H&E stain unmixing techniques when estimating relative stain concentrations. In: Polish Conference on Biocybernetics and Biomedical Engineering, PCBBE 2017. AISC, vol. 647, pp. 87–97. Springer (2017)

6. Korzynska, A., Roszkowiak, L., Pijanowska, D., Kozlowski, W., Markiewicz, T.: The influence of the microscope lamp filament colour temperature on the process of digital images of histological slides acquisition standardization. Diagn. Pathol. **9**(1), S13 (2014)

7. Kowal, M., Filipczuk, P., Obuchowicz, A., Korbicz, J., Monczak, R.: Computer-aided diagnosis of breast cancer based on fine needle biopsy microscopic images. Comput. Biol. Med. **43**(10), 1563–1572 (2013)

8. Kowal, M., Korbicz, J.: Marked point process for nuclei detection in breast cancer microscopic images. In: Polish Conference on Biocybernetics and Biomedical Engineering, PCBBE 2017. AISC, vol. 647, pp. 230–241. Springer (2018)

9. Kowal, M., Skobel, M., Nowicki, N.: The feature selection problem in computer-assisted cytology. Int. J. Appl. Math. Comput. Sci. **28**(4), 759–770 (2018)

10. Kumar, N., Verma, R., Sharma, S., Bhargava, S., Vahadane, A., Sethi, A.: A dataset and a technique for generalized nuclear segmentation for computational pathology. IEEE Transact. Med. Imaging **36**(7), 1550–1560 (2017)

11. Li, J., Speier, W., Ho, K.C., Sarma, K.V., Gertych, A., Knudsen, B.S., Arnold, C.W.: An em-based semi-supervised deep learning approach for semantic segmentation of histopathological images from radical prostatectomies. Comput. Med. Imaging Graph. **69**, 125–133 (2018)

12. Li, X., Plataniotis, K.N.: A complete color normalization approach to histopathology images using color cues computed from saturation-weighted statistics. IEEE Transact. Biomed. Eng. **62**(7), 1862–1873 (2015)

13. Macenko, M., Niethammer, M., Marron, J., Borland, D., Woosley, J.T., Guan, X., Schmitt, C., Thomas, N.E.: A method for normalizing histology slides for quantitative analysis. In: IEEE International Symposium on Biomedical Imaging, ISBI 2009, pp. 1107–1110. IEEE (2009)

14. Mazurek, P., Oszutowska-Mazurek, D.: From the Slit-Island method to the Ising model: analysis of irregular grayscale objects. Int. J. Appl. Math. Comput. Sci. **24**(1), 49–63 (2014)

15. Nurzynska, K.: Optimal parameter search for colour normalization aiding cell nuclei segmentation. CCIS, vol. 928, pp. 349–360. Springer (2018)

16. Onder, D., Zengin, S., Sarioglu, S.: A review on color normalization and color deconvolution methods in histopathology. Appl. Immunohistochem. Mol. Morphol. **22**(10), 713–719 (2014)

17. Oszutowska-Mazurek, D., Mazurek, P., Parafiniuk, M., Stachowicz, A.: Method-induced errors in fractal analysis of lung microscopic images segmented with the use of histaenn (histogram-based autoencoder neural network). Appl. Sci. **8**(12), 2356 (2018)
18. Piorkowski, A., Gertych, A.: Color normalization approach to adjust nuclei segmentation in images of hematoxylin and eosin stained tissue. In: Information Technology in Biomedicine. ITIB 2018. AISC, vol. 762, pp. 393–406. Springer (2019)
19. Reinhard, E., Adhikhmin, M., Gooch, B., Shirley, P.: Color transfer between images. IEEE Comput. Graph. Appl. **21**(5), 34–41 (2001)
20. Roszkowiak, L., Korzyńska, A., Siemion, K., Pijanowska, D.: The influence of object refining in digital pathology. In: Image Processing and Communications Challenges 10. IP&C 2018. AISC, vol. 892, pp. 55–62. Springer (2019)
21. Ruifrok, A.C., Johnston, D.A., et al.: Quantification of histochemical staining by color deconvolution. Anal. Quant. Cytol. Histol. **23**(4), 291–299 (2001)
22. Salvi, M., Molinari, F.: Multi-tissue and multi-scale approach for nuclei segmentation in H&E stained images. Biomed. Eng. online **17**(1), 89 (2018)
23. Starosolski, R.: New simple and efficient color space transformations for lossless image compression. J. Vis. Commun. Image Represent. **25**(5), 1056–1063 (2014)
24. Starosolski, R.: Human visual system inspired color space transform in lossy JPEG 2000 and JPEG XR compression. In: International Conference: Beyond Databases, Architectures and Structures, pp. 564–575. Springer (2017)
25. Tosta, T.A.A., Neves, L.A., do Nascimento, M.Z., Segmentation methods of h&e-stained histological images of lymphoma: Segmentation methods of H&E-stained histological images of lymphoma: a review. Inf. Med. **9**, 35–43 (2017)

Algorithm for Finding Minimal and Quaziminimal st-Cuts in Graph

Andrey Grishkevich$^{(\boxtimes)}$

Kazimierz Wielki University, Chodkiewicza 30, 85-064 Bydgoszcz, Poland
`grishkev.gmai@gmail.com`

Abstract. This paper presents an algorithm for enumerating the set of minimal edge st-cuts of an oriented graph based on search for the set of undecomposable cuts in a graph and on the entire set of minimum cuts synthesis build upon a subset of solely undecomposable cuts in the distributive lattice of minimum cuts. The method of listing the quasi-minimal (following the minimal, closest to minimal) cuts of the graph is considered.

1 Introduction

Algorithms for the minimum cuts (maximum flows) search, which are considered to be the classical problem of combinatorial analysis, are well studied theoretically. The development of computer programs and their parallel-based counterparts that are highly efficient on high-dimensional graphs [1–3] has stimulated application of minimal cuts search algorithms in vision and imaging technology such as image and video segmentation, co-segmentation, stereo vision, and multiview reconstruction [4,5]. The network cuts can also be used for the assessment of the complex network reliability. Under different circumstances, including cases when element reliabilities are high, cut-based reliability evaluation can provide more accurate results [6].

The number of cuts of a graph is exponential with respect to the number of vertices of the graph; therefore, their enumeration and usage is rather demanding, especially for large graphs. In real applications not all of the cuts are used, but rather minimal cuts and certain cuts that are close to minimal [7], or even a disjoint cut from this set [8].

In this paper, it is proposed to reduce the search on a graph by searching only for a certain subset of undecomposable minimal cuts, the number of which is limited by the number of vertices of the graph (reducing the complexity of the algorithm by reducing the complexity of the search in the graph, what is very important for large graphs). The entire set of minimal cuts or some of its subsets can be synthesized from a subset of undecomposable cuts based on algebraic operations in the distributive lattice of minimal cuts. This approach can also be used to search for quaziminimal cuts, treating them as minimal cuts in the modified graph.

M. Choraś and R. S. Choraś (Eds.): IP&C 2019, AISC 1062, pp. 65–72, 2020.
https://doi.org/10.1007/978-3-030-31254-1_9

2   Distributive Lattice of Minimal Edge st-Cuts of an Oriented Graph

Let $G(V, E)$ be a connected directed graph, where $V = \{v\}$ the set of vertices of the graph, $E = \{e = (i, j) : i, j \in V\}$ the set of directed edges of the graph. Let us select two vertices the source s and the sink t $(s, t \in V, s \neq t)$ in the graph G. Let A, B $(A \cap B = \emptyset)$ be some subsets of the vertex set. Let us denote

$$(A, B) = \{(i, j) : (i, j) \in E, i \in A, j \in B\}$$

the set of oriented arcs leading from $i \in A$ to $j \in B$. Additionally let us suppose that, firstly, there is at most one oriented arc $(i, j) \in E$ and one oriented arc $(j, i) \in E$ between any two vertices $i, j \in V$, and, secondly, there are no loops (that is, arcs of the form $(i, i) \in E$).

The cut [9], separating the vertices s, t of the graph G, is the set of arcs

$$r = (R, \bar{R}) \subseteq E,$$

where

$$R \cap \bar{R} = \emptyset, R \cup \bar{R} = V, s \in R, t \in \bar{R}.$$

The set of all such cuts is denoted by $\mathcal{R}$. Each edge $e \in E$ of the graph G is associated with a nonnegative number $c(e) \geq 0$, which is called the weight (throughput) of the edge. The throughput (weight) of the cut $r \in \mathcal{R}$ is determined by

$$c(r) = c(R, \bar{R}) = \sum_{e \in (R, \bar{R})} c(e). \tag{1}$$

In the set of cuts $\mathcal{R}$ of graph G a subset of minimal cuts (cuts of minimal weight) can be distinguished with respect to the weight function c

$$\mathcal{M}_{min,c} = \{m : m = \arg\min_{r \in \mathcal{R}} c(r)\}. \tag{2}$$

On set $\mathcal{M}_{min,c}$ binary operations $\vee, \wedge$ are defined. For any

$$m_i = (M_i, \overline{M_i}) \in \mathcal{M}_{min,c}, \ i = 1, 2,$$

we put

$$m_1 \vee m_2 = (M_1 \cup M_2, \overline{M_1 \cup M_2}),$$

$$m_1 \wedge m_2 = (M_1 \cap M_2, \overline{M_1 \cap M_2}).$$

Known [10]

$$m_1 \vee m_2, m_1 \wedge m_2 \in \mathcal{M}_{min,c}.$$

The set of the minimal cuts with defined on it operations is a distributive lattice $< \mathcal{M}_{min,c}; \vee, \wedge >$ [11].

The minimal cut $p \in \mathcal{M}_{min,c}$ of a distributive lattice is undecomposable ($\vee$-undecomposable), if for any relation $p = m_1 \vee m_2$ follows $p = m_1$ or $p = m_2$ [12].

Theorem 1. *Let $v \in V$, $\mathcal{M}_v = \{m = (M, \overline{M}) : m \in \mathcal{M}_{min,c}, v \in M\}$. If $\mathcal{M}_v \neq \emptyset$, then*

$$p = \bigwedge_{m \in \mathcal{M}_v} m \tag{3}$$

is a undecomposable cut.

We shall designate as $\mathcal{P}_c$ the set of undecomposable cuts of the lattice $< \mathcal{M}_{min,c}; \vee, \wedge >$. It is obvious, that $\mathcal{P}_c$ is a partially ordered set and it is a subset of partially ordered set $\mathcal{M}_{min,c}$. The set of distributive lattice of the minimal cuts in graph can be analytically described [11]

$$\mathcal{M}_{min,c} = \bigcup_{A \in \mathcal{A}(\mathcal{P}_c)} \bigvee_{a \in A} a, \tag{4}$$

where $\mathcal{A}(\mathcal{P}_c)$ - set of antichains A of partially ordered set $\mathcal{P}_c = \bigcup_i P_i$.
The following estimates

$$|\mathcal{P}_c| \leq |V| - 1,$$

$$|\mathcal{P}_c| \leq |\mathcal{M}_{min,c}| \leq 2^{|\mathcal{P}_c|}$$

are valid for the presentation (4).

3   Algorithm for Enumeration of Minimal Edge st-Cuts of a Graph in a Distributive Lattice of Minimum Cuts

The specified representation (4) shown above, forms a basis for a new original decomposition approach to finding minimal cuts in graph. It consist first, from searching of only undecomposable minimal cuts in graph and secondly, from synthesis in distributive lattice of minimal cuts of all set of minimal cuts on partially ordered subset of undecomposable cuts. The offered approach allows to reduce search in graph (number of graph's connection checks) due to allocation of only subsets of undecomposable minimal cuts.

The following is introduced: a set of labelled vertices M, flow f from source s to sink t in arc-node form, the current value of the capacity of the edge $\delta(e)$.

The algorithm is a stage in singling the set of undecomposable minimum cuts $\mathcal{P} \subseteq \mathcal{M}$ of graph $G(V, E)$.

Step 1. Place $\forall e \in E$ $\delta(e) = c(e)$, $\mathcal{P} = \emptyset$.
Step 2. Construct maximum flow.
Step 3. Put flow f in arc-path form [9] and obtain the set of chain

$$\mathcal{D} = \{D_i : i = 1, 2, \ldots, l\}.$$

Step 4. If $\mathcal{D} = \emptyset$, then go to step 5. Otherwise, select chain $D_i = D_i(A_i, U_i)$, $D_i \in \mathcal{D}$. Place $\mathcal{D} = \mathcal{D} \setminus D_i$.
Step 5. Apply the Ford, Fulkerson labelling procedure [10]. If $t \in M$, then for $\forall u \in U_i$ place $\delta(u) = c(u)$ and go to step 4. Otherwise, we obtain cut $m = (M, \overline{M})$.

Step 6. Remember minimum cut

$$P_i = P_i \bigcup (M, \overline{M}).$$

For $u = U_i \bigcap (M, \overline{M})$, place $\delta(u) = \infty$. Go to step 5.

Step 7. Place

$$\mathcal{M} = \mathcal{P} = \bigcup_{i=1}^{l} P_i.$$

In the circle consisting of steps 5–6, the labelled set constructed during the previous iteration may be used. This has been omitted in writing the algorithm to simplify the text. Steps 5–6 can be performed in parallel.

The algorithm proposed for singling out the undecomposable cut set for the directed graph $G(V, E)$ has a time complexity $O(\sigma(|V|, |E|) + l|E|)$, where $\sigma(|V|, |E|)$ is the time consuming tendency of step 2 (of the procedure for finding maximum flow); $l|E|$ is the complexity of steps 3 and 4–6.

Representation of the flow in arc-path form may be done in such a way so that the inequality $l \leq |E|$ is satisfied.

The correctness of the undecomposable minimal cuts sets search algorithm follows from the Theorem 1.

The following algorithm is a stage in the synthesis of cut set $\mathcal{M} \backslash \mathcal{P}$.

Step 8. Find

$$\circ = \bigwedge_{p \in \mathcal{P}} p,$$

the zero element of the minimum cut distributive lattice. Develop the set

$$\mathcal{P} \backslash \circ = \bigcup_{i=1}^{i=d} C_i$$

in chain C_i, where

$$\forall i \neq j \Rightarrow C_i \bigcap C_j = \emptyset;$$

$$\forall p_1, p_2 \in C_i \Rightarrow p_1 < p_2 \text{ or } p_1 > p_2.$$

Step 9. Construct set

$$\mathcal{T} = \{\alpha : \alpha = \{p_1, p_2, \ldots, p_{d'}\};$$

$$\forall i \ p_i \in \mathcal{P}; 2 \leq d' \leq d;$$

$$\forall i \neq j \ \exists i' \neq j' \text{ are such that } p_i \in C_{i'}; \ p_j \in C_{j'}\}.$$

Step 10. If $\mathcal{T} = \emptyset$, then go to step 13. Otherwise, select $\alpha \in \mathcal{T}$. Place $\mathcal{T} = \mathcal{T} \backslash \alpha$.

Step 11. If α is not an anti-chain ($\exists i \neq j$ are such that $p_i < p_j$ or $p_i > p_j$), then go to step 10.

Step 12. Construct minimum cut

$$m = \bigvee_{p \in \alpha} p.$$

Remember

$$\mathcal{M} = \mathcal{M} \bigcup m.$$

Go to step 10.

Step 13. End.

Evidently, $l \geq d = d(\mathcal{P})$, where $d(\mathcal{P})$ is the number of the Dilworth set $\mathcal{P}$.

Step 9–12 of the algorithm are exhaustive searches in set $\mathcal{T}$. This permits a time complexity to be obtained in the synthesis algorithm for the cut set $\mathcal{M} \backslash \mathcal{P}$ as $O(d|\mathcal{T}|)$.

The search may be reduced by employing, for example, the following condition for reducing the extent of the exhaustive search in set $\mathcal{T}$ when determining the set of anti-chains set $\mathcal{P}$. Let $\alpha = \{b_1, b_2\} \in \mathcal{T}$, $b_1 \in P_1'$, $b_2 \in P_2'$, $b_1 \leq b_2$. Then

(1) $\forall \beta = \{b_1, b_3\}$, $b_3 \in P_2'$, $b_3 \geq b_2$ and $\forall \omega \in \mathcal{T}$, if $\beta \in \omega$, then ω is not an anti-chain;
(2) $\forall \gamma = \{b_4, b_2\}$, $b_4 \in P_1'$, $b_4 \leq b_1$ and $\forall \omega \in \mathcal{T}$, if $\gamma \in \omega$, then ω is not an anti-chain.

4   Enumeration of Quasiminimal st-Cuts of an Oriented Graph

The set of the quasiminimal (following the minimal, closest to minimal) cuts of the graph

$$\mathcal{M}_{min+1,c} = \{\arg \min_{r \in \mathcal{R} \backslash \mathcal{M}_{min,c}} c(r)\}.$$

We are required to list the quasiminimal cuts based on the set of minimal cuts.

Let set

$$S = \{u : u \in U\} \subseteq U,$$

for any

$$m \in \mathcal{M}_{min,c} \Rightarrow m \cap S \neq \emptyset.$$

Consider the function

$$c_S(u) \rightarrow \mathbb{R}^+,$$

which we define as follows

$$c_S(u) = \begin{cases} \infty, & \text{if } u \in S, \\ c(u), & \text{if } u \in U \backslash S. \end{cases} \tag{5}$$

The constructed weight function prohibits all minimal cuts relative to the weight function c, since at least one edge of the specified section has weight ∞. For given S, c_S the set of minimal cuts in the graph G is denoted by $\mathcal{M}_{min,c_S}$.

Consider the set of all possible sets

$$\mathcal{S} = \{S\}. \tag{6}$$

As $\mathcal{S}$ various subsets of set edges

$$E^* = \{e : e \in p, p \in \mathcal{P}\} \tag{7}$$

can be chosen.

The set of quasiminimal cuts is

$$\mathcal{M}_{min+1,c} = \{\arg \min_{r \in \bigcup_{S \in \mathcal{S}} \mathcal{M}_{min,c_S}} c(r)\}. \tag{8}$$

A quasiminimal cut with respect to the weight function c (1) is a minimal cut with respect to the modified weight function c_S (5). Therefore, the algorithm for enumeration of minimal cuts (2) in the distributive lattice of minimal cuts (4) can be used to list quasiminimal cuts (8).

5 Example

Let us list the minimal and quasiminimal cuts of the graph $G(V, E)$ (benchmark directed networks № 10 [6]), where $V = \{s, 1, 2, 3, 4, 5, t\}$, $E = \{1 = (s, 1), 2 = (1, 3), 3 = (1, t), 4 = (s, 2), 5 = (2, 3), 6 = (3, t), 7 = (8, 4), 8 = (2, 4), 9 = (2, 5), 10 = (5, 3), 11 = (5, t), 12 = (4, 5)\}$, $c(e) = 1 \ \forall e \in E$ (Fig. 1).

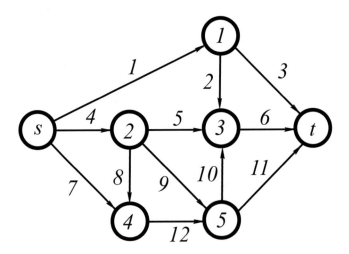

Fig. 1. Graph $G(V, E)$

For the graph presented, three-element cuts are minimal, and four-element cuts are quasiminimal.

Intermediate results for determining the minimum and quasiminimal cuts of the graph presented:

$f(e) = 1, e \in \{1, 3, 4, 5, 6, 7, 11, 12\} \subseteq E,$
$f(e) = 0, e \in \{2, 8, 9, 10\} \subseteq E$ (step 2, maximum flow, $\max(f) = 3$);
$A_1 = \{(s, 1, t)\}, U_1 = \{1, 3\}\},$
$A_2 = \{(s, 2, 3, t)\}, U_2 = \{4, 5, 6\}\},$
$A_3 = \{(s, 4, 5, t)\}, U_3 = \{7, 12, 11\}\},$
$\mathcal{D} = \{D_1(A_1, U_1), D_2(A_2, U_2), D_3(A_3, U_3)\}$ (step 3, flow in arc-path form);
$P_1 = \{(1, 4, 7), (3, 6, 11)\},$
$P_2 = \{(1, 4, 7), (1, 6, 11)\},$
$P_3 = \{(1, 4, 7), (1, 4, 12), (1, 6, 11)\}$ (step 4–6, chain undecomposable minimal cuts);
$\mathcal{P}_c = \mathcal{M}_{min,c} = \{(1, 4, 7), (3, 6, 11), (1, 6, 11), (1, 4, 12)\}$ (step 7, set undecomposable cuts; step 8–12, set minimal cuts);
$E^* = \{1, 3, 4, 6, 7, 11, 12\}$ (Eq. (7), edges of the graph included in the minimum cuts);
$S_1 = \{1, 11\},$
$S_2 = \{7, 12, 6\},$
$S_3 = \{4, 6\},$
$\mathcal{S} = \{S_1, S_2, S_3, ...\}$ (Eq. (6), subsets of set E^*);
$\mathcal{M}_{min, c_{S_1}} = \{(3, 2, 4, 7), (3, 2, 4, 12), (3, 6, 9, 12)\},$
$\mathcal{M}_{min, c_{S_2}} = \{(1, 4, 10, 11), (1, 5, 10, 11)\},$
$\mathcal{M}_{min, c_{S_3}} = \{(1, 5, 9, 12), (1, 5, 10, 11)\},$
$\mathcal{M}_{min+1, c} = \mathcal{M}_{min, c_{S_1}} \bigcup \mathcal{M}_{min, c_{S_2}} \bigcup \mathcal{M}_{min, c_{S_3}}$
$\quad = \{(3, 2, 4, 7), (3, 2, 4, 12), (3, 6, 9, 12),$
$\quad\quad (1, 4, 10, 11), (1, 5, 10, 11), (1, 5, 9, 12)\}$ (Eq. (8), set quasiminimal cuts).

Based on the proposed algorithm, a set of minimal $\mathcal{M}_{min,c}$ and quasiminimal $\mathcal{M}_{min+1,c}$ cuts was constructed for the graph presented on Fig. 1.

6 Conclusions and Future Work

A method for a compact description of the minimal cuts set by a subset of undecomposable minimal cuts, the number of which does not exceed the number of vertices of the graph, is obtained. An efficient algorithm for the search of solely undecomposable minimal graph cuts of complexity

$$O(\max\{\sigma(|V|, |E|), |E|^2\})$$

is proposed. The algorithm of synthesis minimal and quasiminimal cuts based on a subset of undecomposable cuts is considered.

It is of interest to study the conditions that make it possible to reduce the brute-force during the minimal cuts set synthesis for a partially ordered subset of undecomposable minimal cuts for large Dilworth numbers $(d(\mathcal{P}) > 3)$. Another area of research is to obtain the minimum aggregate of the sets $\{S_i\}$, that allow to describe all the quasiminimal cuts.

References

1. Otsukia, K., Kobayashib, Y., Murota, K.: Improved max-flow min-cut algorithms in a circular disk failure model with application to a road network. Eur. J. Oper. Res. **248**, 396–403 (2016)
2. Fishbain, B., Hochbaum, D.S., Mueller, S.: A competitive study of the pseudoflow algorithm for the minimum s-t cut problem in vision applications. J. Real-Time Image Proc. **11**(3), 589–609 (2016)
3. Gianinazzi, L., Kalvoda, P., Palma, A., et al.: Communication-avoiding parallel minimum cuts and connected components. In: ACM Conference Principles and Practice of Parallel Programming 2018, PPoPP 2018 (2018). http://www.unixer.de/~htor/publications
4. Sun, W., Dong, E.: Kullback-Leibler distance and graph cuts based active contour model for local segmentation. Biomed. Sig. Process. Control **52**, 120–127 (2019)
5. Liu, Z., Song, Y.Q., Sheng, V.S., et al.: Liver CT sequence segmentation based with improved U-Net and graph cut. Expert Syst. Appl. **126**, 54–63 (2019)
6. Mishra, R., Saifi, M.A., Chaturvedi, S.K.: Enumeration of minimal cutsets for directed networks with comparative reliability study for paths or cuts. Qual. Reliab. Engng. Int. **32**, 555–565 (2016)
7. Emadi, A., Afrakhte, H.: A novel and fast algorithm for locating minimal cuts up to second order of undirected graphs with multiple sources and sinks. Electr. Power Energy Syst. **62**, 95–102 (2014)
8. Liu, W., Li, J.: An improved cut-based recursive decomposition algorithm for reliability analysis of networks. Earthq. Eng. Eng. Vib. **11**, 1–10 (2012)
9. Ford, L., Fullkerson, D.: Flows in Networks. Princeton University Press, Princeton (1962)
10. Hu, T.: Integer Programming and Network Flows. Addison-Wesley, Reading (1969)
11. Grishkevich, A.A.: Combinatory Methods of Research of Extreme Structures of Mathematical Models of Electric Circuits and Systems. Publishing House JuUrGu, Chelyabinsk (2004)
12. Aigner, M.: Combinatorial Theory. Springer, Heidelberg (1979)

The Influence of the Number of Uses of the Edges of a Reference Graph on the Transmission Properties of the Network Described by the Graph

Beata Marciniak[(⊠)] [iD], Sławomir Bujnowski [iD], Tomasz Marciniak [iD], and Zbigniew Lutowski [iD]

Faculty of Telecommunications, Computer Science and Electrical Engineering, UTP University of Science and Technology, Al. prof. S. Kaliskiego 7, 85-796 Bydgoszcz, Poland
{Beata.Marciniak,Slawomir.Bujnowski,Tomasz.Marciniak, Zbigniew.Lutowski}@utp.edu.pl

Abstract. This paper presents the ability to evaluate the transmission properties of ICT networks modelled by Reference Graphs. To determine the features, the number of uses of individual graph edges in the paths connecting the graph nodes is used. The way of determining the parameter is discussed, and the formulas describing it for the graphs of any degree and number of nodes in their diameter function is given.

Keywords: ICT network · Regular graph · Probability

1 Introduction

The ICT systems designed to transmit data between their users have to provide an adequate quality, speed and reliability [1] While designing such systems it is important not only to choose the hardware installed in the nodes, but also to select the network connections configuration [2–4]. Normally, telecommunication network have a ring structure and optical fibres are used as a transmission medium [5,6]. To lower the investment costs associated with the construction and utilisation of networks, a standard of node equipment is introduced, which make it possible to model the systems with regular graphs [7]. Regularity also plays an important role in the operation of the communication system, eg it allows for more effective detection of anomalies [8]. Their structure can be described with graphs [3,9] where vertices are constituted by telecommunication nodes, and the edges connecting the vertices – by independent transmission channels. The basic parameters influencing the transmission properties of a network are the diameter and the average path length of graphs describing their topologies [10,11]. To evaluate the connection structures objectively, a notion of referential graphs has been defined. They are called Reference Graphs (RG) [12,13].

© Springer Nature Switzerland AG 2020
M. Choraś and R. S. Choraś (Eds.): IP&C 2019, AISC 1062, pp. 73–81, 2020.
https://doi.org/10.1007/978-3-030-31254-1_10

Definition 1. *Reference Graphs are regular structures of a predetermined number of nodes, where a diameter and an average path length determined on the basis of a random source node have equal, theoretically calculated, lower length limits.*

The study showed that despite equal basic values of Reference Graphs consisting of the same number of nodes (i.e. The diameter and average path length), in some cases the transmission parameters are different (the transmission parameter means the probability of rejecting the call for realisation). An example can be eight-node, third degree Reference Graphs RG (8,3) (Figs. 1(a) and (b)), whose diameter is 2 and the average path length is 1.57.

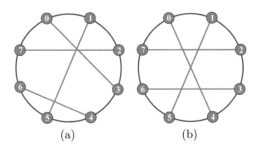

(a)   (b)

Fig. 1. Examples of eight-node Reference Graphs.

Figure 2 shows the results of simulations of the above mentioned graphs referring to the probability of rejecting the call for realisation in the function of traffic generated in the network nodes. The tests were run with a simulator designed by the authors of this paper.

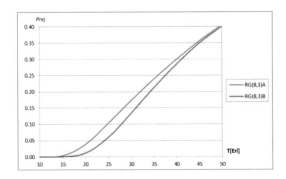

Fig. 2. The chart of the probability of rejecting a service call in the function of the intensity of the generated traffic.

The verification of the reason of the differences of the transmission parameters was completed theoretically by determining the number of uses of specific graph edges to transmit information with the assumption that the graphs are not directed, that is the transmission is possible in two directions. The results are presented in Table 1.

Table 1. The number of uses of individual edges of the analysed graphs.

Graph A				Graph B			
Edge	l		L_e	Edge	l		L_e
	1	2			1	2	
0-1	2	8	10	0-1	2	8	10
0-3	2	8	10	0-4	2	8	10
0-7	2	8	10	0-7	2	8	10
1-2	2	8	10	1-2	2	8	10
1-5	2	8	10	1-5	2	8	10
2-3	2	8	10	2-3	2	8	10
2-7	2	8	10	2-6	2	8	10
3-4	2	8	10	3-4	2	8	10
4-5	2	4	6	3-6	2	8	10
4-6	2	4	6	4-5	2	8	10
5-6	2	4	6	5-6	2	8	10
6-7	2	8	10	6-7	2	8	10

where l means the length of a minimum length path containing the chosen edge, L_e - the number of occurrences of a given edge in the minimum length paths connecting the graph nodes. The minimum length path is the shortest path connecting any two nodes included in the graph.

Table 1 shows that the distribution of use of specific edges in the analysed graphs are different. To explain the reason for this, the example shown in Fig. 3 was used.

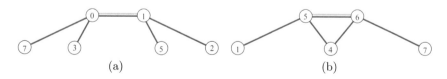

(a)   (b)

Fig. 3. The use of the example edges in minimum length paths included in the analysed graphs.

Figure 3 shows that the number of paths of length 1 is 1 in both cases, and the number of uses of edges **0-1** in minimum length paths $l = 2$ in graph A is 4 (Fig. 3(a)), whereas in graph B of edges **5-6-2** (Fig. 3(b)). It results from the fact that the path connecting the edges **4-5-6** is no a minimum length path. The total numbers of occurrences of analysed edges in minimum length paths are equal 5 and 3 respectively, and when both transmission directions are taken into consideration - 10 and 6. It is consistent with the results presented in Table 1.

More complex structures have also been analysed and their examples are presented in Fig. 4.

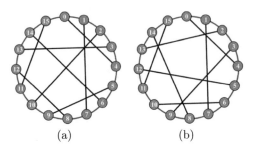

(a)   (b)

Fig. 4. Analysed RG(16,3) graphs.

The graphs have a diameter 3 and the average path length in both cases is $d_{av} = 2.2$. The chart (Fig. 5) shows the simulation results.

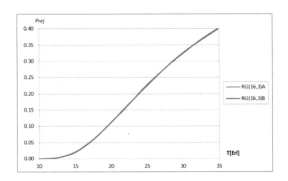

Fig. 5. Simulation results.

Although the graphs are not isomorphic, the measurement results are analogical. The analysis of the totals of uses of all the edges showed that they are equal in both cases and their value is $L = 696$.

The conclusion arises: If the total of uses of all edges in the minimum length paths is equal, the compared RG graphs have the same transmission properties. It was also determined that the number of uses of edges in the paths of length 1 and 2 is the same for every edge - 2 and 8 respectively. The number for the paths of length 3 is different and its values are 12, 18 and 24.

Therefore another conclusion is possible: To evaluate the transmission properties of a network modelled by Reference Graphs it is enough to analyse the distribution of uses of individual edges in the paths of the length equal to the graph's diameter.

2 The Establishment of the Maximum Number of Uses of the Edges of Reference Graphs in Minimum Length Paths

The presented analysis refers to tree structures created on the basis of Reference Graphs. It was concluded that it is impossible to define the maximum number of uses of any edge of the analysed RG graphs on the basis of their diameter and node degree.

The adopted way of conduct is illustrated with the example shown in Fig. 6 where for a virtual graph, whose degree of nodes other than leaves is $d(V) = 3$, a maximum number L_{max} of uses of edged 0-1 was determined in the function of the minimum length paths (for simplicity's sake only one transmission direction was analysed).

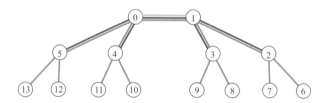

Fig. 6. The illustration of the method of determining the L_{max} value.

Figure 6 shows that if the path length l, which includes edge **0-1**, equals 1 (green) then it is used once; if $l = 2$, the edge occurs *four* time in paths **0-1-2**, **0-1-3**, 5-**0-1**, 4-**0-1** (red); if $l = 3$, the edge occurs *twelve* times (purple): **0-1-2-6**, **0-1-2-7**, **0-1-3-8**, **0-1-3-9**, 4-**0-1**-2, 4-**0-1**-3, 5-**0-1**-2, 5-**0-1**-3, 13-5-**0-1**, 12-5-**0-1**, 11-4-**0-1**, 10-4-**0-1**.

The sum of the numbers of uses of the edge **0-1** is the value L_{max} typical for a tree of a give node degree and path length. In the analysed case it would be: when $l = 1$ then $L_{max} = 1$; if $l = 2$ then $L_{max} = 5$; if $l = 3$ then $L_{max} = 17$. To determine the number of paths of a given length L_l which include the chosen path, for any node degree and graph diameter using the discussed mode of operation, an algorithm has been developed. Its simplified plan is presented in Fig. 7. The arrows mark the paths formed by five edges which include edge a-b and the values written by the nodes are the number of nodes connected by these paths.

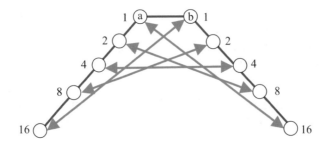

Fig. 7. The rule of determining the L_l value for third degree nodes when the minimum path length is 5.

In general, for any node degree, the number of uses of a selected edge in the function of the length of paths connecting the nodes of the graph is defined by the formula (1).

$$L_{maxl} = l \cdot (d(V) - 1)^{l-1} \tag{1}$$

where l - length of path, $d(V)$ - degree of nodes.

Table 2 shows examples of calculated distributions of use of graph edges in the function of path length and node degree.

Table 2. Distributions of use of graph edges in the function of path length and node degree.

$d(V)$	l	L_{maxl}	$d(V)$	l	L_{maxl}
3	1	1	5	1	1
	2	4		2	8
	3	12		3	48
	4	32		4	256
	5	80		5	1280
	l	$l \cdot 2^{l-1}$		l	$l \cdot 4^{l-1}$
4	1	1	6	1	1
	2	6		2	10
	3	27		3	75
	4	108		4	500
	5	405		5	3125
	l	$l \cdot 3^{l-1}$		l	$l \cdot 5^{l-1}$

Using the results given in Table 2, the total and maximum number of uses of the selected edge L_{max} were given in the function of the size of the diameter of a subgraph constituting a tree created by a Reference Graph with a presupposed node degree. The results of the calculations are presented in Table 3.

Table 3. The maximum number of uses of a selected edge of graph L_{max} in the function of the size of its diameter $(D(G))$ with a presupposed node degree $(d(V))$.

$d(V)$	$D(G)$	$L_{maxD(G)}$	$d(V)$	$D(G)$	$L_{maxD(G)}$
3	1	1	5	1	1
	2	5		7	8
	3	17		34	48
	4	49		142	256
	5	129		547	1280
	k	$2^k(k-1)+1$		k	$k \cdot 3^{k-1} + L_{max(k-1)}$
4	1	1	6	1	1
	9	6		2	11
	57	27		3	86
	313	108		4	586
	1593	405		5	3711
	k	$k \cdot 4^{k-1} + L_{max(k-1)}$		k	$k \cdot 5^{k-1} + L_{max(k-1)}$

In general, the maximum total number of uses of a given edge of a regular degree graph $d(V) > 3$ in the function of its diameter is determined by the recurrence formula (2).

$$L_{maxD(G)} = D(G) \cdot (d(V) - 1)^{k-1} + L_{max(D(G)-1)} \qquad (2)$$

where $D(G))$ - diameter of graph.

The analysis of the results made it possible to define a general formula (3) for polynomials describing the distribution of use of individual edges in the function of the graph's diameter and the degree of its nodes. It allows the determination of the maximum number of uses of a chosen edge included in the paths connecting the nodes.

$$\begin{aligned}
L_{maxD(G)d(V)} = {} & D(G) \cdot (d(V) - 1)^{D(G)-1} + \\
& + (D(G) - 1) \cdot (d(V) - 1)^{D(G)-2} + \\
& + \cdots + 2 \cdot (d(V) - 1) + 1
\end{aligned} \qquad (3)$$

Table 4 shows examples of polynomials describing the distribution of uses of individual graph edges in the function of their diameter for the node degree $d(V) > 2$.

The calculation result obtained thanks to the above polynomials are shown in Table 5.

Table 4. The number of uses of individual edges of the analysed graphs.

$D(G)$	Polynomials
2	$2(d(V) - 1) + 1$
3	$3(d(V) - 1)^2 + 2(d(V) - 1) + 1$
4	$4(d(V) - 1)^3 + 3(d(V) - 1)^2 + 2(d(V) - 1) + 1$
5	$5(d(V) - 1)^4 + 4(d(V) - 1)^3 + 3(d(V) - 1)^2 + 2(d(V) - 1) + 1$
6	$6(d(V) - 1)^5 + 5(d(V) - 1)^4 + 4(d(V) - 1)^3 + 3(d(V) - 1)^2 + 2(d(V) - 1) + 1$

Table 5. The maximum number of uses of edges of regular graphs in the function of their degree and diameter.

$D(G)$	$d(V)$					
	3	4	5	6	7	8
2	5	7	9	11	13	15
3	17	34	57	86	121	162
4	49	142	313	586	985	1534
5	126	547	1593	3711	7465	13539
6	321	2005	7737	22461	54121	114381
	L_{max}					

3   Summary and Conclusions

The article presented the results of the analysis of network structures described by Reference Graphs with the emphasis on the influence on the number of uses of edges constituting the graphs on their transmission properties. Examples were given, explaining the source of differences in these properties in spite of equal values of base parameters: diameter and the average path length. It was concluded that to evaluate the transmission properties of a network modelled by Reference Graphs it is enough to analyse the distribution of uses of individual edges in the paths of the length equal to the graph's diameter. The larger the value of the determined sum, the better the transmission properties of the compared graphs. When the values are equal, the graphs have the same transmission properties. On the basis of the theoretical analysis it was possible to determine the maximum number of uses of any edge of RG graphs in the function of their diameter and node degree, and to describe the dependencies with mathematical formulas.

References

1. Coffman, K.G., Odlyzko, A.M.: Growth of the Internet, Optical Fiber Telecommunications IV B: Systems and Impairments, pp. 17–56 (2002)
2. Pedersen, J.M., Riaz, T.M., Dubalski, B., Madsen, O.B.: A comparison of network planning strategies. In: 10th International Conference on Advanced Communication Technology, pp. 702–707. ICACT, IEEE (2008)
3. Kotsis, G.: Interconnection topologies and routing for parallel processing systems. ACPC, Technical Report Series, ACPC/TR92-19 (1992)
4. Xu, J.: Topological structure and analysis of interconnection networks. Springer, Heidelberg (2010)
5. Ramaswami, R., Sivarajan, K.N., Sasaki, G.H.: Optical Network: A practical Perspective. Morgan/Kaufmann Elesevier, Burlington (2010)
6. Ledziński, D., Śmigiel, S., Zabłudowski, L.: Analyzing methods of network topologies based on chordal rings. Turk. J. Electr. Eng. Comput. Sci. **25**(6), 1–14 (2017)
7. Camarero, C., Martinez, C., Beivide, R.: L-networks: a topological model for regular 2D interconnection networks. IEEE Transact. Comput. **62**(7), 1362–1375 (2013)
8. Andrysiak, T., Saganowski, L., Kiedrowski, P.: Anomaly detection in smart metering infrastructure with the use of time series analysis. J. Sens. **2017**, 1–15 (2017)
9. Graham, R., Knuth, D., Patashnik, O.: Concrete matematics, 2nd edn. Addison-Wesley, Reading (1994)
10. Yebra, J.L.A., Fiol, M.A., Morillo, P., Alegre, I.: The diameter of undirected graphs associated to plane tessellations. ArsCombinatoria **20B**, 159–172 (1985)
11. Bujnowski, S., Dubalski, B., Zabłudowski, A.: Analysis of chordal rings. In: Mathematical techniques and problems in telecommunications, pp. 257–279. Centro International de Matematica, Tomar, PT (2003)
12. Ledziński, D., Marciniak, B., Śrutek, M.: Referential graphs. Telecommun. Electr. **17**, 37–73 (2013). Scientific Journal 262
13. Ledziński, D.: The optimalization of the transmission network topology described with regular graph. PhD Thesis, UTP University of Science and Technology, Bydgoszcz (in Polish) (2014)

Imbalanced Data Classification Using Weighted Voting Ensemble

Lin Lu and Michał Woźniak[(✉)] [iD]

Department of Systems and Computer Networks, Faculty of Electronics,
Wroclaw University of Science and Technology, Wrocław, Poland
{242208,michal.wozniak}@pwr.edu.pl

Abstract. Imbalanced data classification is still remaining thje important topic and during the past decades, plenty of works are devoted
to this field of study. More and more real-life based imbalanced class
problems inspired researchers to come up with new solutions with better performance. Various techniques are employed such as data handling
approaches, algorithm-level approaches, active learning approaches, and
kernel-based methods to enumerate only a few. This work aims at applying a novel dynamic selection methods on imbalanced data classification
problems. The experiments carried out on several benchmark datasets
confirm its pretty high performance.

Keywords: Imbalanced learning · Classifier ensemble ·
Dynamic selection

1 Introduction

In many practical decision tasks the distribution of classes might not be equal
or balanced. For example, in a data set that is collected data for a rare disease
diagnosis, the number of positive samples (i.e., related to a particular disorder)
is many times less than the amount of negative observations.

Recently, a lot of researches have been devoted to imbalanced data classification. This topic has become an interest because it can be applied to many practical scenarios in industry, security and medical field. The base issue in imbalanced
data classification is that usually standard classifiers are devoted to normal class
distributions or similar costs on misclassification. Thus it is more likely for standard classifiers to make incorrect recognition on imbalanced distributed data,
particularly the minority class. Thus some technologies are developed to solve
classification difficulties in imbalanced learning. Data preprocessing method is
one of the main categories in this field. Applying data preprocessing methods is
aiming at changing the distribution of the original data set and balancing the
majority class and minority class. One of the advantages of such techniques is
that it can be applied to the existing recognition tools and the data handling
model can be modified according to the goals of users.

The main contributions of this work are as follows:

© Springer Nature Switzerland AG 2020
M. Choraś and R. S. Choraś (Eds.): IP&C 2019, AISC 1062, pp. 82–91, 2020.
https://doi.org/10.1007/978-3-030-31254-1_11

- Proposing a novel ensemble method using weighted voting rule and data preprocessing called *Weighted Voting Ensemble* (WVE).
- Experimental evaluation of the proposed algorithm backed-up by the statistical tests.

2 Related Works

This work focuses on employing classifier ensemble technique to imbalanced data recognition. Then let us firstly present the classifier selection techniques and then a few main approaches to imbalanced data classification.

2.1 Classifier Selection

The one of most promising approaches is classifier selection [12], where a selection phase may be static or dynamic. The dynamic selection is preferred over static when a more locally accurate classifier or ensemble is needed on an unknown object. Both of them can be applied to provide a single classifier or an ensemble from the pool.

Generally, the selection is done when a new sample is given to classify. Then the local competence of a classifier is evaluated. Usually, to measure it, the performance of a classifier in a neighborhood of the classifying object is evaluated There are different approaches to measure a classifier local competence. The individual based approaches take a particular classifier performance (usually accuracy) into account, while group based approaches focus on interactions among classifiers. The interesting taxonomy of different dynamic selection approaches has been proposed by Britto et al. [12].

Ranking-based method referred as DCS-Rank was proposed by Sabourin et al. [13]. In this method, various classifier parameters are taken on given example, namely, distance to winner, distance to non-winner and distance ratio. Also the author defined a classifier "success" variable. They are used to estimate the average uncertainty in the decision which is resolved by observing classifier parameters. The mutual information is used to indicate which classifier is correct for a so-called meta-pattern space in the training data set which is formed by those classifier parameters. Later, a NN classifier determines the nearest neighbors of an unknown example, thus an optimal classifier is selected based on the rank of classifiers on the neighborhood. A modified ranking methods was proposed by Woods et al. [17]. It is a simplified method with computing classifier competence by calculating the correct recognition in the neighborhood of an unknown sample. The classifier with the most amount of correct recognition is chosen as the most competent classifier.

Accuracy-based methods - a base classifier competence is simply represented by its accuracy or percentage of correct recognition. The accuracy here can be overall local accuracy (OLA) or local class accuracy (LCA). The difference between these two approaches is that OLA is the classifier accuracy on all local region that is close to the unknown pattern in given feature space, where LCA is the accuracy that is calculated only on examples which have same class label as a classifier predicted the unknown pattern. Both of these variations are proposed by Woods et al. [17]. Rather than simply estimating a classifier competence by calculating the accuracy, Giacinto [6] proposed probabilistic based methods named *A Priori* and *A Posteriori*. Similar to accuracy based methods, k-nearest neighbors are determined when given an unknown pattern, the probability for a classifier to correctly classify the unknown pattern is defined by the percentage of its correct recognition on its neighborhood. To deal with the uncertainty of the neighborhood size, the probabilities are weighted by the Euclidean distances between the unknown pattern and its neighbor. Another often used probabilistic-based approach is proposed by Woloszynski et al. [16]. This approach takes into consideration the probability of a random classifier performance using uniform distribution on a L class classification problem.

Behavior-based methods has been first introduced by Huang et al. [9], who also proposed a classifier combination approach called Behavior Knowledge Space. Later Giacinto et al. [7] proposed another classifier selection method derived from this concept, referred as DCS-MCB. In this method, a neighborhood of unknown pattern is first determined in the validation set based on k-NN rule, it is also the region of competence of the unknown pattern that all base classifiers will be measured competence on. Next step, the similarity of classifier behaviors on unknown pattern and its neighbors is measured to indicate similar neighbors to the unknown pattern. Then local accuracy of each classier is computed on selected neighbors as classifier competence, the classifier with highest OLA is selected to classify the unknown pattern.

Oracle-based approach is used by two popular classification techniques, k-Nearest Oracle-Eliminate (KNORA-E) and k-Nearest Oracle Union (KNORA-U) [10]. The kernel of these methods is to find a local Oracle that is achieving full recognition of all examples that are included in the competence area of an unknown pattern. In the KNORA-Eliminate technique, a classifier is included in the ensemble if it fully recognized all samples from the local competence region, later the ensemble will be used to give a final prediction. In the case that none of the classifiers from the pool achieved full recognition, the area of competence will be narrowed and previous steps are repeated until there will be an Oracle ensemble. In the KNORA-Union technique, a classifier is included as long as it recognize at least one sample correctly from the competence area.

DS-KNN is a method proposed by Santana et al. in [14]. In DS-KNN, ensemble is chosen based on accuracy and diversity among the classifiers. Both of these classifier competence measures are computed on the local region that is defined by k-NN rule on a unknown pattern. First the most accurate classifiers are taken from the base classifier pool, then the most diverse classifiers among them are taken to construct the ensemble.

Meta-learning was proposed by Cruz et al. [5]. The key to use meta-learning method is to consider classifier ensemble selection as a different classification problem, which can be called meta-problem. In a meta-problem, input is called meta-features, which are represented by different standards used to estimate base classifier competence. A meta-classifier determines whether a base classifier is qualified enough to make decision output on given unknown pattern. The advantage of meta-learning technique is that it use five different criteria as measure of classifier competence, as proposed in the paper, the system can still have good performance even some of these criterion might not work because of problems with local regions or feature space.

It is worth also mentioning the work by Zyblewski et al. [18], where simple classifier selection techniques has been applied to the imbalance stream data classification.

2.2   Imbalanced Data Classification

According to Branco et al. [3], methods that are used to sway class distributions are divided into three main categories, namely, stratified sampling, synthesizing sampling and hybrid methods. It can also be divided into over-sampling methods and under-sampling methods or hybrid methods, based on whether samples are added or removed during the data handling phase. For example, widely used methods in applications are random over-sampling, random under-sampling, synthetic sampling with clustering techniques and synthetic sampling with data removing after generation.

Random undersampling (RUS) removes a random set of negative class samples from the original training data set. A main drawback of using random under-sampling method is that it may remove the potential informative sample data and cause problems during recognition phase. Seiffert et al. [15] proposed a hybrid algorithm using random under-sampling technique and boosting technique called RUSBoost. Like typical random under-sampling method, RUSBoost removes samples from training data set with majority class label. Unlike to typical RUS, RUSBoost uses resampling training set as the boosting iteration model, in each iteration the training set is resampled according to sample weights assigned. RUSBoost is a method that improves standard RUS method with less information lost during resampling procedure. Under-sampling techniques can also be combined with data cleaning techniques. The advantage of using these methods is that it removes potential noise samples or overlapping samples from the training set. The removal of potential noise samples can be achieved by using Tomek links. The removal can be of both classes or only the majority class. One of the implementation of this technology is in [2]. Cluster Centroids is an under-sampling method that is based one clustering method. It replaces the original majority samples in the training set with cluster centroids of them. Centroids of each cluster are determined by k-Means algorithm. After applying k-Means algorithm, the new determined cluster centroids are seen as the new majority part of training set. Clustering method using k-Means algorithm is simple and easy to implement, but such methods might not be a good option when comes to recognizing outlier examples.

The simplest oversampling technique is Random oversampling (ROS) which selects randomly minority class samples and replicates them. SMOTE [4] is a synthesizing sampling method where operations are taken on the original training data set to generate extra minority class training data. The synthesizing takes place along the line connecting the current sample and its k minority class nearest neighbors.

An adaptive synthetic sampling (ADASYN) method was proposed by He et al. [8] motivated by SMOTE and some of its variants. In this method, synthetic samples are created not only to solve the imbalanced distribution of original data, but also to focus more on the samples that are hard-to-learn for classifiers. In ADASYN algorithm, the distribution of minority class examples is changed adaptively according to the their difficulty level of learning. ADASYN achieved a good performance on improving imbalanced classification according to the original testing results from the authors.

To solve the drawbacks of SMOTE method, Batista et al. [2] proposed to apply Tomek links to resampled data set as a data cleaning strategy. Usually Tomek links are used to clean majority class examples, but in this case both classes from the links are removed. By applying Tomek links one can get a resampled data set with more clearly defined class clusters.

Another variant of SMOTE with data cleaning method is SMOTEENN [1], in this method, also samples from both classes are removed after applying SMOTE as synthesizing method. More samples are removed in this method than in SMOTETomek. The removal happens on evert sample that is incorrectly classified by three nearest neighbors of it.

3   Proposed Algorithm

Weighted Voting Ensemble (WVE) is dedicated only for imbalanced data classification problems and it is a variant of a standard weighted voting classifier ensemble. The kernel of this method is to make more competent estimators having stronger decision supports. The support degree is counted based on the amount of samples a classifier correctly recognized on a nearest neighbors. To fit in imbalanced data problems, the support degrees can be various on original samples and generated samples. Given an unknown sample x, k-nearest neighbors are determined on validation set. For each neighbor sample, the attribution of support degree can be represented as

$$d_{i,j} = \begin{cases} 2 & \text{if } c_i \text{ labels } N_j \text{ correctly and } N_j \text{ is an original sample} \\ 1 & \text{if } c_i \text{ labels } N_j \text{ correctly and } N_j \text{ is a generated sample} \\ 0 & \text{otherwise} \end{cases} \quad (1)$$

The total support degree of classifier c_i can be represented as

$$d_i = \sum_{j=1}^{k} d_{i,j} \quad (2)$$

The pseudocode of WVE is shown in Algorithm 1.

Input pool of classifiers C, datasets Tr and Te, neighborhood size k
Output label prediction p of unknown sample x
for each testing sample x in Te **do**
  for each classifier c_i in C **do**
   $a_i \leftarrow$ the number of correctly labeled samples in the neighborhood of x by c_i
   $b_i \leftarrow$ the number of correctly labeled synthetic samples in the neighborhood of x by c_i
   classifier c_i weight $d_i \leftarrow \frac{2a_i+b_i}{\sum_i(2a_i+b_i)}$
  end for
  calculate predictions $p \leftarrow \sum_i c_i \cdot d_i$
end for

Algorithm 1. *Weighted Voting Method* for imbalanced data classification.

4 Experiments

The experiments aim at the proposed method performance testing on different benchmark data sets (Table 1).

Table 1. Data sets used in this work.

Data set name	Total examples	Imbalanced ratio
yeast1	1484	2.46
vehicle0	846	3.25
segment0	2308	6.02
page-blocks0	5472	8.79
yeast4	1484	28.1
winequality	1599	29.17
yeast6	1484	41.4
poker	1485	58.4

4.1 Set-Up

All experiments are implemented using Python 3.6. All datasets come from KEEL (Knowledge Extraction based on Evolutionary Learning) data set repository. All data sets are binary class imbalanced datasets. All datasets are divided into training set, validation set and test set. Wilcoxon test has been used as the statistical tool to confirm differences between methods.

Dynamic selection methods that are applied in the experiments are:

- k-Nearest Oracle-Union technique (KNU)
- k-Nearest Oracle-Eliminate technique (KNE)
- Dynamic Ensemble Selection performance (DSP)

- Meta learning for dynamic ensemble selection (MDS)
- Weighted Voting Ensemble method (WVE)

Data preprocessing methods that are used to resample training data are:

- RandomOverSampler (ROS): Random over-sampling.
- ClusterCentroids (CC): Under-sampling by generating centroids based on clustering methods.
- SMOTE (SM): Oversampling using SMOTE.
- SMOTEENN (SMEN): Combine over- and under-sampling using SMOTE and Edited Nearest Neighbours.
- ADASYN (ADS): Over-sampling using Adaptive Synthetic technique.

4.2   Result Analysis

Experiment 1 - Comparing WVE with Other Dynamic Classifier Selections.
Tables 2 and 3 show results achieved by Wilcoxon tests on precision and recall. In this test, the dynamic selection method WVE is compared to each other methods on precision, recall, F-measure and G-mean. The null hypothesis in these tests are that WVE is an identical classification model to other solutions.

Table 2. Results obtained by the Wilcoxon test for WVE according to precision.

VS	R^+	R^-	Exact P-value	Asymptotic P-value
KNU	630.0	150.0	0.000526	0.00079
KNE	615.0	165.0	0.0012464	0.00165
DSP	672.0	148.0	0.0002482	0.000418
MDS	666.0	154.0	0.0003572	0.000565

Table 3. Results obtained by the Wilcoxon test for WVE according to recall.

VS	R^+	R^-	Exact P-value	Asymptotic P-value
KNU	532.5	287.5	0.10147	0.097709
KNE	571.0	249.0	0.0299	0.029673
DSP	511.0	269.0	0.09288	0.089974
MDS	565.0	255.0	0.03684	0.036607

According to the test results, Wilcoxon procedure on precision shows all p-value are less than 0.05, thus it rejects all null hypotheses with WVE against other solutions. Wilcoxon procedure on recall rejects the hypotheses between WVE against KNE and MDS. On the measure of F-measure and G-mean, there are no statistically significant differences between WVE and other solutions.

Experiment 2 - Comparing Different Preprocessing Techniques Using by WVE
Table 4 shows the results achieved by Wilcoxon test on recall measurement. In
the test, data preprocessing method is compared to other over-sampling methods
on recall score.

Table 4. Results obtained by the Wilcoxon test for algorithm CC.

VS	R^+	R^-	Exact P-value	Asymptotic P-value
SMEN	522.0	298.0	0.13508	0.126626
SM	608.5	211.5	0.006807	0.007278
ADS	625.5	194.5	0.003127	0.003578
ROS	682.0	98.0	0.00001353	0.000043

Detailed results of the all experiments may be found on the article's webpage[1].

Observations. Different dynamic selection methods were implemented, also
comparison between these methods was performed. According to the analysis
procedures, they are performed at a significance level of $\alpha = 0.05$.

A pairwise comparison between WVE against all other dynamic selection
methods was performed during the test procedure. Considering precision measure, WVE method is significantly different from all other selection methods.
It is achieving a better performance at classification precision than all other
methods. Also it achieves a much lower average rank at precision measure, thus
WVE is good at doing imbalanced data classification problems. When considering recall measure, it is considered significantly different from KNE and MDS
methods, on the other hand, statistically equivalent to KNU and DSP methods
on performance. Comparison among different data preprocessing methods are
also performed in experiments. According to the test, under-sampling method
CC is significantly different from SM, ADS, ROS methods on recall score, other
methods are not showing big differences on results. Therefore, it is considered
that under-sampling method CC is doing a different job from other over-sampling
methods on data preprocessing.

Lessons Learnt. It is observed that when the size of data set is small, the
performance of classifiers are not always promising enough. The connections
between size of a training set and classification performance level is like to be that
small amount of minority class examples in training set might lead to not enough
learning on minority class. It is worse when the minority class and majority class
are having overlapping spaces.

When a data set is having much higher imbalanced ratio, under-sampling
method CC is having better performance than over-sampling methods because

---

[1] https://github.com/velvet37/des_and_imb.git.

synthetic examples are created around original minority class samples, which is possible to cause overlapping of minority class and majority class in training set when too much data was generated. Thus it is possible to cause misclassification by recognizing a positive sample as a negative one instead.

The statistical analysis of different dynamic ensemble methods shows that Weighted Voting Ensemble shows much better performance than other methods on precision measure but not so competent on other measures, this can be explained by "no free lunch" theorem. There is not a perfect criterion which measures classifier competence so that there will be a single classifier or ensemble that is competent for all classification problems. It is also observed that each ensemble is having its own competence area. Thus when a classification problem which is more requiring on precision performance, Weighted Voting Ensemble will be a good choice.

5   Conclusions

In this work, a novel dynamic ensemble selection method called Weighted Voting Ensemble applied to imbalanced data classification has been presented. The proposed algorithm has been evaluated on the different datasets and several preprocessing method have also been used. Statistically, all data preprocessing methods are working well with Weighted Voting Ensemble method, the proposed method in this work. It improves the statistical results and performance. It is also worth noting that Weighted Voting Ensemble is the only method that is significantly different from all other methods. Thus for future research direction, this method can be extended by using a bootstrapping technique on the original data set first, then embed with data preprocessing methods. Different base classifiers may also be tested to find a better ensemble, e.g., as proposed by Ksieniewicz and Wozniak [11], where feature selection techniques has been applied to ensure a high diversity of the ensemble. Also more tests on different data sets are needed for deeper analysis. In general, SMOTE and SMOTEENN methods are providing good performance over all. When the data set is higher imbalanced, ClusterCentroids is also a good option to improve the classifier performance. It is also a good future research direction to focus on the structure of a data set and find out more about where the difficulty of learning locates.

All methods that are compared and analyzed were having satisfying performance, they should be discussed in deeper way in the future.

Acknowledgement. This work was supported by the Polish National Science Centre under the grant No. 2017/27/B/ST6/01325 as well as by the statutory funds of the Department of Systems and Computer Networks, Faculty of Electronics, Wroclaw University of Science and Technology.

References

1. de Almeida Prado Alves Batista, G.E., Bazzan, A.L.C., Monard, M.C.: Balancing training data for automated annotation of keywords: a case study. Rev.a Tecnologia da Informação **3**(2), 15–20 (2003)

2. Batista, G.E.A.P.A., Prati, R.C., Monard, M.C.: A study of the behavior of several methods for balancing machine learning training data. SIGKDD Explor. Newsl. **6**(1), 20–29 (2004)
3. Branco, P., Torgo, L., Ribeiro, R.P.: A survey of predictive modeling on imbalanced domains. ACM Comput. Surv. **49**(2), 31:1–31:50 (2016)
4. Chawla, N.V., Bowyer, K.W., Hall, L.O., Kegelmeyer, W.P.: Smote: Synthetic minority over-sampling technique. J. Artif. Int. Res. **16**(1), 321–357 (2002)
5. Cruz, R.M., Sabourin, R., Cavalcanti, G.D., Ren, T.I.: Meta-des: a dynamic ensemble selection framework using meta-learning. Pattern Recognit. **48**(5), 1925–1935 (2015)
6. Giacinto, G., Roli, F.: Methods for dynamic classifier selection. In: Proceedings 10th International Conference on Image Analysis and Processing, pp. 659–664, September 1999
7. Giacinto, G., Roli, F.: Dynamic classifier selection based on multiple classifier behaviour. Pattern Recognit. **34**, 1879–1881 (2002)
8. He, H., Bai, Y., Garcia, E.A., Li, S.: ADASYN: Adaptive synthetic sampling approach for imbalanced learning. In: 2008 IEEE International Joint Conference on Neural Networks (IEEE World Congress on Computational Intelligence), pp. 1322–1328, June 2008
9. Huang, Y.S., Suen, C.Y.: A method of combining multiple experts for the recognition of unconstrained handwritten numerals. IEEE Transact. Pattern Anal. Mach. Intell. **17**(1), 90–94 (1995)
10. Ko, A.H., Sabourin, R., Alceu Souza Britto, J.: From dynamic classifier selection to dynamic ensemble selection. Pattern Recognit. **41**(5), 1718–1731 (2008)
11. Ksieniewicz, P., Woźniak, M.: Imbalanced data classification based on feature selection techniques. In: Yin, H., Camacho, D., Novais, P., Tallón-Ballesteros, A.J. (eds.) Intelligent Data Engineering and Automated Learning - IDEAL 2018, pp. 296–303. Springer International Publishing, Cham (2018)
12. Kuncheva, L.I.: Combining Pattern Classifiers: Methods and Algorithms. Wiley-Interscience, New York (2004)
13. Sabourin, M., Mitiche, A., Thomas, D., Nagy, G.: Classifier combination for hand-printed digit recognition. In: Proceedings of 2nd International Conference on Document Analysis and Recognition, ICDAR 1993, pp. 163–166, October 1993
14. Santana, A., Soares, R.G.F., Canuto, A.M.P., de Souto, M.C.P.: A dynamic classifier selection method to build ensembles using accuracy and diversity. In: 2006 Ninth Brazilian Symposium on Neural Networks, SBRN 2006, pp. 36–41, October 2006
15. Seiffert, C., Khoshgoftaar, T.M., Van Hulse, J., Napolitano, A.: Rusboost: a hybrid approach to alleviating class imbalance. IEEE Transact. Syst. Man Cybern. Part A Syst. Hum. **40**(1), 185–197 (2010)
16. Woloszynski, T., Kurzynski, M., Podsiadlo, P., Stachowiak, G.W.: A measure of competence based on random classification for dynamic ensemble selection. Inf. Fusion **13**(3), 207–213 (2012)
17. Woods, K., Bowyer, K., Kegelmeyer, W.P.: Combination of multiple classifiers using local accuracy estimates. In: Proceedings CVPR IEEE Computer Society Conference on Computer Vision and Pattern Recognition, pp. 391–396, June 1996
18. Zyblewski, P., Ksieniewicz, P., Woźniak, M.: Classifier selection for highly imbalanced data streams with minority driven ensemble. In: Rutkowski, L., Scherer, R., Korytkowski, M., Pedrycz, W., Tadeusiewicz, R., Zurada, J.M. (eds.) Artificial Intelligence and Soft Computing, pp. 626–635. Springer, Cham (2019)

Evaluation of the MRI Images Matching Using Normalized Mutual Information Method and Preprocessing Techniques

Paweł Bzowski[1,2](✉) ⓘ, Damian Borys[1,2] ⓘ, Wiesław Guz[3,4] ⓘ,
Rafał Obuchowicz[5] ⓘ, and Adam Piórkowski[6] ⓘ

[1] PET Diagnostics Department,
Maria Sklodowska Curie Institute - Oncology Center Gliwice Branch,
Gliwice, Poland
pawel.bzowski@io.gliwice.pl

[2] Biotechnology Centre, Silesian University of Technology, Gliwice, Poland
[3] Department of Radiology, Diagnostic Imaging and Nuclear Medicine,
Institute of Experimental and Clinical Medicine, Faculty of Medicine,
University of Rzeszow, Rzeszow, Poland
[4] Department of Electroradiology,
Institute of Nursing and Health Sciences, Faculty of Medicine,
University of Rzeszow, Rzeszow, Poland
[5] Department of Diagnostic Imaging,
Jagiellonian University Medical College,
19 Kopernika Street, 31-501 Cracow, Poland
[6] Department of Biocybernetics and Biomedical Engineering,
AGH University of Science and Technology,
al. Mickiewicza 30, 30-059 Cracow, Poland

Abstract. One of the common methods for medical diagnosis is Magnetic Resonance Imaging (MRI), a safe, non-invasive method. During each imaging session a patient's position may be different, therefore comparison of two sequences can become difficult. The primary goal of this work is preparation of an optimal algorithm for co-registration of T1 and T2 weighted MRI images. To adjust co-registration sensitivity, different preprocessing methods to perform normalizations and edge detection were used. The obtained results allow to increase quality of the co-registration process.

Keywords: Magnetic Resonance Imaging · Co-registration ·
Matching · T1 and T2 weighted images

1 Introduction

Magnetic Resonance Imaging is one of the most popular diagnostic methods due to its sensitivity and lack of ionizing radiation. During an MRI exam it is essential for the subject to remain in the same position between sequences,

© Springer Nature Switzerland AG 2020
M. Choraś and R. S. Choraś (Eds.): IP&C 2019, AISC 1062, pp. 92–100, 2020.
https://doi.org/10.1007/978-3-030-31254-1_12

otherwise each MRI series may not be complementary to the rest of the study. Such situations could affect the quality of imaging and therefore the value of the diagnosis. Attempting to match images achieved from different protocols can also be a challenge due to the different signals obtained for each tissue. In such cases similarity methods such as SSD (sum of squared differences) will not work properly. To solve this problem, a method such as MI (mutual information) can be used. The aim of the presented study was to find a method to reduce differences between MRI images obtained from different protocols used in the same study. The presented algorithm uses the rigid method based on affine transformation with mutual information as a quality factor. This study is a continuation of work presented at ITiB Conference in Kamień Śląski, 2018: Rigid and Non-rigid Registration Algorithm Evaluation in MRI for Breast Cancer Therapy Monitoring [2].

2   Materials

This study was conducted in accordance with good medical practice guidelines and was approved by the Ethics Committee of Jagiellonian University (permission no. 1072.6120.16.2017) and complied following the Declaration of Helsinki and good medical practice. Informed written consent for participation was obtained from the legal guardian of each examined child. We included images of the left hand of seven healthy boys (age range 9 to 15 years), including two 9-year-olds, three 12-year-olds, and two 15-year-olds.

MR studies containing uncorrectable motion artifacts or that were technically imperfect were rejected. A 1.5-T system (GE Optima 360, Chicago, IL, USA) with a dedicated 4-channel wrist coil was used. T1 and T2 coronal plane sequences were applied. Images of the wrist area with radius and ulna bones were acquired.

The algorithm was prepared using MATLAB 2016b (MathWorks) software with Image Processing Toolbox. During the tests we focused on the wrist area instead of the entire hand, because this region is the most important in this study. This method also allowed to decrease calculation time and avoid differences in the image due to changed finger positions. All runs were performed on a personal computer with an Intel i7 processor, 16 GB RAM memory, fast SSD disc and Nvidia GeForce GTX 1060.

3   Methods

The main goal of this work was to compare results of registration with and without preprocessing using different methods. For this purpose Local Normalize [12], MATLAB Edge function (Sobel filter) and Statistic Dominance Algorithm (SDA) [4,7] were chosen. As an example pair of MRI wrist images T1 and T2 sequences (Fig. 1(a) and (b)) were used.

At first MATLAB Edge (Sobel), an element of the MATLAB image processing toolbox was used. This function is based on the Sobel method to detect the

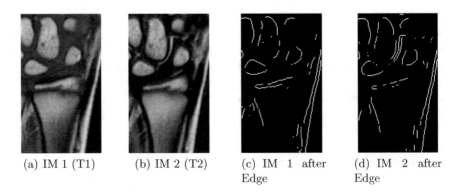

(a) IM 1 (T1)   (b) IM 2 (T2)   (c) IM 1 after Edge   (d) IM 2 after Edge

Fig. 1. Original images and images after edge detection.

edges. It is one of the most popular methods to detect edges mainly due to its simplicity. The results of using this method are presented in Fig. 1(c) and (d).

Next the Local Normalize [12] was used. This is an open source code for MATLAB. This algorithm was made by Guangeli Xiong and uploaded to the File Exchange platform. This type of normalization is intended to uniformize the mean and variance of an image in the surrounding area (Eq. 1). The result of this algorithm is presented in Fig. 2(a) and (b).

$$g(x, y) = \frac{f(x, y) - m_f(x, y)}{\sigma_f(x, y)} \tag{1}$$

where:

  f(x,y) - original image
  m_f(x,y) - estimation of local mean of f(x,y)
  σ_f(x,y) - estimation of local variance
  g(x,y) - output image

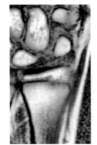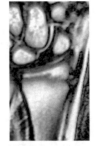

(a) IM 1 after Local Normalization   (b) IM 2 after Local Normalization   (c) IM 1 after SDA   (d) IM 2 after SDA

Fig. 2. Images 1 and 2 after local normalization and after SDA.

The final method was normalization using Statistical Dominance Algorithm (SDA). The algorithm was proposed in [7]. The main idea is to determine for each pixel the number of neighboring points in a specific area. The resulting value is the number of pixels with values higher than the pixel in the center. Using this method it is possible to find the result of the dominance of this center point over the neighboring points and allow to classify these points. The result of this method is presented in Fig. 2(c) and (d) for radius R = 50.

The next step was image registration. The algorithm consists of the main program to perform rotations and displacement and an additional part to calculate similarity of images. The main part is based on affine transformation. Using different d_x, d_y and α factors in Eq. (2), it is possible to obtain new coordinates for each pixel in an image.

$$\begin{bmatrix} x_2 \\ y_2 \\ 0 \end{bmatrix} = \begin{bmatrix} 1 & 0 & d_x \\ 0 & 1 & d_y \\ 0 & 0 & 1 \end{bmatrix} \begin{bmatrix} cos(\alpha) & -sin(\alpha) & 0 \\ sin(\alpha) & cos(\alpha) & 0 \\ 0 & 0 & 1 \end{bmatrix} \begin{bmatrix} x_1 \\ y_1 \\ 0 \end{bmatrix} \tag{2}$$

where:

x_1 and y_1 - coordinates of image before transformation
x_2 and y_2 - coordinates of image after transformation
d_x and d_y - translation in x and y direction
α - rotation(angle)

To compute the similarity of two images, the mutual information method could be used [3,9]. This method is based on the probability theory and information theory. It shows how one set of variables is dependent on a second set of variables. These calculations are based on the entropy of each data set. The entropy of any variable is a function which attempts to characterize the random variable. This value not only shows all possibilities of the value but provides some information about frequency.

To calculate mutual information, the entropy of each data set (images) should be calculated as in the Eq. (3). In the next step the joint histogram of two data sets should be computed (4). During the last step the mutual information is calculated by subtracting entropies of each image from the common entropy (5). If the mutual information value is higher, this indicates the images are more similar. Another way is to use normalized mutual information, which is better for comparing results from different sets (6).

$$H(A) = -\sum_{i}^{n} p(a_i) * log(p(a_i)) \tag{3}$$

$$H(A, B) = -\sum_{i}^{n} p(a_i, b_i) * log(p(a_i, b_i)) \tag{4}$$

where:

a_i, b_i - pixel value in each matrix element (for data set A and B)
$p(a_i)$, $p(b_i)$ - marginal distribution of elements a_i and b_i

$$MI = H(A, B) - H(A) - H(B). \tag{5}$$

$$NMI = 2 * \frac{H(A, B)}{H(A) + H(B)} \tag{6}$$

In each iteration the values of displacement and rotation are changed. Next, using the Eq. 2 coordinates of new pixels could be obtained. Using the interpolation method the new image could be reconstructed. In the last step the new Mutual Information value could be calculated. If this value was the highest, then the displacement and rotation values were saved. Using the interpolation method and new coordinates, the new image could be reconstructed. The obtained images could then be analyzed due to their similarity.

4   Results

After registration, the obtained values of displacement and rotation were transferred to the original image. To obtain the new image, the old image must be interpolated with new values. Next, each new image was compared with the reference image and the mutual information value for each pair were calculated. In Table 1 are presented displacements and rotations for presented examples and the mutual information values are presented in Table 2. Using a different preprocessing method obtained values of rotations and displacements that were not the same, however, these values do not differ significantly. In each case the MI value was higher than before the registration method. That means that images after registration were more similar to the reference image than before registration.

To present dissimilarities before and after image registration the images were binarized using the Otsu method, which is an automatic method for binarization. Images without preprocessing are presented on Fig. 3(b), after local normalization on the Fig. 4(b), after edge detection Fig. 4(a) and after SDA Fig. 4(c).

For the group of patients, the processing time of each method was calculated. Time began at the start of preprocessing of the image and ended after the registration process with the saving of a new image. In Table 3 the execution time for different preprocessing methods are presented.

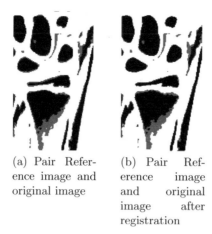

(a) Pair Reference image and original image

(b) Pair Reference image and original image after registration

Fig. 3. Difference between reference image and image before and after registration (Without preprocessing).

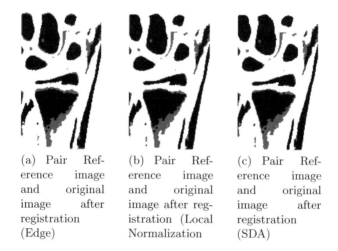

(a) Pair Reference image and original image after registration (Edge)

(b) Pair Reference image and original image after registration (Local Normalization

(c) Pair Reference image and original image after registration (SDA)

Fig. 4. Differences between reference image and image after registration.

The presented processing time results (Table 3) show that in each case all of the preprocessing methods increase calculation speed. Very good results were obtained using the Edge method for preprocessing. For each result, the computation time was almost two times less than registration without preprocessing.

Table 1. Values of obtained rotations (degrees) and displacements (pixels) for each method.

Name	Method	x translation	y translation	Rotation
No. 1	Without preprocessing	0	1	0,1
No. 1	Local	0	1	−0,4
No. 1	Edge	0	2	0,4
No. 1	SDA	0	1	0
No. 2	Without preprocessing	−2	1	0,5
No. 2	Local	−2	3	1,6
No. 2	Edge	−2	2	1
No. 2	SDA	−2	1	0
No. 3	Without preprocessing	−3	−3	0,4
No. 3	Local	−4	−4	0,6
No. 3	Edge	−4	−4	−0,3
No. 3	SDA	−3	−3	0
No. 4	Without preprocessing	0	−1	0
No. 4	Local	0	−1	0,1
No. 4	Edge	0	−1	0,3
No. 4	SDA	0	−1	0
No. 5	Without preprocessing	0	−1	−0,7
No. 5	Local	1	−1	−1,3
No. 5	Edge	1	−1	−0,3
No. 5	SDA	0	−1	0

Table 2. Values of normalized mutual information for images pairs.

Name	Original	Without preprocessing	Local normalization	Edge	SDA
No. 1	0,275	0,285	0,284	0,270	0,285
No. 2	0,326	0,357	0,340	0,354	0,355
No. 3	0,173	0,197	0,196	0,195	0,195
No. 4	0,331	0,352	0,349	0,348	0,352
No. 5	0,292	0,299	0,290	0,289	0,296

Table 3. Table with processing time for each method and different images set (time in seconds).

Images set	Without preprocessing	SDA	Local normalization	Edge
No. 1	303,1	280,1	280,1	165,9
No. 2	261,9	255,9	253,5	153,1
No. 3	221,5	225,8	236,7	143,2
No. 4	222,8	213,2	235,7	146,0
No. 5	221,0	212,5	237,7	145,5

5   Discussion and Conclusions

With the rapid growth of utilization of MR and ongoing improvement of techniques utilized in MR image acquisition the idea of use of MR for the estimation of patient age is emerging. However, the long period for the acquisition and dependence of the image quality from the overall conditions of the acquisition results in a serious limitation of this technique. A highly significant problem can also be the patient changing position during the exam. In this case comparison of two sequences can be difficult. Different preprocessing methods can affect the registration process. The presented figures and MI values shows that all images were correctly registered using these methods. With each preprocessing method the differences between values of displacements and rotation are small and are not significant. The processing time trials prove that using edge detection as a preprocessing method can increase computation speed. The presented method could be useful for correctly matching bone images of different MRI sequences.

Further work will mainly focus on finding the proper methods of segmentation which can be used to find optimal images co-registration [10,11]. The results obtained should be compared with known X-rays images and segmentation can be tested for methods described for other modalities [1,5,6,8,13].

References

1. Bielecka, M., Bielecki, A., Korkosz, M., Skomorowski, M., Wojciechowski, W., Zieliński, B.: Application of shape description methodology to hand radiographs interpretation, vol. 6374, pp. 11–18 (2010)
2. Bzowski, P., Danch-Wierzchowska, M., Psiuk-Maksymowicz, K., Panek, R., Borys, D.: Rigid and non-rigid registration algorithm evaluation in MRI for breast cancer therapy monitoring, vol. 762, pp. 150–159 (2019)
3. D'Agostino, E., Maes, F., Vandermeulen, D., Suetens, P.: A viscous fluid model for multimodal non-rigid image registration using mutual information. Med. Image Anal. **7**(4), 565–575 (2003)
4. Kociołek, M., Piórkowski, A., Obuchowicz, R., Kamiński, P., Strzelecki, M.: Lytic region recognition in hip radiograms by means of statistical dominance transform. In: ICCVG. LNCS, vol. 11114, pp. 349–360. Springer (2018)
5. Nurzynska, K., Smolka, B.: Segmentation of finger joint synovitis in ultrasound images. In: 2016 IEEE ICCE, pp. 335–340. IEEE (2016)
6. Pietka, E., Pospiech, S., Gertych, A., Cao, F., Huang, H., Gilsanz, V.: Computer automated approach to the extraction of epiphyseal regions in hand radiographs. J. Digit. Imaging **14**(4), 165–172 (2001)
7. Piorkowski, A.: A statistical dominance algorithm for edge detection and segmentation of medical images. In: Information Technologies in Medicine. AISC, vol. 471, pp. 3–14. Springer (2016)
8. Tadeusiewicz, R., Ogiela, M.R.: Picture languages in automatic radiological palm interpretation. Int. J. Appl. Math. Comput. Sci. **15**, 305–312 (2005)
9. Viola, P.: Alignment by maximization of mutual information. A.I. Technical Report No. 1548 (1995)

10. Włodarczyk, J., Czaplicka, K., Tabor, Z., Wojciechowski, W., Urbanik, A.: Segmentation of bones in magnetic resonance images of the wrist. Int. J. Comput. Assist. Radiol. Surg. **10**(4), 419–431 (2015)
11. Włodarczyk, J., Wojciechowski, W., Czaplicka, K., Urbanik, A., Tabor, Z.: Fast automated segmentation of wrist bones in magnetic resonance images. Comput. Biol. Med. **65**, 44–53 (2015)
12. Xiong, G.: Local Normalization (2005). www.mathworks.com/matlabcentral/fileexchange/8303-local-normalization
13. Zieliński, B., Skomorowski, M., Wojciechowski, W., Korkosz, M., Sprężak, K.: Computer aided erosions and osteophytes detection based on hand radiographs. Pattern Recogn. **48**(7), 2304–2317 (2015)

Remote Heart Rate Monitoring Using a Multi-band Camera

Piotr Garbat$^{(\boxtimes)}$ ⓘ and Agata Olszewska

Institiute of Microelectronics and Optoelectronics,
Warsaw University of Technology,
Nowowiejska 15/19 St., Warsaw, Poland
p.garbat@elka.pw.edu.pl
http://www.imio.pw.edu.pl

Abstract. Video based heart rate estimation based on the PPG technique is a remote optical technique which allows us to determine the heart and breath rates through the intensity of motion variations. This paper proposes a new monitoring method for estimating the heart rate using a vision system with the VIS and SWIR camera.

Keywords: Heart rate moinitoring · Remote PPG · Multispectral

1 Introduction

Recent advances in computer vision and optics technology have allowed the camera to become a tool for monitoring physiological parameters remotely. The ability to determine vital signs, including the heart rate (HR) and the respiratory rate (RR), with the use of a video camera has recently been reported in [1–3]. The task of measuring skin blood perfusion and heart rate measurements from facial images became the subject of many publications. The majority of research is focused on how to decrease the influence of head movement and facial expressions during the examination. Overall, three basic steps in Imaging Photo Plethysmography (IPPG) may be differentiated: extraction of the intensity fluctuation signal, determination of the PPG signal, and heart rate estimation. Depending on the final application, this process may be implemented in many different ways. The first step is abstracting the most valuable, in terms of further PPG signal analysis, set of points from the face image. The most common methods of Regions of Interest (ROI) detection are based on automatic face localisation and finding edges. For more precise positioning of the information, facial features detection algorithms are used. In the next step, unwanted low and high frequency noises, caused by illumination changes and by head position variations respectively, are eliminated. This is accomplished by using a bandpass filter [4] or by signal detrending algorithms [5] in combination with a low pass filter [6]. For a multidimensional signal (as RGB + SWIR) a signal dimension reduction is performed by mixing the channels linearly. It is mostly done by Blind

© Springer Nature Switzerland AG 2020
M. Choraś and R. S. Choraś (Eds.): IP&C 2019, AISC 1062, pp. 101–107, 2020.
https://doi.org/10.1007/978-3-030-31254-1_13

Source Separation (BSS) techniques, such as ICA [7] or PCA [5]. Different types of methods, for instance CHROM [9] and POS [13], based on transformation of skin colour space have been proposed. The third group of solutions, including SB [2], enables the independent suppression of multiple motion-frequencies by signal decomposition into independent frequency bands. The HR estimation can be realized using a frequency analysis. The most ubiquitous algorithms are based on Fast Fourier Transform [4,5,7]. The approach proposed in this work enables estimation of heart rate using a multiband camera.

2   Heart Rate Monitoring

The proposed system consists of two cameras: a depth camera with RGB image and additional SWIR (Short Wavelenght Infra Red) camera. Figure 1 provides a general scheme of the system for extracting the blood pulse from a four-band image captured by a vision system, and examples of images captured for two sub-bands.

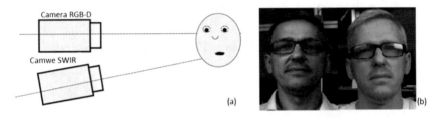

Fig. 1. The four-band camera system for blood-pulse monitoring (a). A sample RGB image with landmarks and SWIR channel frame (b).

The image fusion algorithm is based on transformations between the cameras' coordinates systems supported by depth maps from a 3D camera. The reconstructed model is then used to carry data through different images. All steps are performed automatically. The only process, when human intervention is needed, is calibration. The PPG signal is evaluated from a multiband face image (see Fig. 1). The video signals were analyzed by custom software written in Python and C++.

The face and facial landmark detection algorithm, used in this work, is based on Deep Alignment Network (DAN) [14], a robust face alignment method based on a deep neural network architecture. The facial landmark detector finds the x- and y- coordinates of facial landmarks on the participant's face in multiple stages process. Each stage improves the locations of the facial landmarks estimated by the previous stage.

Based on these points the ROI is selected. Next, the spatial averages of the four colour channels (red, green, blue, narrow red) pixel values within the resulting ROI are calculated for each frame (Fig. 2). The raw traces are detrended using three techniques:

– based on empirical mode decomposition EMD [12], and the ICA applied to recover source signals from the observations,
– based on POS method,
– based on SB algorithm.

This allows for removing only very low-frequency components of the signal, without damaging the high-frequency information. Then the heart rate is determined by finding the maximum frequency peak in the FFT spectrum. The spectrum is filtered using a Butterworth filter with low- and high-frequency cutoffs at 0.8 and 2.5 Hz, respectively, corresponding to acceptable pulse frequencies.

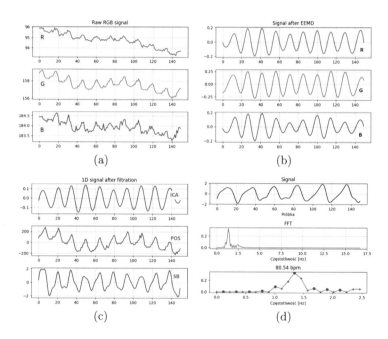

Fig. 2. Pulse signal processing: (a) raw signals from three RGB channels, (b) signals filtered and detrended, (c) reconstructed HR signal from three different methods, (d) reconstructed signal and HR value.

3   Experiment

To evaluate the performance of the proposed system, a series of experiments have been performed, in which vital signs of 6 persons have been detected. The reference heart rate (HR) was measured by the contact PPG device. The obtained results show that experimentally computed heart rate is well correlated with the reference measurements (Figs. 3 and 5). All examined patients were taken into account. The impact of an additional SWIR channel on the quality of the

Table 1. The comparison of correlation coefficients for all band combination

	ICA	POS	SB
RGB	0.62	0.81	0.88
SWIR	0.10	–	–
RGB+SWIR	0.69	0.84	0.92

pulse signal determination and determining the HR value was investigated. The Table 1 provides comparison of the correlation coefficient for individual bands.

The improvement of the correlation after addition of the SWIR channel to the RGB signal is clearly noticeable. Pearson correlation coefficient is equal to 0.88 for RGB and 0.92 for RGB+SWIR. Differences between the experimental and the reference examinations were calculated. The resulted Bland-Altman plots with mean error value and 95% limits of agreement are presented on Figs. 4 and 6.

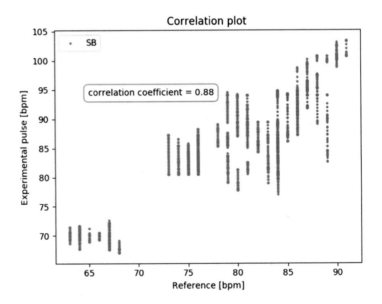

Fig. 3. Pulse rate: correlation between experimental and reference measurements for connected bands RGB.

4   Summary

We presented remote monitoring of heart rates using a multiband camera system. The obtained results are comparable with similar approaches presented in [4,6,7]. We can clearly notice improvements in the obtained results for a signal with

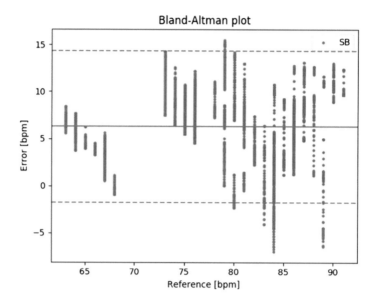

Fig. 4. Pulse rate: Bland-Altman error plot for connected bands RGB.

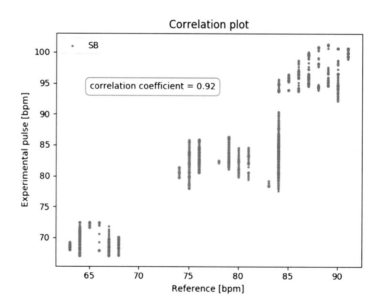

Fig. 5. Pulse rate: correlation between experimental and reference measurements for connected bands RGB and SWIR.

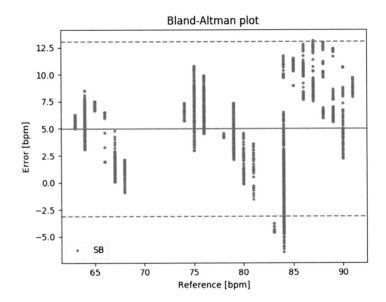

Fig. 6. Pulse rate: Bland-Altman error plot for connected bands RGB and SWIR.

an additional SWIR band. In the best case scenario, the correlation coefficient between the camera and the reference contact finger measurements equals 0.92, which is an acceptable value for the video pulse monitoring method. Setup was resistant to small patient's movements.

Acknowledgement. This work was supported by the National Centre for Research and Development OPTO-SPARE, grant agreement PBS3/B9/41/2015.

References

1. Hassan, M.A., Malik, A.S., Fofi, D., Saad, N., Karasfi, B., Ali, Y.S., Meriaudeau, F.: Heart rate estimation using facial video: a review. Biomed. Sig. Process. Control **38**, 346–360 (2017)
2. Wang, W., den Brinker, A.C., Stuijk, S., de Haan, G.: Robust heart rate from fitness videos. Inst. Phys. Eng. Med. Physiol. Meas. **38**, 1023–1044 (2017)
3. Rouast, P.V., Adam, M.T.P., Chiong, R., Lux, E.: Remote heart rate measurement using low-cost RGB face video: a technical literature review. Front. Comput. Sci. **1**, 1–15 (2016)
4. Poh, M.-Z., McDuff, D.J., Picard, R.W.: Non-contact, automated cardiac pulse measurements using video imaging and blind source separation. Opt. Express **18**, 10762–10774 (2010)
5. Tarvainen, M.P., Ranta-aho, P.O., Karjalainen, P.A.: An advanced detrending method with application to HRV analysis. IEEE Trans. Biomed. Eng. **49**(2), 172–175 (2002)
6. Verkruysse, W., Svaasand, L.O., Stuart, N.J.: Remote plethysmographic imaging using ambient light. Opt. Express **16**, 21434–21445 (2008)

7. Verkruysse, W., Svaasand, L.O., Nelson, J.S.: Remote plethysmographic imaging using ambient light. Opt. Express **16**, 21434–21445 (2008)
8. Lewandowska, M., Rumiński, J., Kocejko, T., Nowak, J.: Measuring pulse rate with a webcam - a non-contact method for evaluating cardiac activity. In: 2011 Federated Conference on Computer Science and Information Systems (FedCSIS), Szczecin, pp. 405–410 (2011)
9. de Haan, G., Jeanne, V.: Robust pulse rate from chrominance-based rPPG. IEEE Trans. Biomed. Eng. **60**(10), 2878–2886 (2013)
10. Procházka, A., Schätz, M., Vyřata, O., Valiř, M.: Microsoft kinect visual and depth sensors for breathing and heart rate analysis. Sensors **16**(7), 996 (2016)
11. Kowalski, M., Naruniec, J.: Face alignment using k-cluster regression forests with weighted splitting. IEEE Sig. Process. Lett. **11**(23), 1567–1571 (2016)
12. Huang, N.E., Shen, Z., Long, S.R., Wu, M.C., Shih, H.H., Zheng, Q., Yen, N.-C., Tung, C.C., Liu, H.H.: The empirical mode decomposition and the Hilbert spectrum for nonlinear and nonstationary time series analysis. Proc. Roy. Soc. Lond. A **454**, 903–995 (1998)
13. Wang, W., Brinker, A., Stuijk, S., Haan, G.: Algorithmic principles of remote PPG. IEEE Trans. Biomed. Eng. **64**(7), 1479–1491 (2017)
14. Kowalski, M., Naruniec, J., Trzcinski, T.: Deep alignment network: a convolutional neural network for robust face alignment. In: CVPRW (2017)

Initial Research on Fruit Classification Methods Using Deep Neural Networks

Zbigniew Nasarzewski[1(✉)] and Piotr Garbat[2]

[1] Institute of Radioelectronics and Multimedia Technology,
Warsaw University of Technology, Warsaw, Poland
z.nasarzewski@ire.pw.edu.pl
[2] Institiute of Microelectronics and Optoelectronics,
Warsaw University of Technology, Warsaw, Poland
p.garbat@elka.pw.edu.pl

Abstract. In the following paper, the methods from deep learning area, with the classical method of machine learning, have been compared. The fine-tuned networks: VGG, ResNet and MobileNet, as well as SVM, have been put together. Models of neural networks were trained using own dataset of fruit photos taken in laboratory conditions with the help of high-speed industrial cameras.

Keywords: Fruit classification · Object recognition ·
Image recognition · Image processing · Machine learning ·
Deep learning · Deep neural network · SVM classifcator ·
Image processing applications

1 Introduction

The study is part of the research carried out as part of a project aiming at the implementation of an industrial device for automatic optical sorting of fruit with the possibility of automatic separation of full-value and damaged fruit with defects. The effect of industrial research will be the development of a complete detection system and obtaining a ready solution in the form of a prototype ready to work in conditions close to reality. As part of the development work, the machine's construction, control systems and optimization will be performed by optical, pneumatic and mechanical systems. The sorting machine for cherries, blueberries, strawberries and cranberries will work with an approval for use in the food industry.

The test stand will be equipped with matrices illuminated with special lamps and a set of cameras that will take a set of photos analyzed by a computer located in the device, linked with a controller and display. After examining and determining where the fruit is damaged or if it is of low quality, the next step will be to remove it by means of pneumatic nozzles. The machine's capacity is about 3–3.5 tons of fruit per hour. Accuracy sorting min. 90% and wrong fruit

© Springer Nature Switzerland AG 2020
M. Choraś and R. S. Choraś (Eds.): IP&C 2019, AISC 1062, pp. 108–113, 2020.
https://doi.org/10.1007/978-3-030-31254-1_14

classification error below 3%. The result of the project will be a technological sorter in the final form ready to be implemented in the target line.

The biggest challenge in the research is to find the optimal method for quickly distinguishing small objects - a minimum size of 10 mm. Research focused on the most difficult fruit, i.e. blueberries and their classification based on deep neural network and machine learning methods.

2   Releated Solutions

In recent years we have been experiencing big impacts of new solutions in the field of machine learning. The vast majority focus on the recognition and detection of objects in high resolution images. Successfully available solutions on the road fine-tuning [1] can be adapted to a different category of problems.

VGG architecture is noteworthy and adapted to this category of problem because it reinforced the notion that convolutional neural networks have to have a deep network of layers in order for this hierarchical representation of visual data to work [2].

Deep residual networks (ResNet), provided the breakthrough idea of identity mappings in order to enable training of very deep convolutional neural networks. This network architecture set new records in classification, detection, and localization [3].

Mobile solutions are also available, which can be used on devices with less computing power, characterized by high efficiency and simple architecture. MobileNets are small, low-latency, low-power models parameterized to meet the resource constraints of a variety of use cases [4].

Machine learning technique performance in terms of use have been confirmed in areas of small objects recognition and few outstanding achievements obtained using deep learning [5]. There are also studies and dedicated datasets that allow you to train the network on characteristic collections of fruits [6]. There are also studies that describe the detection of fruit - blueberry fruit - subject to the following description [7].

3   General Concept

Market standards accept two approaches to the implementation of the camera-data acquisition system. It is possible to mount cameras allowing observation of fruit in flight or a stable moving system on the belt (see Fig. 1). After preliminary tests (details outside the area of interest of this article), the second solution was selected as the starting point for further research.

Data acquisition for the needs of the research was carried out based on an industrial cam as well as a standard RGB full-HD cam. Dataset based on blueberry fruit has been divided into two general categories: (a) "NOK" category - containing spoiled and dried lower class fruit as well as organic scrap (leaves, fragments of twigs, seeds) and inorganic (plastic, pieces of metal, pieces of paper) and (b) "OK" category - height quality, fresh fruit. The fruit was subjected to

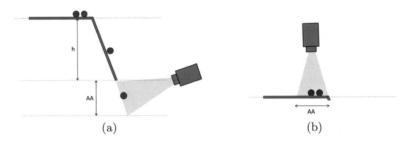

Fig. 1. Two research systems considered: (a) free fall of the fruit (b) tape with a horizontal sliding fruit system under the camera.

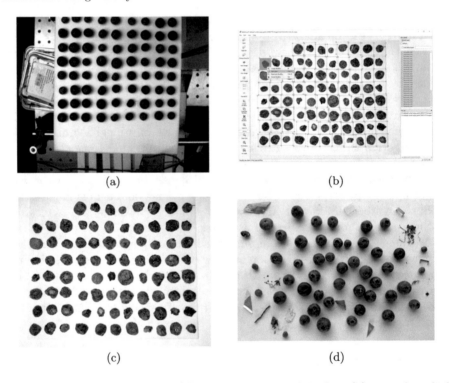

Fig. 2. Objects used for training: (a) raw photography of the fruit (b) manual marked poor quality fruit (c) clean picture of 'NOK' set of data (d) test picture with good quality fruit - 'OK' with 'NOK' objects.

manual marking by the operator, photographed waste was marked automatically (see Fig. 2).

4   Initial Analaysis

The created dataset was used to train SVM as well as selected deep neural network models.

4.1   Traditional Machine Learning

The proposed traditional solution basis of a histogram and HOG descriptor and an SVM classifier. The detection process is realised for two types of descriptors: a two dimensional histogram (8 × 8 bins) and HOG descriptor. In the next step a linear SVM (Support Vector Machine) classifier was instructed to classify good or bad fruit. The method demonstrated high detection speed and high accuracy (over 99%), more details Table 1. The ROC curve was shown on Fig. 3.

Table 1. Results of SVM classification.

Solution	Accuracy	Precision	ROC_AUC
Histogram 2D (8 × 8 bins)	0.975	0.988	0.996
HoG	0.967	0.945	0.994
Histogram 2D + HoG	0.992	0.981	0.998

4.2   Deep Learning

In order to carry out the experiments, pre-trained, based on the ImageNet dataset network was used. The implementations available in the Tensorflow package with the help of Keras were also used. Pre-trained network architectures were fine-tuned, previously trained top-layer were reduced, the base weights

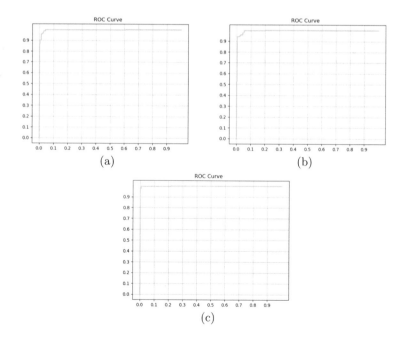

Fig. 3. ROC curves for three type of descriptors: (a) Histogram 2D, (b) HoG, (c) Both Histogram 2D and HoG.

were frozen, and the new top-layer retrained - Table 2 lists the parameters of the network. Dataset consists of 2700 images with resolution of 80 × 80 pixels each (divided into training, validation and test subset). Train set were divided into three groups: train (2216 images), validation (553 images) with general split: 0.2 and test data (40 images) all belonging to 2 classes.

Table 2. Deep neural networks and trainable params.

DNN	Total params	Trainable params
VGG16	134,268,738	8,194
MobileNet	3,230,914	2,050
ResNet50	23,591,810	4,098

Among the networks tested, the highest prediction speed characterizes the VGG16 network, which maintains the highest efficiency, followed immediately by MobileNet and ResNet50. The set parameters below (Table 3).

Table 3. Deep neural networks speed and efficiency of prediction on test data.

DNN	Prediction	True negative	False positive	False negative	True positive
VGG16	17 ms	20	0	0	20
MobileNet	30 ms	20	0	6	14
ResNet50	70 ms	17	3	0	20

Below are the charts presenting the accuracy and loss of binary cross entropy over the 300 epochs of training (see Figs. 4, 5 and 6).

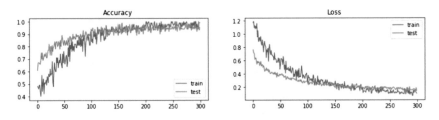

Fig. 4. VGG16 accuracy and loss.

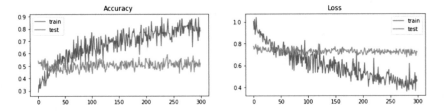

Fig. 5. MobileNet accuracy and loss.

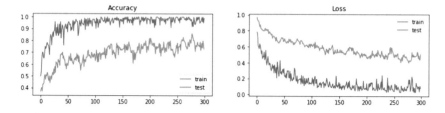

Fig. 6. ResNet50 accuracy and loss.

5 Summary

This study compared two alternative methods for detecting blueberries. The first one based on a Histogram and HOG descriptor and an SVM classifier, the second on deep learning classifier. Both methods demonstrated high detection accuracy 90% for our datasets. Early experiences showed that classical methods are faster than deep learning methods, however, by optimizing the complexity of the network architecture, the speed of prediction can be obtained - which is the subject of further work and the results will be presented in the extension of the article.

Acknowledgement. This work was supported by the National Centre for Research and Development, project: Operational Programme Intelligent Development 2014–2020, edition: 5/1.1.1/2017.

References

1. Thrun, S.: Is learning the n-th thing any easier than learning the first? (1996)
2. Simonyan, K., Zisserman, A.: Very deep convolutional networks for large-scale image recognition. In: ICLR (2015)
3. He, K., Zhang, X., Ren, S., Sun, J.: Deep residual learning for image recognition (2015)
4. Howard, A.G., Zhu, M., Chen, B., Kalenichenko, D., Wang, W., Weyand, T., Andreetto, M., Adam, H.: MobileNets: efficient convolutional neural networks for mobile vision applications (2017)
5. Sa, I., Ge, Z., Dayoub, F., Upcroft, B., Perez, T., McCool, C: DeepFruits: a fruit detection system using deep neural networks (2016)
6. Muresan, H., Oltean, M.: Fruit recognition from images using deep learning (2018)
7. Tan, K., Lee, W.S., Gan, H., Wang, S.: Recognising blueberry fruit of different maturity using histogram oriented gradients and colour features in outdoor scenes (2018)

3D Optical Reconstruction of Building Interiors for Game Development

Mariusz Szwoch[1(✉)] and Dariusz Bartoszewski[2]

[1] Gdańsk University of Technology, Gdańsk, Poland
`szwoch@eti.pg.edu.pl`
[2] Forever Entertainment s.a., Gdynia, Poland
`dariusz.bartoszewski@forever-entertainment.com`

Abstract. This paper presents current possibilities of using stereophotogrammetry to scan interiors for video games. This technology offers great possibilities for effective room scanning based on popular cameras, however, it also has some limitations in the implementation of a fully automatic image processing. As the result of the STERIO project, the Sterio Reconstruction tool has been created that offers semi-automatic 3D reconstruction of building interiors. The proposed technology can be used, among others, for modeling rooms for video games which action takes place in locations reflecting real interiors. The original, large-scale experiment has been carried out in order to validate the created technology in real conditions. The goal of the experiment was to develop a prototype 3D video game by a small development studio using created solutions. The experiment showed great possibilities of the developed tool and the technology used.

Keywords: Image processing · Stereo-photogrammetry ·
3D scanning · Video games · Interiors reconstruction

1 Introduction

Three-dimensional vision plays an extremely important role in the everyday life of a human being, allowing, among others, for orientation in the field, avoiding obstacles, efficient movement and precise handling. For many years, it has also found numerous applications in digital systems. Among the many fields of applications, one can mention here simultaneous localization and mapping (SLAM) systems for robotics and autonomous vehicles, vehicle support systems (active cruise control, obstacle detection systems, and road analysis), optical quality control, 3D scanning of static and animated objects.

The latter solutions can be used for the development of modern video games that require hard use of high quality 3D assets for both game characters and the

This work was created within the Sectoral Programme GAMEINN within the Operational Programme Smart Growth 2014–2020 under the contract no POIR.01.02.00-00-0140/16.

scenery. In the traditional approach, all such assets are created by 3D artists or bought from a store. The first approach is a very labor-intensive task, while the second one does not guarantee uniqueness of bought assets, what is easily recognized by players. In case when game concept requires that the action takes place in a location that represents a real world place, traditional, handmade modeling seems to be non-creative, yet time and labor consuming task that could be replaced by automatic or at least semi-automatic software solutions. Our studies and large-scale experiments have shown that automation of room reconstruction significantly reduces the time and cost of modeling the 3D scene [1]. Moreover, it also supports rapid prototyping giving preliminary object models, which is a very important element of agile development process and often determines the further direction of the game development. For those reasons, it is very important to accelerate the process of modeling 3D scenes that are meant to reflect real objects. Three-dimensional scanning has already been successfully used in the development of several commercial games, such as The Vanishing of Ethan Carter by The Astronauts in 2014, Star Wars Battlefront by EA DICE in 2015, and Get Even by The Farm 51 in 2017.

There are available several technologies allowing for reconstruction of 3D objects, including photogrammetry [2,3], laser based devices, e.g. light detection and ranging (LIDAR) [4] and time-of-flight (ToF) technologies, and finally depth sensors [5,6]. Unfortunately, all of the mentioned solutions have some drawbacks that limit their usage in video game industry, especially by small and medium developer teams. Laser based scanners are very precise as to geometry scanning but are very expensive and do not support textured models. ToF sensors are cheaper but have limited resolution and also do not provide texture information. In turn, RGB-D depth sensors, based mostly on the structured infrared light patterns, like Microsoft Kinect or Intel RealSense, can provide complete 3D models of usually enough quality but are limited to small distances of a few meters. Finally, photogrammetry methods can provide high quality textured models but require numerous input data and high computational and memory resources. Moreover, it often fails to fully reconstruct objects with flat, uniform surfaces and poorly illuminated ones.

In this paper, the main concept of the STERIO project is presented, which aims to overcome problems of above approaches by using photogrammetry and stereo-photogrammetry methods for 3D scanning of building interiors for game development purposes at reasonable financial costs. The performed experiments proved that the proposed approach is faster than traditional one and allows to provide good quality models that can be used in video games.

2 Background

Three-dimensional video games are the most spectacular part of the market as they allow to present reliable game worlds that provide deep immersion experience. 3D games dominate primarily on the platforms of personal computers and video consoles, but they become an essential and growing part of games for mobile and browser platforms. Because the action of many 3D games takes place in worlds that resemble, reflect or even simulate our world, the creation of realistic models and scenes has become a very important task of 3D graphics, animators, and level designers. In general, scenes can be divided into outdoor and indoor ones with different characteristics, requirements and limitations. External scenarios usually include a wide area with diversified terrain, vegetation cover and numerous objects, e.g. buildings, vehicles, animals, opponents, etc. It requires, among other things, paying special attention to the scene rendering performance, which in some cases may require the use of simpler objects with low number of vertices (low-poly). In turn, indoor scenes have a very limited horizon, which allows the player to register the details of rooms, which usually enforces the use of more accurate models with more vertices (mid- and high-poly).

In the case of manual modeling of objects, high-poly models are usually created, which can be easily converted to simpler ones tailored to the needs of the designer. However, manual approach is time-consuming and requires high imaginary, design, and modelling skills. That is why different methods, algorithms, and tools supporting the 3D scanning have been developed recently. In general, they can be divided into *active* methods that use some additional light source for measurements and *passive* ones, that are based on the image analysis of the existing, non-enriched scene. Active methods may use the measurement of flight time or phase difference (ToF) of a laser beam reflecting from the object's surface (e.g. Microsoft Kinect v.2) or use triangulation method based on lasers (LIDAR) or a structural light patterns (e.g. Microsoft Kinect v.1). Precise active scanners are usually very expensive and often scan only geometry of the models. In turn, passive methods generally use triangulation on a set of images taken by calibrated stereo-camera set (stereo-photogrammetry) or by different, independent and not calibrated cameras (multi-view stereo, MVS). The first approach demands more effort to properly take pictures but provides higher precision and can deal with many problematic issues.

3D scanning technology has been developed for the last decades and used in different fields, such as preserving objects significant for a cultural heritage [7], reconstruction of buildings for 3D maps, character modeling for 3D animations etc. Initially, most efforts focused on scanning small and medium sized 3D objects using relatively cheap cameras, 3D scanners, and depth sensors based on structural-light technology [8]. Unfortunately, due to their limitations, such 3D scanners cannot be used for scanning bigger objects such as buildings and their interiors.

Modeling of larger objects and whole 3D locations could be carried out mainly using expensive laser scanners which are available to larger development studios [9, 10]. Laser based technics do not provide color information, unless hybrid laser-camera systems are used. With a smaller game budget, the modeling of larger assets remained the domain of 3D graphic designers. It is only recently that methods allowing the use of cameras for scanning buildings, their interiors, and even vegetation have attracted a lot of interest [5, 7–9].

3D scans can be also obtained with use of standard, optical cameras which are very accessible and relatively cheap. In general, there are two kinds of methods for acquiring 3D scans with cameras, which are *photogrammetry* and *stereo-photogrammetry*. The photogrammetric approach bases on a series of usually unrelated photos taken from around a scanned object. Special algorithms, such as scale-invariant feature transform (SIFT) or speeded up robust features (SURF) are used to detect the characteristic points in those images [11]. Matching algorithms, such as random sample consensus (RANSAC), are then used to find the correspondence between images pair. On this base, algorithms, such as structure from motion (SfM), allow to reconstruct a 3D point cloud representing the scanned object. This approach is most popular nowadays as it allows for the reconstruction of buildings, plants, and other objects based on uncorrelated images possible taken using different cameras at different time, e.g. as is the case in social media services. This approach is also used in a professional tools, such as Agisoft Photoscan [12]. The greatest disadvantage of this approach is the computational power needed to process big sets of images.

Another technique which allows to retrieve a 3D view of objects is stereo-photogrammetry that belongs to the same class of technologies as stereo vision [3]. In this method a rigid set of two cameras is used called a stereo-camera. As the relative position of cameras is fixed, it is possible to reconstruct partial point cloud of the scanned object directly from a stereo-image, simplifying the calculations time. This can be an essential advantage over SfM in some application area such as rapid modeling and 3D scenes prototyping in video games. Although, this approach requires using a tripod, it can become another advantage in some situations, such as insufficient illumination or hard-to-process surfaces, when it is possible to take several photos from exactly the same location in different lighting conditions [13].

3   The STERIO Project

The goal of the STERIO project was to develop a low-cost technology for making 3D scans of building interiors based on images from standard photo cameras [13]. The results of the project developed by Forever Entertainment s.a. allow for faithful recreation of real world scenery in the virtual world of video games. Applying the proposed technology also allows for more effective and less costly location generation for games which action takes place in real locations. Such approach will have a big advantage over the indirectly competitive 3D laser-scanning technology as it produces a similar effect with fewer resources and shorter production time. The technology is intended for use in small- and medium-sized game development studios.

3.1   The STERIO Data Set

During the project, a comprehensive collection of over 10,000 photos of 26 various interiors were taken. These photos, single and stereo pairs, were used to develop, test and verify the algorithms and tools being developed. For each location, a different number of data subsets were prepared, consisting of images taken under different lighting conditions with different exposition parameters, camera location and orientation points, additional visual markers displayed, and other experiment setting. Some datasets are accompanied by additional images of a special chessboard pattern for cameras calibration. Sample images from the STERIO dataset are presented in Fig. 1.

(a)   (b)

Fig. 1. Sample images from the STERIO dataset: the restaurant's cellar (a), the university hall (b).

3.2   Sterio Reconstruction Tool

The main effect of the project is the Sterio Reconstruction tool that allows designer to perform a full process of processing the set of input images of the object into its 3D model. The tool uses image processing algorithms based on open source OpenMVG and openMVS libraries [14]. This tool offers a user-friendly interface that allows full control over the reconstruction process, starting from selection of input photos, through configuration of parameters of individual processing stages, saving of intermediate data, and presentation of results (Fig. 2). SR offers also some manual and semi-automatic tools allowing for modifying intermediate point clouds, flattening regions etc.

The whole processing pipeline consist of several steps including (Fig. 3):

– collecting photos, initial processing, and determining characteristic points (landmarks);
– searching for pairs of pictures covering a common fragment of a room;
– calculating relative and global camera positions and orientations for each analyzed pair;
– building local sparse point clouds and fitting them into the global sparse point cloud;
– building global dense point cloud;
– building object's mesh;
– building textures;
– decimating and retopologizing the model.

Fig. 2. Example screen of the Sterio Reconstruction tool.

Fig. 3. Visualization of main steps of image processing in the Sterio Reconstruction tool: calculating camera positions and orientations (a), building global sparse (b) and dense (c) point cloud, building object's mesh (d), decimating the mesh (e), retopologizing the model with textures (f).

4   Experiments

The goal of the experiment was to validate the usefulness of the developed technology and tools in the video game industry. A large-scale experiment was carried out, using them in the development of a prototype video game, which takes place in locations based on real places. Within the experiment, complete design work was carried out that allowed to create a prototype survival horror game "The Mansion", containing one complete level. Particular attention was paid to the design of the game world, especially the layout of the title mansion, allowing for the correct placement of chambers and rooms created using the developed solution (Fig. 4a). As part of the task, a complete scenography was created, consisting of a reconstructed mansion, additional rooms and an open area with small architecture. A set of several simple logic puzzles allows players to explore the area and assess the quality of its components.

(a)   (b)

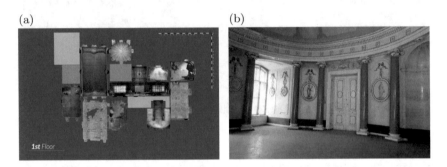

Fig. 4. The designed 1st floorplan for The Mansion game (a), example photography of the Bożków Mansion (The Pillar Room).

In order to create interiors corresponding to the actual location, a complete photographic session was made in accordance with the developed recommendations for a historic mansion from the 16th century located in the village of Bożków in Silesia region (Fig. 4b). Searching for the target location revealed a big threat to this type of projects, which is the lack of consent or exorbitant financial expectations of property owners. All photos were taken in accordance to the previously developed recommendations for image acquisition for 3D room scanning [15]. These recommendations advise how to cover the room with photographs, properly balancing the completeness of the cover guaranteeing the completeness of the final model and the number of photographs deciding on the processing time and the required memory. Additional criteria set out the rules for exposure of photographs, as well as cases in which you can use individual photos to speed up the work, and when you need to use a stereophoto set to get better pictures in dark rooms or flat surfaces with no landmarks. On such surfaces special graphical pattern was displayed that gives the best results according to previous tests and experiments [1].

According to the designed plan of the rooms, several chambers were selected from the mansion in Bożków for full reconstruction and use in the prototype game. For each room an initial photo set was chosen for the Sterio Reconstruction tool according to the experience of the level designer. After a full reconstruction of the room, the model was inspected for its completeness. In the case of significant space shortages to the initial set of photographs, additional photos of a given part of the room were added.

Each fully reconstructed room underwent further processing aimed at simplifying the mesh of the model. At this stage, the built-in Sterio Reconstruction tool was used to flatten the selected surfaces, such as walls, ceilings, floors or their fragments. This operation allows for a radical reduction in the number of mesh tops. However, it causes some problems with the continuity of the grid at the border of the simplified area, which makes it necessary to leave a certain margin allowing its smooth connection with the environment. After each change of the mesh geometry one should also remember about the need to convert the texture mapping (retopologization).

In the next stage, the model is automatically simplified (decimation) to the extent that can be used in the game. Unfortunately, modern decimation algorithms still do not work optimally. If the level of reduction is too high, it often causes different artifacts, such as lack of edge and corner retention, highly irregular surface division, disappearance of details in important places, etc. Therefore, each reconstructed model was subjected to final processing by 3D graphics, whose task was to eliminate any imperfections of the model and its optimization and adaptation to the needs of the game. An exemplary appearance of the models of the reconstructed rooms is shown in Fig. 5.

(a)   (b)

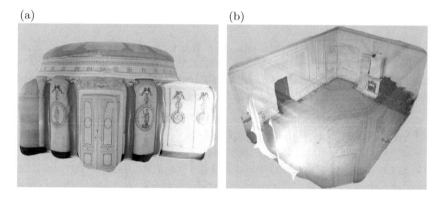

Fig. 5. An exemplary appearance of the models of the reconstructed rooms: The Pillar Room (a), The White Room (b).

Finally, optimized models were imported into Unity 3D engine and put together, creating a plan of the first floor of the title mansion. In the level design process the models were combined with other room models, hand-modeled due to their nature, preventing them from 3D scanning (the orangery). All rooms have been adequately lit, enriched with detail models, interaction elements and additional special effects, e.g. floating dust, moon glow, mirror reflections, etc. An example of the appearance of rooms in the game is shown in Fig. 6.

The entire process of reconstruction of the room was assessed by the level designer and 3D graphic designer in terms of suitability in the process of designing the rooms in the actual course of production. The overall rating was positive. The level designer emphasized the advantages of quickly receiving room prototypes, omitting the detailed concept sketches required in the traditional production process. In turn, 3D graphic designer emphasized the generally good quality

Fig. 6. An exemplary appearance of the rooms during the gameplay.

of reconstructed models and their compatibility with 3D modelling tools like 3ds Max, Blender 3D and ZBrush. Although models still required a relatively large amount of work to eliminate all shortcomings, the process of creating models was faster than in the traditional approach.

5   Conclusion and Future Work

The experiment carried out with the use of the Sterio Reconstruction tool showed the high suitability of the photogrammetric approach for room reconstruction. Tools of this type can be valuable elements of the game design and development process, whose locations are based on real places.

The solutions proposed within the project result in greater efficiency in the reconstruction of rooms [15]. You can mention here about the set of recommendations regarding taking of photographs extremely important due to the high costs of potential repetition and photo session. Comprehensive collection of photographs of diverse rooms is a great base for testing and developing algorithms [13]. The proposed method for displaying a special graphic patterns with an LCD projector in rooms with poor lighting or flat surfaces allows to effectively scan these rooms using stereo-photogrammetry [1].

The conducted research also indicated some limitations of the photogrammetric approach. One of the most important limitations is very large computing and memory power requirements for image matching algorithms, creation and joining of point clouds. This significantly limits the possibility of using typical workstations, even with usage of the power of graphics cards (GPGPU). On such devices one should use a smaller set of pictures, which does not guarantee full coverage of the reconstructed room and thus completeness of the obtained model mesh that must be corrected manually. The solution to the problem may be the use of expensive, specialized workstations or the use of a cloud computing model.

Another problem is the correct decimation of the model grid. It is a well-known problem of combining simple and complex walls in 3D modeling, for which a completely satisfying solution has not been developed so far. This limits cooperation in some way with the proposed semi-automatic options for inviting flat surfaces, enforcing the need to leave a certain margin of unprocessed vertices. One solution may be development of decimation methods that keep edges and corners.

Another group of problems is related to the availability and condition of scanned rooms, which should also be taken into account in this type of projects. During the project there were unexpected problems with finding a manor house that could be formally used to implement the game. In many situations, some obstacles were: legal problems (lack of owner), lack of consent for temporary removal of furniture or absurdly high costs of granting permission for use in the production of the game.

Future work includes further development of photogrammetric image processing algorithms and the Sterio Reconstruction tool to overcome all mentioned problems and to maximize the automation level of the entire process. In particular, the efficiency of cloud computing and high-performance workstations will be examined. Advanced algorithms for flattening the surface and decimation will also be developed.

References

1. Szwoch, M., Kaczmarek, A.L., Bartoszewski, D.: Using stereo-photogrammetry for interiors reconstruction in 3D game development. In: Choraś, M., Choraś, R. (eds.) Image Processing and Communications Challenges 10, IP&C 2018. Advances in Intelligent Systems and Computing, vol. 892. Springer, Cham (2019)
2. Mikhail, E.M., Bethel, J.S., McGlone, J.C.: Introduction to Modern Photogrammetry. Wiley, New Jersey (2011)
3. Kaczmarek, A.L.: Stereo vision with Equal Baseline Multiple Camera Set (EBMCS) for obtaining depth maps of plants. Comput. Electron. Agric. **135**, 23–37 (2017)
4. Musialski, P., Wonka, P., Aliaga, D.G., Wimmer, M., Van Gool, L., Purgathofer, W.: A survey of urban reconstruction. Comput. Graph. Forum **32**(6), 146–177 (2013)
5. Escorcia, V., Dávila, M.A., Golparvar-Fard, M., Niebles, J.C.: Automated vision-based recognition of construction worker actions for building interior construction operations using RGBD cameras. In: Construction Research Congress (2012)
6. Szwoch, M., Pieniążek, P.: Detection of face position and orientation using depth data. In: The 8th International Conference on Image Processing and Communications; Image Processing and Communications Challenges 7 Book Series: Advances in Intelligent Systems and Computing, vol. 389, pp. 239–251. Springer, Cham (2016)

7. Balsa-Barreiro, J., Fritsch, D.: Generation of visually aesthetic and detailed 3D models of historical cities by using laser scanning and digital photogrammetry. Digit. Appl. Archaeol. Cult. Herit. **8**, 57–64 (2017)
8. Ha, H., Oh, T.H., Kweon, I.S.: A multi-view structured-light system for highly accurate 3D modeling In: International Conference on 3D Vision (3DV), pp. 118–126. IEEE (2015)
9. Kazmi, W., Foix, S., Alenya, G.: Plant leaf imaging using time of flight camera under sunlight, shadow and room conditions. In: International Symposium on Robotic and Sensors Environments (ROSE), pp. 192–197. IEEE (2012)
10. Tse, R.O., Gold, C., Kidner, D.: 3D city modelling from LIDAR data. In: van Oosterom, P., Zlatanova, S., Penninga, F., Fendel, E.M. (eds.) Advances in 3D Geoinformation Systems. Lecture Notes in Geoinformation and Cartography. Springer, Heidelberg (2008)
11. Bay, H., Tuytelaars, T., Van Gool, L.: SURF: speeded up robust features. In: Leonardis, A., Bischof, H., Pinz, A. (eds.) Computer Vision – ECCV 2006. ECCV 2006. Lecture Notes in Computer Science, vol. 3951. Springer, Heidelberg (2006)
12. Agisoft PhotoScan user Manual (2014). https://www.agisoft.com/pdf/photoscan-pro_1_4_en.pdf. Accessed 10 July 2019
13. Szwoch, M., Kaczmarek, A.L., Bartoszewski, D.: STERIO - reconstruction of 3D scenery for video games using stereo-photogrammetry. In: Napieralski, P., Wojciechowski, A. (eds.) Computer Game Innovations in Monograph of the Lodz University of Technology. Łódź University of Technology, Łódź (2017)
14. Kaczmarek, A.L., Szwoch, M., Bartoszewski, D.: Using open source libraries for obtaining 3D scans of building interiors. In: Borzemski, L., Świątek, J., Wilimowska, Z. (eds.) Information Systems Architecture and Technology: Proceedings of 39th International Conference on Information Systems Architecture and Technology - ISAT 2018. ISAT 2018. Advances in Intelligent Systems and Computing, vol. 852. Springer, Cham (2019)
15. Szwoch, M., Kaczmarek, A.L., Bartoszewski, D.: Recommendations for image acquisition for 3D room scanning. J. Appl. Comput. Sci. **26**(2), 45–55 (2018)

A Simplified Classification of Electronic Integrated Circuits Packages Based on Shape Descriptors

Kamil Maliński and Krzysztof Okarma[(✉)] [iD]

Department of Signal Processing and Multimedia Engineering,
Faculty of Electrical Engineering,
West Pomeranian University of Technology in Szczecin,
26 Kwietnia 10, 71-126 Szczecin, Poland
okarma@zut.edu.pl

Abstract. Automatic classification of electronic elements based on image analysis may be useful for verification of a proper selection of electronic integrated circuits previously assembled in the Printed Circuit Boards (PCBs), as well as for an automated sorting of such elements in robotic systems supporting their production. Although modern high speed pick and place machines dedicated for small surface-mount devices (SMD), utilising very small packages, largely replaced the through-hole assembling technology in many industrial application, there are still some types of applications where such technology is less suitable, e.g. due to thermal, mechanical or power constraints. Hence, a proper classification of electronic elements in dual in-line packages (DIP) using shape analysis may be an important element for the combination with further steps based on the recognition of alphanumerical markings. Some experimental results obtained for selected shape descriptions are presented in this paper, which are promising also for natural images.

Keywords: Classification · DIP · Shape analysis · Electronic packages

1 Introduction

Automatic classification of electronic components according to their package shape, such as popular dual in-line packages (DIP) containing various elements, such as transistors, switches, LEDs, resistors, etc., may be useful in semi-automatic assembling systems or prototyping of electronic circuits, in which the application of surface-mount devices (SMD) is not possible. Assuming unknown location of unsorted individual packages, similar methods may be useful in robotic applications, as well as for diagnostic purposes. Possible applications of such image based recognition might be related to checking the connections as well as the type of electronic components mounted on the printed circuit board (PCB).

© Springer Nature Switzerland AG 2020
M. Choraś and R. S. Choraś (Eds.): IP&C 2019, AISC 1062, pp. 125–133, 2020.
https://doi.org/10.1007/978-3-030-31254-1_16

Although an automatic classification of electronic components is not very popular topic of scientific research, its practical usefulness may be confirmed e.g. by the artificial intelligence system introduced in 2017 by Fujitsu company[1], utilizing template pattern matching during inspecting parts for misalignment. Some other attempts to recognition of electronic elements are typically based on the analysis of hand-drawn circuit diagrams [2,9] or analysis of transfer function [1] without computer vision methods.

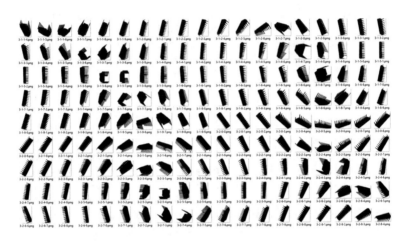

Fig. 1. A fragment of the set of binary images obtained assuming various views of an exemplary DIP.

Assuming the potentially limited visibility of alphanumerical markings, an important element of such systems, useful also for initial limiting the number of checked types of elements, is the recognition of the integrated circuit (IC) package type. This task may be conducted applying some of the shape analysis methods, assuming previous binarization of the image captured by the camera. Nevertheless, for unknown relative position (and rotation) of the individual IC packages, their proper recognition may be troublesome due to similarity of shapes obtained for various DIPs for different angles of observations. Therefore, an appropriate combination of shape descriptors would be necessary to ensure the proper classification regardless of the relative position of the analysed DIP.

The main purpose of research presented in the paper is the analysis of possible application of various simple shape descriptors and parameters during the initial stage of classification of IC packages without the use of text recognition. The first part of the conducted experiments consists of the classification of binary DIP images obtained from the intentionally prepared database containing the synthetic images of the STL 3D models captured assuming different views. After extensive experiments some of the most relevant shape descriptors were selected

---

[1] AI-Enabled Image Recognition System to Revolutionize the Manufacturing Line, Available online: https://journal.jp.fujitsu.com/en/2017/04/19/01/.

applying the Principal Component Analysis (PCA) method. Finally, some additional experiments were made using natural images as described in the respective sections of the paper.

2 Database Preparation

Preparation of the image database used in experiments was made using Javascript and PHP scripts. Initially, the STL 3D model was loaded and its screenshot was captured for a specified location of a virtual camera, after manual positioning on the canvas. Next, the captured image was cropped to ensure it contains only a single object and binarized for a faster further shape analysis. After calculation of shape parameters all data was saved in MySQL database. Each model was rotated by 18° around the X axis. After the rotation by 180° it was rotated around the Y axis, and after 180° around the Z axis. It was assumed that the 180° angle should be enough due to the symmetry of the shapes. After capturing such images, partially illustrated in Fig. 1 for an exemplary IC package, the process was repeated starting from taking a first screenshot of the next DIP. Finally, the database was created, which contains 1000 images for each package, and the set of all measured values.

3 Proposed Method and Experimental Verification

3.1 Shape Descriptors

Considering the exemplary binary images illustrated in Fig. 1, it can be easily noticed that some of the simplest shape parameters, such as e.g. Feret's coefficient defined as the ratio of two Feret diameters, or just the area or the perimeter of an object, would not be sufficient for a reliable classification of IC package images. Therefore, after the application of PCA algorithm, the following shape descriptors were selected as the most representative ones, combining good separation of classes with their simplicity. Hence, in further experiments the following shape descriptors were considered:

- aspect ratio (width/height of the bounding rectangle),
- relative area (number of object's pixels/area of the bounding rectangle),
- circularity (area of the bounding rectangle/(longRadius$^2 \cdot \pi$),
- roundness (perimeter/(2 $\cdot \pi \cdot$ longRadius),
- center of mass (average coordinates of the pixels in the object),
- shape signature.

The shape signature was calculated in the following way:

- for each pixel of the perimeter the distance from the centre of mass and the angle were determined,
- all values were normalized dividing each value by the largest distance,
- the results were sorted according to the angle (clockwise),

Fig. 2. Images of various IC packages used in experiments.

Fig. 3. Images of various random shapes used in experiments.

– the results were divided into 72 bins (5° each) and the average of the distances in each bin was calculated.

The images of various IC packages used in experiments are presented in Fig. 2, whereas Fig. 3 illustrates some other random shapes used for the initial verification of DIP shapes recognition, similarly as in typical General Shape Analysis approaches [3,4].

3.2   Shape Similarity Calculation

The analysis of test images loaded by the user can be made using a dedicated testing website. The uploaded photo is converted into the binary image using the classical Otsu method [7] and the shape parameters described above are calculated. Next, the shape descriptors are compared with those stored in the database trying to find the best k matches for aspect ratio, area, circularity, roundness and centre of mass. For these parameters each similarity factor is calculated by dividing the smaller of the two compared values by the larger one and then multiplying by 100 to express the similarity as the percentage. The first-check score is the average of these percentages.

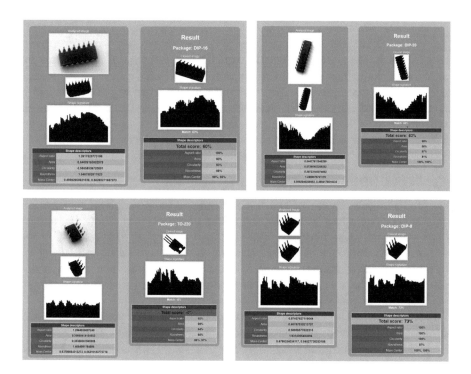

Fig. 4. Illustration of some exemplary results of classification.

In the second stage, the similarity factors between the shape descriptors for each result are determined - each element in the shape signature of the analysed image is compared with each element of the signature from the database. Since the value tends to stay below 10, the match percentage is calculated using the following formula:

$$Match\ percentage = 10 \cdot (10 - AD)\,, \tag{1}$$

where AD is the aggregated distance of the shape signature elements. The procedure is then repeated with the shape signatures shifted from -3 to 3. Each

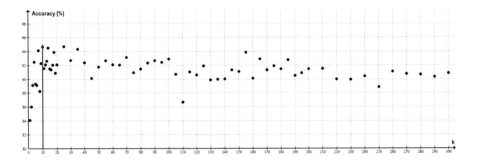

Fig. 5. Illustration of dependency of classification accuracy on the number of selected images using basic shape descriptors.

entry in the database differs from the next by 18°, and the shape signature has the 5° resolution, so by shifting the signature by 15° each side the best possible match can be found undoubtedly. Finally, the match percentage is multiplied by the average of all the others to calculate the highest score and choose it as the result. Some exemplary results of classification are shown in Fig. 4, where some problems caused by the presence of shadows can be noticed.

3.3 Experimental Verification

One of the disadvantages of the initial version of the proposed method was relatively high computational cost, hence the appropriate choice of the number of images chosen using basic shape descriptors (denoted as k) is crucial. To find the appropriate number of images k, some experiments were made using the synthetic reference "perfect" images. As illustrated in Fig. 5, the most reasonable choice seems to be $k = 10$ as too many selected images cause the necessity of calculation of their shape signatures.

Assuming $k = 10$ images selected using basic shape parameters, the verification of the system for real photos (containing only a single IC package after assumed segmentation) was conducted. For each image in the database, the algorithm checks, whether the recognized package is correct and then the accuracy can be determined, as shown in Table 1.

Another investigated problem was related to the choice of the best threshold for the detection of the integrated circuits among the other shapes, shown in Fig. 3. The goal of this experiment was to find the threshold of certainty, that allows to determine, whether or not, the image actually contains an IC package. Since the chosen images were going to be further analysed, the aim was to minimize false negatives, because false positives can be dealt with during the later stages of processing. The overall accuracy of 77% can be achieved for the threshold equal to 52, as shown in Fig. 6, whereas the number of false negatives is less than 10%. The detailed results are shown in Table 2, where TP, TN, FP and FN denote the numbers of true positives, true negatives, false positives and false negatives respectively (considering IC package images as "positives").

Table 1. Experimental recognition results obtained for real photos of IC packages.

Package type	Number of images	Correct	Incorrect	Accuracy
DIP-8	45	40	5	88.89%
DIP-16	45	34	11	75.56%
DIP-20	45	36	9	80.00%
TO-220	32	31	1	96.88%
All	167	141	26	84.43%

Table 2. Experimental results of detection of the IC packages among the other shapes.

	Total	IC packages	others	TP+TN	FP+FN	FP	FN
Number	365	167	198	282	83	51	32
Percentage	100	45.75	54.25	**77.26**	22.74	13.92	8.77

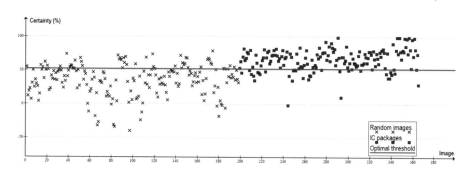

Fig. 6. Illustration of the experimentally selected threshold for the detection of IC packages.

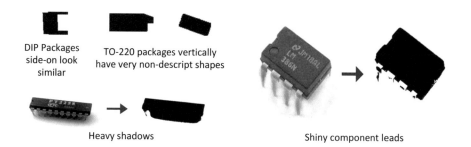

DIP Packages side-on look similar

TO-220 packages vertically have very non-descript shapes

Heavy shadows

Shiny component leads

Fig. 7. Illustration of the exemplary problems occurring during the recognition.

Analysing the obtained results, some of the most troublesome images can be found in the database, where the main sources of problems can be noticed for some specific angles as well as in the presence of heavy shadows or shiny component leads, being troublesome for global binarization. The illustration of such exemplary issues is shown in Fig. 7.

4   Conclusions and Future Work

Presented system allows for the detection of IC package images as well as its reliable recognition utilizing relatively simple shape descriptors. Its potential applications and extensions may be considered not only in electronics and robotics but also - in a modified version - in some other areas of industry, where the recognition of similar object shapes plays an important role. In contrast to General Shape Analysis, the presented approach is based on the two-stage combination of some simple shape parameters, therefore a natural direction of its further development might be the use of neural networks or some other machine learning methods to improve the classification stage.

One of its limitations, which should also be considered in further research, is the influence of lighting conditions, since the results of global binarization in the presence of heavy shadows have a significant impact on the results of further shape analysis. One of the promising directions of experiments towards the solution of this challenge may be the application of adaptive binarization, e.g. Niblack [6] or Sauvola [8] algorithms. Nevertheless, their computational complexity is much higher in comparison to global thresholding, hence a reasonable compromise might be the application of region based approach [5].

References

1. Barrah, E.M., Ahdid, R., Safi, S., Malaoui, A.: New technique to determination of electronic circuits for semiconductor components by recognizing nyquist curve. J. Comput. Sci. Appl. **3**(4), 100–104 (2015). https://doi.org/10.12691/jcsa-3-4-3
2. Edwards, B., Chandran, V.: Machine recognition of hand-drawn circuit diagrams. In: Proceedings of the IEEE International Conference on Acoustics, Speech, and Signal Processing (ICASSP), vol. 6, pp. 3618–3621, June 2000. https://doi.org/10.1109/ICASSP.2000.860185
3. Forczmański, P., Frejlichowski, D.: Robust stamps detection and classification by means of general shape analysis. In: Bolc, L., Tadeusiewicz, R., Chmielewski, L.J., Wojciechowski, K. (eds.) Computer Vision and Graphics. LNCS, vol. 6374, pp. 360–367. Springer, Heidelberg (2010). https://doi.org/10.1007/978-3-642-15910-7_41
4. Frejlichowski, D.: An experimental comparison of seven shape descriptors in the general shape analysis problem. In: Campilho, A., Kamel, M. (eds.) Image Analysis and Recognition. LNCS, vol. 6111, pp. 294–305. Springer, Heidelberg (2010). https://doi.org/10.1007/978-3-642-13772-3_30
5. Michalak, H., Okarma, K.: Fast adaptive image binarization using the region based approach. In: Silhavy, R. (ed.) Artificial Intelligence and Algorithms in Intelligent Systems. AISC, vol. 764, pp. 79–90. Springer, Heidelberg (2019). https://doi.org/10.1007/978-3-319-91189-2_9

6. Niblack, W.: An Introduction to Digital Image Processing. Prentice Hall, Englewood Cliffs (1986)
7. Otsu, N.: A threshold selection method from gray-level histograms. IEEE Trans. Syst. Man Cybern. **9**(1), 62–66 (1979). https://doi.org/10.1109/TSMC.1979.4310076
8. Sauvola, J., Pietikäinen, M.: Adaptive document image binarization. Pattern Recogn. **33**(2), 225–236 (2000). https://doi.org/10.1016/S0031-3203(99)00055-2
9. Valois, J.P., Cote, M., Cheriet, M.: Online recognition of sketched electrical diagrams. In: Proceedings of Sixth International Conference on Document Analysis and Recognition (ICDAR), pp. 460–464, September 2001. https://doi.org/10.1109/ICDAR.2001.953832

Impact of ICT Infrastructure on the Processing of Large Raster Datasets

Paweł Kosydor[1,2](✉) ⓘ, Ewa Warchala[3] ⓘ, and Adam Piórkowski[4] ⓘ

[1] Geoscience Department, KGHM Polska Miedz S.A., Lubin, Poland
`pawel.kosydor@kghm.com`
[2] Department of Geology, Geophysics and Environmental Protection,
AGH University of Science and Technology, Cracow, Poland
[3] Surveying Department,
KGHM CUPRUM sp. z.o.o. - Research and Development Centre, Wroclaw, Poland
[4] Department of Biocybernetics and Biomedical Engineering,
AGH University of Science and Technology, Cracow, Poland

Abstract. Applications used for spatial data processing can operate on files stored on servers, which is a commonly used approach. It is known that this method may not be the best fit for processing very large raster files, however, it is difficult to find useful information on the issue. This is the reason the authors of this article carried out research to determine the real impact of ICT infrastructure on processing of such files. The article also aims to verify whether utilizing virtual desktops can help mitigate the identified adverse effects on the process.

Keywords: Computing virtualization · High resolution imagery · Network speed

1 Introduction

The increasing importance of images as a means of sharing spatial information is a phenomenon powered by new photogrammetry and internet technologies [2,3]. These technologies redefined the limits of how much information an image can relay. At the same time, new requirements to the existing ICT (Information and Communication Technology) infrastructure were introduced due to the rapid increase in size of raster datasets and in the number of information recipients. This is not a completely new occurrence, because processing spatial data has always faced the problem of exceeding available resources of current IT systems (e.g. databases).

The currently acquired collections of orthophoto maps are characterized by both high detail and huge sizes. For instance, one of the latest (2014) available orthophoto maps of Poland was developed using a pixel size of 0.1 m. This data set consists of thousands compressed files, each of which has a size of 1 GB. The number of pixels in each of these image files is approximately 2.3 billion and each

© Springer Nature Switzerland AG 2020
M. Choraś and R. S. Choraś (Eds.): IP&C 2019, AISC 1062, pp. 134–141, 2020.
https://doi.org/10.1007/978-3-030-31254-1_17

of those pixels occupies 3 bytes of computer memory. Keeping all the data in memory while processing only one of such files requires 7 GB of RAM (random-access memory). Today, orthophoto maps with a pixel size of 3 cm are available, which translates into about a tenfold increase in demand for RAM memory.

In the area of spatial information systems, a number of technologies have been developed to allow for processing and visualizing very large raster data sets (e.g. Enhanced Compression Wavelet - ECW). However, processing of primary raster data still requires the use of a large amount of memory and generates large data files.

In this test we present the impact of ICT infrastructure on preprocessing of the aforementioned orthophoto data set. This preprocess had to be performed before the final conversion of the data into the ECW format.

This research aims to identify the influence of two layer architecture on processing large spatial data sets. It also aims to verify whether utilizing virtual desktops can help mitigate the identified adverse effects on the process.

2   Preprocess Description

The acquired maps were in a coordinate system with the designation CRS 1992. The target system used another coordinate system with the designation CRS 2000. In order to use this data efficiently, its coordinates had to be reprojected (to obtain all needed spatial data in the same coordinate system). Because of different definitions of these coordinate systems, the results of such data processing must be saved in sheets that differ from the originals. The idea of creating a one sheet map in the target coordinate system is presented in Fig. 1. Blue lines are the borders of the source map sheets and the red line is the border of the result map sheet.

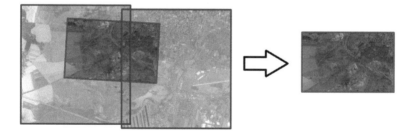

Fig. 1. Reprojection process.

In this case two source map sheets had to be used to create one result map sheet. In the case of reprojection of maps from the "1992" to the "2000" coordinate system the number of source map sheets varied from 1 to 4. It is worth noting that the area of the result sheet is only a fraction of the area of the source sheets. The tests were performed on a 64-bit Windows Operating System. The only application used was GDAL 2.1.0 (Geospatial Data Abstraction Library) with custom batch files. To collect and analyze data on the network traffic and the amount of data written and read from local disk drives a simple application was written in C# language.

The initial version of the preprocess consisted of 3 steps:

- Merging two source map sheets into one map without reprojection of the coordinate system.
- Reprojecting the merged map created in step "1" from the "1992" to the "2000" coordinate system.
- Clipping the reprojected map to the area of a standard "2000" map sheet.

Processing raster files with the algorithm above generated disproportionately high traffic in the computer network and created large temporary files [6]. The algorithm had to be optimized to allow to perform all planned tests. This was done by adding a step "0" - Generating copies of the source sheets reduced to the smallest required area.

3   Test Architecture and Test Variants

All tests were performed in the two layer architecture shown on Fig. 2 on three different Desktops, specifications of which are shown in Table 1.

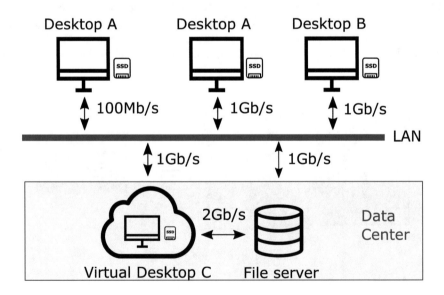

Fig. 2. Test architecture.

Two of those desktops (A and B) were ordinary PCs used in the production environment. Desktop B was used only during tests performed simultaneously

Table 1. Specifications of test desktops.

Desktop	Processor	RAM	LAN Card Speed	Local storage
A	I7-4770 3.40 GHz	8 Gb	1 Gb/s, 100 Mb/s	SSD drive
B	I5-6200U 2.30 GHz	8 Gb	1 Gb/s	SSD drive
C	Xeon 6136 3.00 GHz	128 Gb	2 Gb/2	SSD drive

on two desktops connected to the same network switch. The third computer was a Virtual Desktop hosted on a PowerEdge R740xd server. This server was connected directly to the file server by the internal Data Center Network [1]. To identify the impact of the network on the process, the tests were performed several times in three variants:

- LLL - Source, result and temporary data was stored and processed on local SSD drives.
- NLN - Source and result data was stored on a network shared drive while temporary data was stored on local SSD drives.
- NNN - Source, result and temporary data was stored and processed on a network shared drive.

To simulate the impact of a 100 Mbps transfer speed a few tests were performed on Desktop A with the network card switched to the speed of 100 Mbps [4].

4 Results

Table 2 contains performance results for all the executed variants of the preprocess. Figure 3 shows the durations of the preprocess in all variants. With the exception of the initial version of the preprocess, best results were observed in the variant LLL, in which all data was located on local SSD drives. Duration of the preprocess executed simultaneously on Desktops A and B was 54% longer than the preprocess executed independently on Desktop A only.

Figures 4 and 5 show the usage of resources in selected variants. The greatest impact of the network on preprocess duration time was observed on Desktop A with LAN Card Speed set to 100 Mbps. In Fig. 5, it is easy to see that the amount of data sent through the network by the Virtual Desktop C is less than half the amount of data sent by Desktop A and is nearly equal to the amount of data received.

The smallest ratio of the amount of transferred data to the files' total size was observed when the computer had much available RAM or when the amount of processed data was reduced (Fig. 6). The largest ratios were observed in tests performed on the initial non-optimized algorithm.

Figure 7 shows the ratios of the durations of the preprocesses to the duration of the fastest variant of the preprocess executed on the same Desktop using the same data set and the same network speed.

Table 2. Performance results.

Desktop	Variant	Ethernet speed	Iterations	Network transmission		Local SSD activity		Process duration
				Bytes received	Bytes sent	Bytes read	Bytes written	[seconds]
The modified version of the preprocess								
Raster files with pixel size of 0.1 m								
A	LLL	1 GB/s	9	493 kB	544 kB	10.7 GB	27.8 GB	660
C	LLL	2 GB/s	8	6.1 MB	7.0 MB	na*	na*	385
A	NLN	100 Mb/s	3	861.5 MB	3.1 GB	8.2 GB	26.0 GB	776
A	NLN	1 GB/s	9	860 MB	3.1 GB	10.0 GB	26.4 GB	668
C	NLN	2 GB/s	9	873 MB	3.0 GB	na*	na*	409
A	NNN	100 Mb/s	3	15.7 GB	41.4 GB	4.7 GB	180.6 MB	4 442
A	NNN	1 GB/s	9	14.6 GB	38.7 GB	4.1 GB	53.2 MB	904
C	NNN	2 GB/s	9	11.4 GB	13.8 GB	na*	na*	595
The initial version of the preprocess								
Raster files with pixel size of 0.1 m								
A	LLL	1 GB/s	4	5.5 MB	4.0 MB	73.5 GB	241 GB	3 653
A	NLN	1 GB/s	3	4.7 GB	3.1 GB	68.7 GB	234 GB	3 598
A	NNN	1 GB/s	3	103 GB	288 GB	1.5 GB	254 MB	5 141
The modified version of the preprocess executed simultaneously								
Raster files with pixel size of 0.1 m								
A	NNN	1 GB/s	3	16.8 GB	39.8 GB	3.0 GB	75.9 MB	1 397
B	NNN	1 GB/s	3	20.1 GB	48.0 GB	487 MB	1.0 GB	1 125
The modified version of the preprocess								
Raster files with pixel size of 0.2 m								
A	LLL	1 GB/s	10	130 kB	726 kB	170 MB	2.6 GB	53
A	NLN	1 GB/s	10	268 MB	674 MB	8.2 MB	2.7 GB	59
A	NNN	1 GB/s	10	2.9 GB	4.8 GB	1.0 MB	3.4 MB	127

*It was impossible to log Local SSD disk drive activity on Virtual Desktop C. This was caused by the virtualization mechanism that hid all of the host disk activity [5].

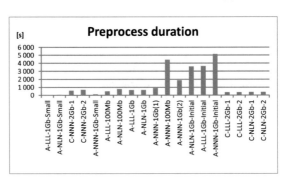

Fig. 3. Duration of all variants of preprocess.

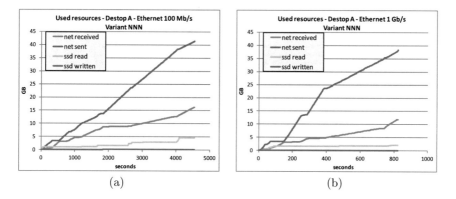

Fig. 4. Resources usage on modified preprocess - Desktop A - variant NNN.

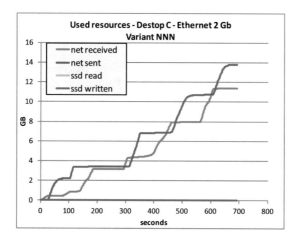

Fig. 5. Resources usage on modified preprocess - Desktop C - variant NNN.

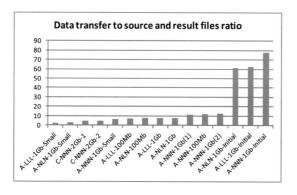

Fig. 6. Ratio of amount of transferred data to files size sum.

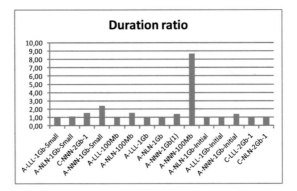

Fig. 7. Ratio of duration of preprocess to LLL (the fastest) variant.

5   Conclusions

Utilizing a Virtual Desktop led to the best results (shortest execution time) for all the variants. In fact, the worst result achieved on Virtual Desktop C was better than the best result achieved on Desktop A. Therefore, the performed tests confirmed the usefulness of this technology in mitigating the adverse effects of LAN Network Data Transfer Speed on spatial data processing.

Processing spatial data located on shared repositories while storing temporary data on local SSD drives took nearly the same amount of time as when all the data was located on local SSD drives. This way of processing large orthophoto maps can reduce the adverse impact of a Network Data Transfer.

The worst impact on the duration of the preprocess was observed when Network Speed was set to 100 Mbps and in the case of using the initial non-optimized algorithm.

Acknowledgement. This work was financed by MNiSW, 0033/DW/2018/02, Doktorat Wdrozeniowy.

References

1. Czerwinski, D.: Influence of the VM manager on private cluster data mining system. In: International Conference on Computer Networks, pp. 47–56. Springer (2014)
2. Krawczyk, A.: A concept for the modernization of underground mining master maps based on the enrichment of data definitions and spatial database technology. In: E3S Web of Conferences, vol. 26, p. 00010. EDP Sciences (2018)
3. Krawczyk, A., Kula, R.: Analiza wariantów modeli obudowy podatno-łukowej wyrobiska górniczego w aspekcie przetwarzania komputerowego. Roczniki Geomatyki **11**, 51–61 (2013)
4. Laskawiec, S., Choraś, M., Kozik, R.: Switching network protocols to improve communication performance in public clouds. In: International Conference on Image Processing and Communications, pp. 224–236. Springer (2018)

5. Munoz-Exposito, J.E., de Prado, R.P., Garcia-Galan, S., Rodriguez-Reche, R., Marchewka, A.: Analysis and real implementation of a cloud infrastructure for computing laboratories virtualization. In: Image Processing and Communications Challenges 7, pp. 275–280. Springer (2016)
6. Przylucki, S., Sierszen, A., Czerwinski, D.: Software defined home network for distribution of the SVC video based on the DASH principles. In: International Conference on Computer Networks, pp. 195–206. Springer (2017)

Gated Recurrent Units for Intrusion Detection

Marek Pawlicki$^{(\boxtimes)}$, Adam Marchewka, Michał Choraś, and Rafał Kozik

UTP University of Science and Technology, Bydgoszcz, Poland
marek.pawlicki@utp.edu.pl

Abstract. As the arms race between the new kinds of attacks and new ways to detect and prevent those attacks continues, better and better algorithms have to be developed to stop the malicious agents dead in their tracks. In this paper, we evaluate the use of one of the youngest additions to the deep learning architectures, the Gated Recurrent Unit for its feasibility in the intrusion detection domain. The network and its performance is evaluated with the use of a well-established benchmark dataset, called NSL-KDD. The experiments, with the accuracy surpassing the average of 98%, proves that GRU is a viable architecture for intrusion detection, achieving results comparable to other state-of-the-art methods.

Keywords: Gated Recurrent Unit · Intrusion detection · Deep learning

1 Introduction

The security of contemporary digital systems is of crucial importance to an immense range of industries, starting from banking and communication, through commerce, law enforcement, medical industries, and even space exploration. Intrusion detection is the course and the measure for monitoring and evaluating the ongoing network traffic for the signs of security breaches. Through the proliferation of computers in every walk of life, cybersecurity gains in importance day by day [10]. Computer networks along with the Internet are now an integral part of the contemporary lifestyle. And so are cyber attacks [8]. According to [1], the massive barrage of cyber threats across all industries is unmanageable. As noticed in the report, criminals are moving towards crypto jacking, the non-consensual use of someone else's computing power to mine cryptocurrencies. In 70 billion security events reported by IBM security task force per day, over half is leveraging the vulnerabilities of operation systems, and nearly a third consists of spearfishing. All of those fall under the umbrella term of intrusions.

This paper is structured as follows: firstly, we delve into a short summary of the state-of-the-art in the use of the gated recurrent unit for intrusion detection. Then the peculiarities of Recurrent Neural Networks in general, and Long Short-Term Memory and Gated Recurrent Unit architectures are illustrated. In the experiments section, the chosen benchmark dataset is introduced, along with

© Springer Nature Switzerland AG 2020
M. Choraś and R. S. Choraś (Eds.): IP&C 2019, AISC 1062, pp. 142–148, 2020.
https://doi.org/10.1007/978-3-030-31254-1_18

the framework and a survey of the evaluation metrics. Finally, the data preprocessing procedure and the experimental setup are outlined. This is followed by the achieved results of our work.

2   Related Work

To the best of our knowledge, the Gated Recurrent Unit has not been thoroughly researched for intrusion detection since its inception in 2014 [4,9]. Evaluates the way Principal Component Analysis improves the results of GRU for Intrusion Detection Systems, achieving, as the authors report, remarkable results.

The use of Variant Gated Recurrent Units (E-GRU) is evaluated in [7] as a preprocessing step in payload aware IDS. The authors notice an improvement over the state-of-the-art methods on a benchmark dataset - ISCX2012. The achieved accuracy reaches 99.9%, however, the memory usage of E-GRU turns out to be 32 times the memory usage of a standard GRU.

The authors of [13] improve the performance of IDS by utilising a Deep Network model with automatic feature extraction. The GRU is used as a feature extractor, which feeds the outputs to a Multi Layer Perceptron (MLP), which then uses a softmax to come up with the final classification. The experiments were performed on KDD99 and NSL-KDD datasets, achieving 99.98% for KDD99 and 99.55% accuracy for NSL-KDD.

3   Recurrent Neural Networks

The usual architectures of neural networks lack the ability to recognise sequential dependencies among data. This makes them underperform on data types that progressively build on previous instances, like time-series data, or text [3]. A different architecture is necessary to handle this kind of data. The answer to this challenge is the Recurrent Neural Network (RNN) - an architecture that features feedback loops. The RNN is capable of learning without relying on the Markov assumption, the premise that, given a present state, all the following states do not depend on the past states [11]. The output from the recent time index T is applied as one of the outputs to time index T+1, or back to itself [6] The unfolded structure of an RNN can be seen on Fig. 1.

Because of this propagation of weights through time the RNN faces its own kind of problem. The mentioned weights are multiplied recursively, thus, if the weights are too small, the subsequent values will be progressively getting smaller and smaller, or, should the weights be sizeable, the final values will approach infinity. This challenge is referred to as the Vanishing Gradient- or the Exploding Gradient problem, respectively [6].

3.1   Long Short-Term Memory and GRU

The vanishing/exploding gradient problem arises whenever the matrices are multiplied repeatedly, destabilising the result. This occurs whenever the RNN is

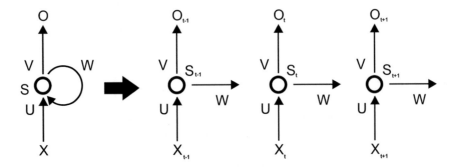

Fig. 1. The unfolded structure of a Recurrent Neural Network.

dealing with a long enough sequence. For short sequences the RNN never even experiences the problem. Thus, one could stipulate that an RNN has a reasonable short-term memory, but underperforms when long term memory is necessary. To address this problem, the Long Short-Term Memory network was introduced. It uses a property called 'cell state' to retain portions of its long-term memory.

The Gated Recurrent Unit (GRU) [4], can be seen as a simplification of the LSTM architecture, following the same basic principle, but not completely mapping one another. The GRU uses a 'reset gate' to partially reset the hidden states of the network [3]. The models attained with the use of a GRU usually have lesser complexity than those of LSTM [6]. The GRU is more efficient than LSTM [3].

4   Experiments

4.1   Dataset

The dataset utilised in this study is a widely recognised intrusion detection benchmark called NSL-KDD. It is an improvement over the KDD'99 dataset, sorting out some of the problems of its predecessors.

- the problem of redundant records in the set has been resolved
- the duplicate records have been eliminated
- the proportions of selected records from different difficulty groups have been improved
- the number of records is reasonable, therefore the researchers are not forced to subsample the dataset, making the results of different research endeavours more comparable.

The NSL-KDD dataset, even though it is not a real-life capture of network traffic, has been so widely adopted by the research community, that it has become the go-to dataset for comparable and consistent evaluation of intrusion detection systems [12].

4.2   TensorFlow and Keras

We utilise TensorFlow, the high-performance, open-source library, made available for python users by the researchers and developers of the Google Brain team. It's continuing mission is to provide adequate support for machine and deep learning. TensorFlow is actually implemented in a wide range of scientific and industrial applications [2].

Keras, an easy-to-use interface to TensorFlow and an array of other machine-learning libraries, allows for an amazing speed of experimentation. Such malleability is achieved via a modular, expandable design. Keras was conceived and developed as an interface rather than an autonomous library, but now it receives the full support of the TensorFlow library. This and makes possible intuitive coding of both machine and deep learning procedures [5].

4.3   Evaluation Metrics

The result of a classification task falls into one of four possible results,

- True Positive (TP) - the number of correct classifications of the intrusion class
- True Negative (TN) - the number of correct classifications of the benign class
- False Positive (FP) - the count of benign instances mistakingly put into the attack class
- False Negative (FN) - represents the attacks that were classified as benign traffic.

From those metrics, a range of other helpful parameters can be estimated, namely the accuracy, precision, recall and the f1-score.

$$Accuracy = \frac{TP + TN}{TP + TN + FN + FP} \tag{1}$$

$$Precision = \frac{TP}{TN + FP} \tag{2}$$

$$Recall = \frac{TP}{TP + FN} \tag{3}$$

$$F1_{Score} = \frac{2 \cdot Precision \cdot Recall}{Precision + Recall} \tag{4}$$

It is worth noticing that Precision and Recall, in a sense work in a pair, where Precision accounts for the number of False Positives (or lack thereof), and high Recall values translate to fewer False Negatives. Thus, one could increase the recall of a classifier to make it more heavy-handed in stopping potential threats, causing that, at the same time, precision will fall, since more false positives will arise.

The f1-score, also called the F-measure, is a metric which combines both precision and recall. In essence, the higher the f1-score, the better the classifier's performance.

4.4   Data Preprocessing and Experimental Setup

The NSL-KDD dataset comes as a mix of categorical and numerical data, so the first step in our course of action to prepare the data is to one-hot encode the categorical values. Then the datapoints are scaled to get the values into the range of 0 to 1. This is followed by a train/test split with a ratio of 9:1. The training data is then used on the GRU for it to come up with the model. Before being fed to the GRU, the data has to be reshaped into a 3D array.

We use a simple GRU architecture with two GRU layers of 50 units and two dropout layers with the dropout of 0.2 intertwined. The activation function of the GRU layers is tanh. This setup is followed by a softmax layer with 21 neurons - matching the number of classes contained in the dataset.

5   Results

The results are encouraging but clearly, need further research. The architecture we have evaluated achieved a weighted average of 97% across the board, getting up to 100% recognition of some of the attack classes.

Table 1. Results for the NSL-KDD dataset using the GRU Neural Network.

Class name	Class number	Precision	Recall	F1-score	Support
back	0	0.94	1.00	0.97	17
buffer_overflow	1	0.00	0.00	0.00	1
ftp_write	2	0.00	0.00	0.00	0
guess_passwd	3	0.00	0.00	0.00	3
imap	4	0.00	0.00	0.00	1
ipsweep	5	0.92	0.94	0.93	72
land	6	0.00	0.00	0.00	0
multihop	7	0.00	0.00	0.00	0
normal	8	1.00	1.00	1.00	808
nmap	9	0.92	0.59	0.72	37
neptune	10	0.98	0.99	0.99	1358
phf	11	0.00	0.00	0.00	0
pod	12	0.00	0.00	0.00	5
portsweep	13	0.93	0.93	0.93	55
rootkit	14	0.00	0.00	0.00	0
satan	15	0.99	0.92	0.95	72
smurf	16	0.98	0.95	0.96	55
spy	17	0.00	0.00	0.00	0
teardrop	18	1.00	1.00	1.00	21
warezclient	19	0.00	0.00	0.00	14
warezmaster	20	0.00	0.00	0.00	0
micro avg	0.98	0.97	0.98	2519	
weighted avg	0.97	0.97	0.97	2519	

Table 2. Class names and attack types in the NSL-KDD dataset.

Attack category	Class name
Probe	ipsweep, mscan, nmap, portsweep, saint, satan
Denial of Service	apache, back, land, mailbomb, neptune, pod, processtable, smurf, teardrop, udpstorm
User-to-root (U2R)	buffer_overflow, loadmodule, perl, rootkit, ps, sqlattack, xterm
Remote-to-Local (R2L)	ftp_write, guess_passwd, imap, multihop

On the other hand, as seen in Table 1 some of the attacks were herded with the other classes. These are, as expected, the underrepresented classes, like '1', '3' and '4'. Since the test dataset was randomly sampled at the time of the train/test split, not all of the attack types made it into the evaluation. The class names along with the attack categories they represent are illustrated in Table 2.

6   Conclusion

In this paper we have performed an initial evaluation of the feasibility of the Gated Recurrent Unit deep learning architecture for the use in the cybersecurity domain. The setup was tested on a benchmark dataset and achieved promising results, although further research is necessary. Further research should compare the results achieved by the GRU with other state-of-the-art methods, along with the computational times. Even though the GRU is one of the lightweight Deep Learning architectures with the ability to find temporal relationships, it is significantly more demanding than traditional machine learning algorithms.

References

1. IBM X-Force Report (2019). https://newsroom.ibm.com/2019-02-26-IBM-X-Force-Report-Ransomware-Doesnt-Pay-in-2018-as-Cybercriminals-Turn-to-Cryptojacking-for-Profit
2. Abadi, M., Agarwal, A., Barham, P., Brevdo, E., Chen, Z., Citro, C., Corrado, G.S., Davis, A., Dean, J., Devin, M., Ghemawat, S., Goodfellow, I., Harp, A., Irving, G., Isard, M., Jia, Y., Jozefowicz, R., Kaiser, L., Kudlur, M., Levenberg, J., Mané, D., Monga, R., Moore, S., Murray, D., Olah, C., Schuster, M., Shlens, J., Steiner, B., Sutskever, I., Talwar, K., Tucker, P., Vanhoucke, V., Vasudevan, V., Viégas, F., Vinyals, O., Warden, P., Wattenberg, M., Wicke, M., Yu, Y., Zheng, X.: TensorFlow: large-scale machine learning on heterogeneous systems (2015). Software available from tensorflow.org. https://www.tensorflow.org/
3. Aggarwal, C.C.: Neural Networks and Deep Learning. Springer, Cham (2018)

4. Cho, K., Van Merriënboer, B., Gulcehre, C., Bahdanau, D., Bougares, F., Schwenk, H., Bengio, Y.: Learning phrase representations using RNN encoder-decoder for statistical machine translation. arXiv preprint arXiv:1406.1078 (2014)
5. Chollet, F., et al.: Keras (2015). https://github.com/fchollet/keras
6. Goyal, P., Pandey, S., Jain, K.: Unfolding recurrent neural networks, pp. 119–168. Apress, Berkeley (2018). https://doi.org/10.1007/978-1-4842-3685-7_3
7. Hao, Y., Sheng, Y., Wang, J.: Variant gated recurrent units with encoders to preprocess packets for payload-aware intrusion detection. IEEE Access **7**, 49985–49998 (2019). https://doi.org/10.1109/ACCESS.2019.2910860
8. Kim, K., Aminanto, M.E., Tanuwidjaja, H.C.: Network Intrusion Detection using Deep Learning, A Feature Learning Approach. Springer, Singapore (2018)
9. Le, T., Kang, H., Kim, H.: The impact of PCA-scale improving GRU performance for intrusion detection. In: 2019 International Conference on Platform Technology and Service (PlatCon), pp. 1–6, January 2019. https://doi.org/10.1109/PlatCon.2019.8668960
10. Maimon, O., Rokach, L.: Data Mining and Knowledge Discovery Handbook, 2nd edn. Springer, Heidelberg (2010)
11. Skansi, S.: Recurrent neural networks, pp. 135–152, January 2018. https://doi.org/10.1007/978-3-319-73004-2_7
12. Tavallaee, M., Bagheri, E., Lu, W., Ghorbani, A.: A detailed analysis of the KDD CUP 99 data set. In: IEEE Symposium on Computational Intelligence for Security and Defense Applications, CISDA, vol. 2, July 2009. https://doi.org/10.1109/CISDA.2009.5356528
13. Xu, C., Shen, J., Du, X., Zhang, F.: An intrusion detection system using a deep neural network with gated recurrent units. IEEE Access **6**, 48697–48707 (2018). https://doi.org/10.1109/ACCESS.2018.2867564

Towards Mobile Palmprint Biometric System with the New Palmprint Database

Agata Giełczyk$^{(\boxtimes)}$ (ID), Karolina Dembińska, Michał Choraś, and Rafał Kozik (ID)

Faculty of Telecommunications, Computer Science and Electrical Engineering,
UTP University of Science and Technology, Bydgoszcz, Poland
agata.gielczyk@utp.edu.pl

Abstract. In this paper, we address the challenge of palmprint-based human verification in the mobile scenario. We propose a novel way of palmprint acquisition that will be used in order to create a new palmprint benchmark dataset. The palmprints are acquired by the handheld devices using an application dedicated to the Android operating system. The application provides the graphical assistant that improves the hand positioning and ROI extraction. It enables acquiring samples of both hands, in unconstrained illumination conditions, with different backgrounds. After the acquisition, each sample is evaluated according to color and rejected if it does not meet the requirements. So far, there are not many mobile biometrics databases acquired in an unconstrained environment and meeting privacy and legal requirements.

Keywords: Palmprint · Mobile biometrics · Sample acquisition · Database evaluation · GDPR

1 Introduction

Biometric systems are considered to be the most reliable as the security measure to protect from identity theft for many years [4]. In May 2018 a new regulation was introduced by the European Union [17], which is also known as the GDPR - the General Data Protection Regulation. Thus, the problem of privacy and data protection has been one of the most challenging issues for research focused on the biometric field. Biometrics is one of the 'special categories of personal data' that can only be used if the data subject has given clear consent [11]. According to the GDPR biometric data are 'personal data resulting from specific technical processing relating to the physical, physiological or behavioural characteristics of a natural person, which allow or confirm the unique identification of that natural person'. The biometric data may be very diversified. We can enumerate some anatomical features: iris, ear, nose, mouth, whole face, fingerprints, knuckles, palmprints and palm veins, some behavioural features: gait, keystroke dynamics, handwritten signature dynamics, voice and others [8]. Each of the enumerated features may be used for identity recognition due to its uniqueness, distinctiveness, permanence and usability. Nevertheless, the following requirements are introduced [16]:

© Springer Nature Switzerland AG 2020
M. Choraś and R. S. Choraś (Eds.): IP&C 2019, AISC 1062, pp. 149–157, 2020.
https://doi.org/10.1007/978-3-030-31254-1_19

- The person may decide which part of his personal data is to be provided.
- The person may declare his consent towards the data collection act.
- The person has to be informed about who is collecting the data, what is the reason for collecting such data and which processes are going to be applied to his data.
- The person may deny the collection of his personal data.

Apart from them, it is extremely important to ensure the biometric data coding or anonymization. Anonymization is a concept in which the link to the subject is permanently broken, while coding does not break the link. The link can be recovered under highly secure conditions.

Among other biometric features, the palmprint is enumerated. This biometric trait has been developed for more than 15 years and provides satisfying results in user verification systems [21]. There are many advantages of using palmprints as a biometric modality such as [9,22]: larger surface in comparison to fingerprints, lower possibility to be damaged, small amounts of dirt do not affect the identification, stable and rich line features, small distortion, easy self-positioning, permanence in time (resilience against aging). Moreover, a palmprint is widely accepted by users [19].

The paper is structured as follows: in Sect. 2 related work and benchmark databases are presented. In Sect. 3 our approach, design and the assumptions for the newly created database are mentioned. Section 4 contains the description of the possible verification schema and provides preliminary results. In Sect. 5 conclusions and future work are mentioned.

2   Related Work

Most research concerning palmprint-based biometrics is conducted using one of the following databases: PolyU [14], IITD [7] and CASIA [1].

PolyU is a database created in China and Hong-Kong. The samples were acquired by a specially designed sensor. In the database there are 7752 images in the BMP format obtained from 389 people. The database is available for any researcher after logging in. IIT Delhi database was created in India, the samples were acquired from the students and researchers from the IIT institute in Delhi, India. Images were taken in a specific environment - in a closed room with artificial illumination. The CASIA database was created in China and contains 5502 images belonging to 312 people. The samples were acquired by a dedicated sensor. The samples of palmprints available in these three benchmark databases are presented in Fig. 1. Apart from the above-mentioned three databases, numerous others have been proposed recently. For example, the dataset described in [15], which was acquired by a dedicated palmprint scanner. However, the presented approach to sample collection is hardly usable in a real-life mobile scenario. That is why, the new, mobile-oriented database has to be introduced. An increasingly high number of researchers have been interested by biometrics in mobile devices. However, this approach has a number of difficulties. Among others it is possible to enumerate two most challenging issues [10]: (1) complex and changing conditions - in a real-life application the verification process is not supervised. Thus, it is impossible to anticipate the background, the illumination and hand pose as well. (2) Mobile devices provide limited processing power and therefore the proposed verification schema has to be optimized and has to guarantee accepted (in most cases real-time) time of verification. In [5] authors add the sensor limitations as a third and the most challenging problem in the mobile scenario. The resolution of the cameras embedded in mobile devices has been steadily increasing, but there are still a very small number of high-end devices.

There are numerous approaches to mobile palmprint-based verification available. In [10] authors describe the sample acquisition step. They proposed a mobile application that ensures the appropriate hand pose, image size and focus that is captured quickly and automatically from a real-time video stream. It also contains a sample pre-assessment step, which uses the YCbCr format for checking if the sample contains skin.

In [13] a mobile approach to palmprint was proposed. However, in this case the authors did not decide to create any application supporting sample acquisition. They only define that the photo's background is dark in order to improve image segmentation. In [2] the authors only focused on the ROI (Region of Interest) extraction step of the biometric verification system. They show how difficult the task may be if the illumination, distance between the hand and the camera or the hand orientation are not supervised. An interesting approach to image acquisition was presented in [12] where a novel assistant technique, named double-line-single-point (DLSP), was proposed. While the sample is captured from the user, two lines and one point are displayed on the screen of the mobile device to help users to locate their hands correctly and accurately. Apart from the graphical assistant, image processing evaluation is performed - the surfaces of the three gaps between fingers are extracted and if their dimensions are appropriate, the enrollment is passed. Another graphical assistant was introduced in [6], where two points are displayed on the screen of a mobile device with iOS. The hand should be placed so that the points will be between the index and middle fingers, and the ring and little fingers.

In [18] a detailed description of a newly created database was mentioned. The authors presented the characteristics of mobile devices used during the dataset creation and the basic assumptions for the database: taking pictures with a complex background and over a wooden surface and providing images taken in various hand positions - the rotation depends only on the user's choice.

3   Proposed Acquisition System

3.1   Software Description

For capturing the samples to the database, a dedicated application was released. It was suitable for Android mobile devices (starting from version Nougat 7.0). We proposed a graphic assistant to ensure proper hand rotation. The graphical assistant is presented in Fig. 2. In the figure the four points are placed: A, B, C

Fig. 1. Sample examples from benchmark database - PolyU, IITD and CASIA (consequently from the left side).

and D. First, we placed two points B and C so that the angle between the BC line segment is close to $30°$. It makes the system more convenient for the user. The real angle is varying in range $<30°, 31°>$ due to the screen dimensions. Then the perpendicular line segments AB and CD are calculated. The final x and y coordinates of the point are expressed with the Eq. 1 for the left hand and Eq. 2 for the right hand, where x - the device screen width, y - the device screen height and point $(0,0)$ is in the upper left corner. Parameters m and n used in the equations are expressed with Eqs. 4 and 3. Thus, the graphical assistant consists of three segment lines: AB, BC and CD. The proper position of the hand is between AB and CD segment lines. The BC has to cross the fingers valleys.

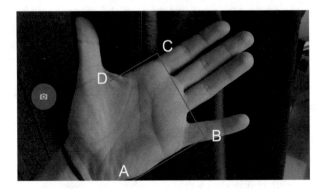

Fig. 2. The graphical assistant ensuring proper hand rotation.

$$A = (x, m) \quad B = \left(\frac{3x}{4}, \frac{y}{3}\right) \quad C = \left(\frac{x}{4}, \frac{y}{2}\right) \quad D = \left(n, \frac{2y}{3}\right) \tag{1}$$

$$A = (x, y - m) \quad B = \left(\frac{3x}{4}, y - \frac{y}{3}\right) \quad C = \left(\frac{x}{4}, y - \frac{y}{2}\right) \quad D = \left(n, y - \frac{2y}{3}\right) \tag{2}$$

$$n = \left(\frac{y}{6} + \frac{3x^2}{4y} \right) \cdot \frac{y}{3x} \tag{3}$$

$$m = \frac{3x^2}{y} + \frac{y}{3} - \frac{9x^2}{4y} \tag{4}$$

In order to ensure high usability, we provide an intuitive graphical user interface (GUI). An example of the screen is presented in Fig. 3. By using our application, it is possible to add a new user, to add the photos and to preview them. During the database creation we used three different average-price devices, their characteristics are presented in Table 1. If the mobile devices possessed by users involved in the database creation are compatible, they were welcomed to add samples taken by their own devices. Such an approach gives us the opportunity to collect data from various devices with various cameras embedded and makes the database independent of the device type.

Fig. 3. The application GUI, from the left side: the list of users, the list of photos for selected user, the photo preview.

3.2 Database Assumptions

For the final version of the database, each user involved will be asked for taking up to 48 photos (2 with a dark background, 2 with a light background, 2 with a complex background - for the right and the left hand - using 3 or 4 devices) in one single session. The exemplary subset for one user is presented in Fig. 4 (top row of images).

Table 1. Mobile devices characteristics.

Device	OS	RAM	Processor	Image size	Sensor resolution	Aperture
Samsung Galaxy A5 2017	Android 8.0 Oreo	3 GB	8×1.9 GHz	4608×3456	16 Mpx	f/1.9
Huawei P10 Lite	Android 7.0 Nougat	3 GB	8×2.1 GHz	3968×2976	12 Mpx	f/2.2
Xiaomi Mi6	Android 8.0 Oreo	6 GB	8×2.45 GHz	4032×3016	12 Mpx	f/1.8

For the final version of the database two sessions are planned at most. Each file will automatically receive an appropriate name using the following scheme **XXXX_YY_B_H_DD_SS.jpg**, where:

- **XXXX** – sequentially numbered ID of the volunteer;
- **YY** – sequentially numbered ID of the sample (starting from 01 for each identity);
- **B** – type of background: D - dark, L - light and C - complex;
- **H** – hand: R - right and L - left;
- **DD** – type of device: D1 - Samsung, D2 - Huawei, D3 - Xiaomi, D4 - other;
- **SS** – session identifier: S1 - first session and S2 - second session.

4   Experiments and Obtained Results

For this research we used the preliminary part of the described database. It contains 50 images taken by three different devices. We have focused only on the pre-processing part of the user identification process. Obviously, after the sample acquisition, the sample needs to be pre-processed. There are two main aims of pre-processing: (1) image enhancement by reducing noise or unwanted details and (2) reducing the size of the image so that make the whole algorithm is less computational time- and resource-demanding. The second aim of pre-processing is often called the ROI extraction. In the previous paper [20] we proposed an ROI extraction algorithm that was successfully used for our research performed on the PolyU and IITD databases. In the first step the segmentation was performed using the thresholding technique. However, this approach is not that promising when the background is light or complex. In Table 2 the percentage of correct segmentation for various backgrounds is presented.

What is striking, we obtained no correct segmentation at all for light background samples. The examples of the correct and incorrect segmentation are presented in Fig. 4. It is remarkable that both PolyU and IITD databases provide the images taken under very good illumination conditions (IITD), preliminary cropped (PolyU) and using dark, uniform background (both). Nevertheless, the newly proposed database contains images taken with dark, light and complex

Table 2. Mobile devices characteristics.

Background	Correct segmentation percentage
Dark	100%
Complex	33%
Light	0%

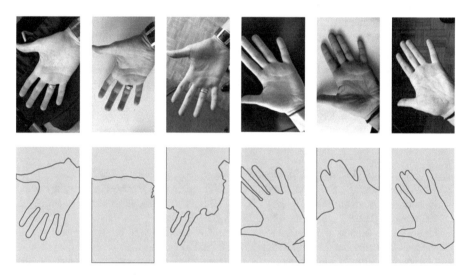

Fig. 4. Examples of a correct and incorrect segmentation for dark, complex and light background; row 1 - original images, row 2 - the result of the segmentation.

backgrounds. Thus, we propose an alternative approach to the ROI extraction using the above described graphical user assistant.

As the hand in the picture is rotated by $30°$, it needs to be turned. The rotation point is the C point from the graphical assistant. Then the ROI (size 450×450) is selected in the specific offset from the C point. In order to speed up the calculations the next step is resizing the ROI to the size 128×128, which is corresponding to our previous works. Then, the sample verification step is performed. In this step we use the expressions 5 [3], where R, G and B are pixel's value for each RGB channel. If all inequalities are true, the pixel is classified as skin. If at least 95% of ROI's pixels are classified as skin, the sample is accepted. Otherwise, the sample is rejected and the process of sample acquisition needs to be repeated. First observations suggest that samples with a dark background are more likely to be accepted, while samples with light or complex background are more often rejected.

$$R > 95; \quad G > 40; \quad B > 20$$
$$max(R, G, B) - min(R, G, B) > 15 \qquad (5)$$
$$|R - G| > 15; \quad R > G; \quad R > B$$

5 Conclusions and Future Work

In this article we have proposed a novel palmprint acquisition system. It utilises a mobile device in order to acquire a palmprint sample. For the system we have developed an application (for Android operating system) that supports hand positioning introduced and described in the article graphical user assistant. We have also proposed a novel ROI extraction algorithm and we have implemented the pixel's color requirements.

The proposed system is used in order to create a new palmprint database - collected in its entirety with mobile devices, using three different background types and not supervised illumination conditions. Therefore, we hope that our database will be interesting and useful for researches interested in mobile palmprint biometrics.

References

1. CASIA. http://biometrics.idealtest.org/index.jsp. Accessed 12 Feb 2019
2. Chai, T., Wang, S., Sun, D.: A palmprint ROI extraction method for mobile devices in complex environment. In: 2016 IEEE 13th International Conference on Signal Processing (ICSP), pp. 1342–1346. IEEE (2016)
3. Choraś, M., Kozik, R.: Contactless palmprint and knuckle biometrics for mobile devices. Pattern Anal. Appl. **15**(1), 73–85 (2012)
4. Clarke, R.: Human identification in information systems: management challenges and public policy issues. Inf. Technol. People **7**(4), 6–37 (1994)
5. Das, A., Galdi, C., Han, H., Ramachandra, R., Dugelay, J.L., Dantcheva, A.: Recent advances in biometric technology for mobile devices. In: 9th IEEE International Conference on Biometrics: Theory, Applications and Systems, BTAS 2018 (2018)
6. Gao, F., Leng, L., Zeng, J.: Palmprint recognition system with double-assistant-point on iOS mobile devices. In: Proceedings of the 29th British Machine Vision Conference, BMVC 2018 (2018)
7. IITD. http://www4.comp.polyu.edu.hk/~csajaykr/database.php. Accessed 12 Feb 2019
8. Ito, K., Aoki, T.: Recent advances in biometric recognition. ITE Trans. Media Technol. Appl. **6**(1), 64–80 (2018)
9. Jia, W., Zhang, B., Lu, J., Zhu, Y., Zhao, Y., Zuo, W., Ling, H.: Palmprint recognition based on complete direction representation. IEEE Trans. Image Process. **26**(9), 4483–4498 (2017)
10. Kim, J.S., Li, G., Son, B., Kim, J.: An empirical study of palmprint recognition for mobile phones. IEEE Trans. Consum. Electron. **61**(3), 311–319 (2015)
11. Kindt, E.: Having yes, using no? About the new legal regime for biometric data. Comput. Law Secur. Rev. **34**(3), 523–538 (2018)
12. Leng, L., Gao, F., Chen, Q., Kim, C.: Palmprint recognition system on mobile devices with double-line-single-point assistance. Pers. Ubiquit. Comput. **22**(1), 93–104 (2018)
13. Moco, N.F., Técnico, I.S., de Telecomunicações, I., Correia, P.L.: Smartphone-based palmprint recognition system. In: 2014 21st International Conference on Telecommunications (ICT), pp. 457–461. IEEE (2014)

14. PolyU. http://www4.comp.polyu.edu.hk/~biometrics. Accessed 12 Feb 2019
15. Qu, X., Zhang, D., Lu, G.: A novel line-scan palmprint acquisition system. IEEE Trans. Syst. Man Cybern. Syst. **46**(11), 1481–1491 (2016)
16. Sanchez-Reillo, R., Ortega-Fernandez, I., Ponce-Hernandez, W., Quiros-Sandoval, H.C.: How to implement EU data protection regulation for R&D in biometrics. Comput. Stand. Interfaces **61**, 89–96 (2019)
17. UE: Regulation 2016/679 of the European parliament and of the council of 27 April 2016 on the protection of natural persons with regard to the processing of personal data and on the free movement of such data. http://eur-lex.europa.eu/eli/reg/2016/679/oj. Accessed 12 Feb 2019
18. Ungureanu, A.S., Thavalengal, S., Cognard, T.E., Costache, C., Corcoran, P.: Unconstrained palmprint as a smartphone biometric. IEEE Trans. Consum. Electron. **63**(3), 334–342 (2017)
19. Verma, S., Chandran, S.: Contactless palmprint verification system using 2-D Gabor filter and principal component analysis. Int. Arab J. Inf. Technol. **16**(1), 23–29 (2019)
20. Wojciechowska, A., Choraś, M., Kozik, R.: Evaluation of the pre-processing methods in image-based palmprint biometrics. In: International Conference on Image Processing and Communications, pp. 43–48. Springer (2017)
21. Zhang, D.D., Kong, W., You, J., Wong, M.: Online palmprint identification. IEEE Trans. Pattern Anal. Mach. Intell. **25**, 1041–1050 (2003)
22. Zhang, D.D., et al.: Palmprint Authentication, vol. 3. Springer, New York (2004)

Vision System for Pit Detection in Cherries

Piotr Garbat[1]([⊠]) [ID], Piotr Sadura[1], Agata Olszewska[1], and Piotr Maciejewski[2]

[1] Institute of Microelectronics and Optoelectronics,
Warsaw University of Technology, Nowowiejska 15/19, St., Warsaw, Poland
p.garbat@elka.pw.edu.pl
[2] Uniwentech, Nowy Rynek 1, Kiernozia, Poland
http://www.imio.pw.edu.pl

Abstract. In the following paper, the results of studies on the impact of lighting configuration on the quality of seed detection in cherries have been presented. The general concept of a vision system for cherry pits detection has been proposed. The two types of light were considered and characterized. The results of cherry classification confirm the effectiveness of the proposed solution and allow to indicate a better solution.

Keywords: Cherry pit detection · Fruits classification ·
Multispectral vision

1 Introduction

The food processing chain has to be monitored at different steps along the way. Recently, food products have been inspected and sorted using cameras RGB, ultraviolet (UV), and infrared (IR) [9]. This recent development allows addressing many problems that are no longer able to be solved using sensor technology (i.e., visual wavelengths). For example, fruits need to be checked during they are harvested, and the same applies when frozen fruits are sorted in production [6,11]. In cherries processing, there are many features, such as pit presence, decay, and insect damage, which can have an impact on the quality of these products [2]. The effectively sorting procedure determines the quality of the final product. The monitoring of the pitting process allows minimizing the risk of presence the unwanted pits in fruits. The cherries with the whole or damaged pit may be removed from the product stream. Modern methods, including computed tomography (CT), x-ray imaging or near-infrared spectroscopy, are being tested for detecting external and internal inhomogeneities within fruits tissue [4,7,10], but the traditional equipment is expensive. In past years, the hyperspectral imaging (HSI) systems have been successfully demonstrated to be the right solution for the inspection of quality parameters in food products [2,5].

The study is part of the project aiming at the implementation of an industrial system for automatic optical sorting of cherry fruits with the possibility of separation of full-value and cherry with the pit. The biggest challenge in the

© Springer Nature Switzerland AG 2020
M. Choraś and R. S. Choraś (Eds.): IP&C 2019, AISC 1062, pp. 158–165, 2020.
https://doi.org/10.1007/978-3-030-31254-1_20

research is to find the optimal method for quick detection of the pit inside the processing fruit that is invisible to the naked eye.

2 Proposed Acquisition System

Market standards accept two approaches to the implementation of the camera-based acquisition system. It is possible to mount cameras allowing observation of fruit in flight or a stable moving system on the belt. Data acquisition for the need of the research was carried out based on an industrial camera for NIR range 740–850 nm and LED light source 835 nm (Fig. 1).

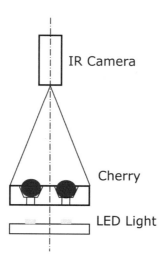

Fig. 1. General concept of vision system for cherry pit detection.

2.1 Direct Light

The first study concerned the directional configuration of light. In this case, LEDs are located in axes of symmetry of holes in the matrix with cherries. Distance between light sources and matrix was adjusted to illuminate whole cherry surface evenly. For small distances (smaller than 30 mm), each hole is illuminated by exactly one LED placed directly under it. As the distance increases, each hole is illuminated by more LEDs (Fig. 2). This phenomenon causes overlapping of light fields shown in Fig. 3.

The cherries without pit are illuminated homogeneously. Irregularities of their surfaces are visible. The cherries differ significantly from others and are exposed less than other ones.

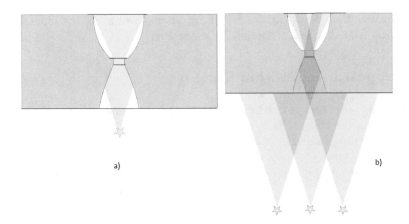

Fig. 2. Difference between illuminating one hole depending on the distance.

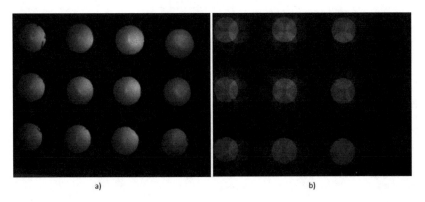

Fig. 3. Distribution of illuminance in small distance and big distance configuration.

The parameters defining the brightness distribution for a single hole were defined Fig. 4. It allows for analyzing the uniformity of cherry luminance. Parameters include: av - average light intensity for whole hole area, avav - ratio of average intensity of half hole area starting from the centre to half radius to av value, contrast - ratio of difference of maximum and minimum intensities to their sum, sym_ang - angular symmetricity, sym_ang_without_centre - sym_ang for area without overexposed centre, sym_rad - radial symmetricity.

2.2 Indirect Light

Another analysed configuration of light is indirect LED sources. In this configuration, LEDs are not located in axes of symmetry of holes in the matrix, but between them. Schema of light propagation is shown in Fig. 5. In this case, the arrangement of LEDs has to illuminate four holes by one diode. Therefore, each hole is illuminated by four LEDS. The use of light in indirect illumination

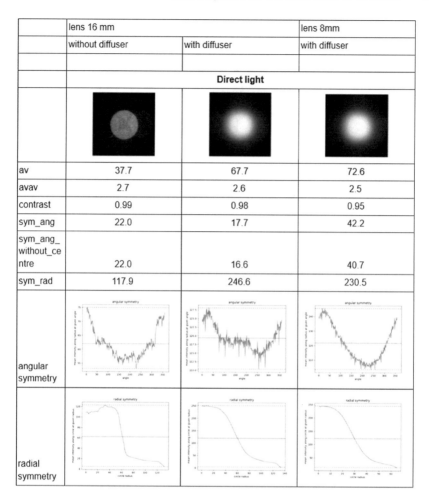

	lens 16 mm		lens 8mm
	without diffuser	with diffuser	with diffuser
		Direct light	
av	37.7	67.7	72.6
avav	2.7	2.6	2.5
contrast	0.99	0.98	0.95
sym_ang	22.0	17.7	42.2
sym_ang_ without_ce ntre	22.0	16.6	40.7
sym_rad	117.9	246.6	230.5
angular symmetry			
radial symmetry			

Fig. 4. Parameters of direct light configuration.

configuration causes light fields overlapping, which can be seen on the diffusing element placed directly on the matrix.

In the case of direct light, the best parameters were obtained for configuration with a 16 mm lens with a diffuser (Fig. 6). For the indirect light application of 8 mm lens with diffuser resulted in best parameters. For both light configurations, the contrast value is above 0.95, which is sufficient for machine vision.

3   Experiment

For two configurations of light - direct and indirect, measurements were performed. The created dataset was used to train the classification model [8]. The proposed solution bases on a histogram, HOG descriptor and an SVM classifier.

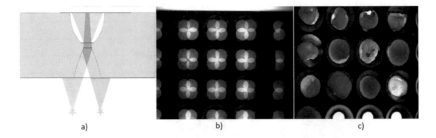

a)   b)   c)

Fig. 5. (a) Schema of light propagation in indirect illuminating, (b) overlapping light fields, (c) cherries in indirect illuminating configuration system.

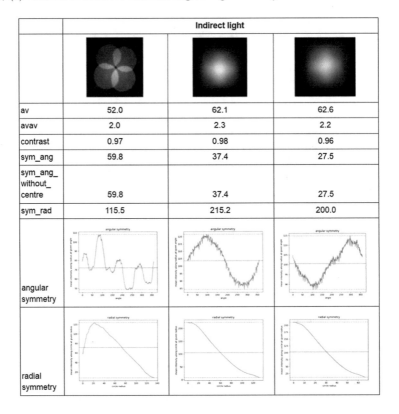

	Indirect light		
av	52.0	62.1	62.6
avav	2.0	2.3	2.2
contrast	0.97	0.98	0.96
sym_ang	59.8	37.4	27.5
sym_ang_ without_ centre	59.8	37.4	27.5
sym_rad	115.5	215.2	200.0
angular symmetry			
radial symmetry			

Fig. 6. Parameters of indirect light configuration.

The detection process is realized for two types of descriptors: a one-dimensional histogram (9 bins) and HOG (Histogram of Gradients) descriptor. In the next step, a linear SVM (support vector machine) classifier was trained to classify fruit with and without pit. For each configuration of light, the training process was repeated. Each measurement consisted of the estimation of following scoring parameters: ROC_AUC, Accuracy, Precision, and ROC curve. The binary dependent variable assumes values: true - for cherry with a pit, false - cherry without a pit. In this case, true positive means correct prediction as pitted cherry for pitted cherry. False-positive is a prediction as pitted cherry, but for cherry without a pit. The main goal of classification is a minimisation of the number of pits in fruits accepted to the next production process (false positives) and false negatives reduction. The data set was made by cropping detected circles with cherries inside from snapshot of whole pit matrix. Each cropped image is resized to dimension 100×100 px. No other image preprocessing methods were performed. 20 sets of 36 positives and negatives were made giving 720 independent observations for each case. The data set was divided into a training and a test part. Size of the test set was 10% of the training set size. The cross-validation was made with the division of training set by 10 parts. The ROC curve was shown in Fig. 7 (Tables 1 and 2).

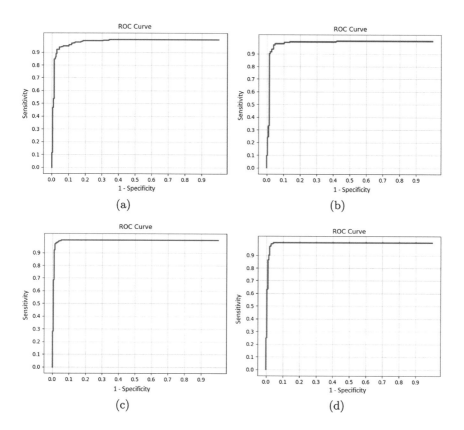

Fig. 7. ROC curves for two light configurations: (a) direct with HoG descriptor, (b) direct with one-dimensional histogram (9 bins) + HoG, (c) indirect with HoG descriptor, (d) indirect with one-dimensional histogram (9 bins) + HoG.

Table 1. Scoring parameters of SVM classification for indirect light configuration.

	Accuracy	Precision	ROC_AUC
Histogram (9 bins)	0.86	0.90	0.95
HoG	0.93	0.92	0.98
Histogram + HoG	0.97	0.96	0.99

Table 2. Scoring parameters of SVM classification for direct light configuration.

	Accuracy	Precision	ROC_AUC
Histogram (9 bins)	0.86	0.91	0.94
HoG	0.95	0.95	0.99
Histogram + HoG	0.98	0.97	1.00

4 Summary

This study compared two alternatives of light for a vision system that detects a pit in cherry. Both configurations enable achieving very good classification results. A small advantage of direct illumination over indirect is observable. It can be related to higher intensity in this case. The best results were obtained for features vector containing information about intensity distribution (Histogram) and gradients distribution (HoG). Both configurations demonstrated high detection speed and accuracy 97% for best features on our data set. The results confirm the effectiveness of the proposed solution. However, the data set consists of very good quality fruit. In industrial case, the obtained results could be a little worse. The final system is going to enrich with additional image modes - polarisation and colour - that allow to achieves higher accuracy.

Acknowledgement. This work was supported by the National Centre for Research and Development, project POIR.01.01.01-00-1045/17.

References

1. Qin, J., Lu, R.: Detection of pits in tart cherries by hyperspectral transmission imaging. Trans. ASAE **48**, 1963–1970 (2005)
2. Siedliska, A., Zubik, M., Baranowski, P., Mazurek, W.: Algorithms for detecting cherry pits on the basis of transmittance mode hyperspectral data. Int. Agrophys. **31**, 539–549 (2017)
3. Donis-Gonzǎlez, I.R., Guyer, D.E., Kavdir, I., Shahriari, D., Pease, A.: Development and applicability of an agarose-based tart cherry phantom for computer tomography imaging. J. Food Meas. Charact. **9**(3), 290–298 (2015)
4. Kawano, S.: Past, present and future near infrared spectroscopy applications for fruit and vegetables. NIR News **27**(1), 7–9 (2016)
5. Lu, Y., Huang, Y., Lu, R.: Innovative hyperspectral imaging-based techniques for quality evaluation of fruits and vegetables: a review. Appl. Sci. **7**(2), 189 (2017)

6. Muresan, H., Oltean, M.: Fruit recognition from images using deep learning (2018)
7. Yeong, T.J., Jern, K.P., Yao, L.K., Hannan, M.A., Hoon, H.T.G.: Applications of photonics in agriculture sector: a review. Open Access Mol. **24**(10), 24 (2019)
8. Meruliya, T., Dhameliya, P., Jainish, P., Dilav, P., Pooja, K., Sapan, N.: Image processing for fruit shape and texture feature extraction - review. Int. J. Comput. Appl. **129**, 30–33 (2015)
9. Bhargava, A., Bansal, A.: Fruits and vegetables quality evaluation using computer vision: a review. J. King Saud Univ. - Comput. Inf. Sci. (2018)
10. Gongal, A.A., Karkee, S., Manoj Zhang, Q., Lewis, K.: Sensors and systems for fruit detection and localization: a review. Cmput. Electron. Agric. **116**, 8–19 (2015)
11. Hameed, K., Chai, D., Rassau, A.: A comprehensive review of fruit and vegetable classification techniques. Image Vis. Comput. **80**, 24–44 (2018)

The Impact of Distortions on the Image Recognition with Histograms of Oriented Gradients

Andrzej Bukała[(✉)], Michał Koziarski, Bogusław Cyganek, Osman Nuri Koç, and Alperen Kara

Department of Electronics, AGH University of Science and Technology, Kraków, Poland
{a.bukala,michal.koziarski,cyganek}@agh.edu.pl

Abstract. While most existing image recognition benchmarks consist of relatively high quality data, in the practical applications images can be affected by various types of distortions. In this paper we experimentally evaluate the extent to which image distortions affect classification based on HOG feature descriptors. In an experimental study based on several benchmark datasets and classification algorithms we evaluate the impact of Gaussian, quantization and salt-and-pepper noise. We examine both known and random types of distortion, and evaluate the possibility of applying distortions on training data and using denoising to mitigate the negative impact of distortions. Although presence of distortions significantly impede classification with the HOG features, in the paper we show how this negative effect can be greatly mitigated in practical realizations. Experimental results underpin our findings.

Keywords: Image processing · Machine learning · Denoising · Sparse features

1 Introduction

Partially due to factors such as the algorithmic advances, as well as an increasing availability of computational resources and data, significant improvements are continuously reported in the field of image recognition. Supposedly, every other scientific paper achieves state-of-the-art results on one of the popular classification benchmarks, which inevitably leads to claims of achieving a superhuman performance by some of the most notable models [22]. However, nearly all of these impressive results are achieved on specially designed benchmark datasets, which contain images of relatively high quality: with limited distortions and to a large extent devoid of artifacts. On the other hand, in the real applications we often face low quality data, affected by various phenomena: noise, which can be caused by the environmental factors and imperfect image acquisition devices; blur, occurring due to the movement of camera or objects; and occlusion, which can happen when the object of interest is partially blocked. In such setting it

© Springer Nature Switzerland AG 2020
M. Choraś and R. S. Choraś (Eds.): IP&C 2019, AISC 1062, pp. 166–178, 2020.
https://doi.org/10.1007/978-3-030-31254-1_21

is important to answer two questions: to what extent distortions can affect the performance of our model in the image recognition task? And what techniques can be applied to successfully mitigate the negative impact of distortions?

Some research on the impact of distortions on various approaches to the image classification can be found in the literature. Khan et al. [17] examined the effect of image deformations on the performance of classification using SIFT and SURF features. Karami et al. [16] also evaluated the impact of distortions on SIFT and SURF, as well as BRIEF and ORB features descriptors, in the image matching task. Dutta et al. [12] examined how distortions affect a commercial face recognition system. Various types of image distortions were also considered in the context of convolutional neural networks [11,15,19,26]. However, to the best of our knowledge previous works did not examine the impact of low image quality on another type of feature descriptors, namely the Histograms of Oriented Gradients (HOG) [8]. HOG features were successfully used in numerous image recognition tasks, with the most notable examples including human detection [2,3,8,18,24,27–29] and face recognition [9,10,25], but also problems such as smile detection [1], traffic sign recognition [23] and handwritten digit recognition [13]. To this day they remain an important approach to the image recognition, especially in the environment, in which the computational resources at our disposal are limited. However, since in such setting we are also more likely to use a low-end image acquisition devices, likely to affect the quality of captured data, it is important to consider the possible impact of introduced distortions on the classification performance.

In this paper we experimentally evaluate the impact of various image distortions on the performance of classification using HOG features. We perform our analysis on several benchmarks in combination with different classification models. We start our analysis with a case in which classifier is trained on undistorted data, but distortions are present during the evaluation. This corresponds to a setting in which either the conditions change during the evaluation phase, or we are forced to train the model on images of a different quality than that observed during the evaluation. Afterwards, we consider the case in which the same type of distortions affects both training and test data. This setting corresponds to either the case in which overall quality of the data is low, both during training and testing, or the one in which we anticipate the presence of distortions during testing, and try to apply artificial distortions as a strategy of mitigating the negative impact of distortions. Finally, in the case of noise we examine the possibility of applying denoising prior to classification while training the classifier on undistorted data, as another strategy of dealing with distortions during the evaluation of the model.

The rest of this paper is organized as follows: in Sect. 2 we describe the HOG feature descriptors. In Sect. 3 we define the distortion models used throughout the paper. In Sect. 4 we describe the conducted experimental study and present the observed results. Finally, in Sect. 5 we present our conclusions.

2   Histograms of Oriented Gradients

Image gradients provide highly discriminative information on image content. In many cases, object recognition based on gradients leads to better results than using only unprocessed intensity signals. It was further observed that collecting local phase of gradient into histograms provides highly discriminative features. Following this idea many object detection and classification methods were proposed which can be divided into two broad groups: global ones, in which histograms of gradients were collected from the whole image(s) [14], and the local ones, in which histograms of gradients were computed only in selected regions of an image [5,6]. The method of Histograms of Oriented Gradients, as proposed by Dalal et al. [8] belongs to the latter group. It is based on evaluating normalized histograms of image gradients, computed on a dense grid with uniformly spaced cells. HOG representation of an image provides several advantages. Gradient structure is very characteristic for local shape, and it provides an easily controllable invariance for local geometric and photometric transformations. HOG is computed by dividing the image into small spatial regions ("cells"), and then for each cell calculating a 1-D histogram of oriented directions or edge orientations over the pixels of the cell. In addition, histogram entries are weighted by the gradient magnitude computed for that entry. Finally, the concatenated histograms of each cell form the HOG descriptor. However, for a better invariance to illumination, shadowing, etc. it is useful to further cross-normalize local responses in larger spatial regions. For this purpose, the cells are grouped together into larger and spatially connected regions, called "blocks". In our case we used the L1-sqrt normalization, given as follows:

$$f = \sqrt{\frac{v}{\|v\|_1 + \epsilon}}$$

where v is a non-normalized vector, containing all histograms in a given block, $\|v\|_1$ is its k-norm, This provides better invariance to illumination, shadowing, etc. Later in this paper we will refer to such normalized descriptor as the HOG descriptor.

3   Image Distortion Models

In this section we provide short characteristics of distortions which were considered in the experiments presented and discussed further on in this paper.

3.1   Gaussian Noise

Gaussian noise is an additive distortion with probability function equal to that of the normal distribution, which is given by:

$$p(x) = \frac{1}{\sigma\sqrt{2\pi}} e^{-\frac{(x-\mu)^2}{2\sigma^2}}$$

(a) $\sigma = 0$ (b) $\sigma = 0.1$ (c) $\sigma = 0.2$

Fig. 1. Gaussian noise applied on example image.

where μ represents mean value and σ the standard deviation. In digital images it is caused mostly by sensor faults due to low illumination or high temperature. Figure 1 shows an effect caused by this type of noise on an image.

(a) $q = 0$ (b) $q = 0.2$ (c) $q = 0.4$

Fig. 2. Quantization noise applied on example image.

3.2   Quantization Noise

Quantization noise arises as a result of quantization of intensity signal to discrete levels of image pixels. It is dependent on bit depth of an image, which limits the number of possible values. This type of noise is modeled by adding a random value η from range:

$$-\frac{1}{2}q \leq \eta \leq +\frac{1}{2}q$$

where q denotes a quantization level. On the other hand, value of η follows an uniform probability distribution p as follows:

$$p(x) = \begin{cases} \frac{1}{x_{max}-x_{min}} & \text{for } x_{min} \leq x \leq x_{max} \\ 0 & \text{otherwise} \end{cases}$$

where x_{max} and x_{min} stand for the maximum and minimum values of the argument x. The random variable η takes the values $\pm\frac{1}{2}q$ with a uniform distribution, where q is a quantization parameter set in the experiments. An example of this type of noise is presented in Fig. 2.

(a) $p = 0$ (b) $p = 0.05$ (c) $p = 0.15$

Fig. 3. Salt-and-Pepper noise applied on example image.

3.3 Salt-and-Pepper Noise

Salt-and-pepper noise is caused by errors in analog-to-digital converters. An image distorted in this way has erroneously bright pixels in dark areas and vice versa. An example is presented in Fig. 3. This type of noise can be modeled by combination of multiplicative and additive components, as follows:

$$\hat{s}(x) = (1 - \mu)s(x) + \mu\beta$$

where $\hat{s}(x)$ and $s(x)$ stand for distorted and pure signal respectively, μ is a random variable with probability $p = Pr(\mu = 1)$ and β is a random variable satisfying the equation $Pr(\beta = s_{max}) = Pr(\beta = s_{min}) = 0.5$.

4   Experimental Study

4.1   Experimental Set-Up

Datasets. Our experiments were conducted using four image databases, as follows. The German Traffic Sign Recognition Benchmark (GTSRB) [23], which is a datasets consisting of around 50,000 images divided into 40 classes. By default image sizes vary from 15×15 to 250×250 pixels. For the purpose of our experiments we resized all images to 32×32. The STL-10 [4] dataset, consisting of 13,000 96×96 images divided into 10 classes, describing objects such as airplanes, birds and cars. The MNIST database of handwritten digits [20] - a dataset consisting of 60,000 train images and 10,000 test images. This dataset is grayscale by default with images of 28×28 pixels in size. For the purpose of our experiment we also used a grayscale version of the FERET [21] database. It contains around 14,000 256×384 images of 2337 classes. FERET database does not sort images into training and test sets by default, so we performed that division randomly, keeping 70%–30% training-test data ratio. In the case of a direct classification, that is the classification on vectorized images, FERET images were downscaled four times. The rest of the datasets were kept in their original sizes. Furthermore, all datasets were normalized to 0–1 range.

Image Distortion Models. In the case of a known distortion intensity, images were distorted using Gaussian noise with standard deviation $\sigma \in \{0.025, 0.05, \ldots, 0.25\}$, Salt-and-Pepper noise with probability of flipping a pixel $p \in \{0.02, 0.04, \ldots, 0.2\}$, quantization noise with range $q \in \{0.05, 0.1, \ldots, 0.5\}$. In the case of a random distortion intensity, the value of a distortion-specific parameter was selected randomly from the considered range, independently for every image.

Histograms of Oriented Gradients were calculated using fixed number of pixels per cell and cells per block. Parameters were chosen for training part of each dataset individually. For the STL-10 database we tested cells of 8×8, 10×10, 12×12 and 16×16 pixels, respectively. For the GTSRB database tested values were 3×3, 4×4, 6×6 and 8×8, 3×3, 4×4 and 6×6 for the MNIST, as well as 24×24, 28×28, 32×32, 40×40, 48×48 for the FERET, respectively. For each setting, we also tested 1×1, 2×2 and 3×3 cells per block values. Each pair of HOG parameters was then used for a grid-search of the classification parameters with the 5-fold cross-validation. From these results, the most memory and time consuming settings were omitted and finally the parameters shown in (Table 1) were used in all tests.

Table 1. HOG parameters chosen for each dataset

Dataset	Pixels per cell	Cells per block
GTSRB	(4, 4)	(2, 2)
STL-10	(16, 16)	(3, 3)
MNIST	(4, 4)	(1, 1)
FERET	(32, 32)	(1, 1)

Classification. For classification we used *Support Vector Machine* (SVM) with a linear kernel. SVM parameters were chosen from range: $C \in \{0.001, 0.1, 1, 10, 100\}$. Grid-search was performed on training set of each dataset with 5-fold cross-validation, both for vectorized images and HOG descriptors classification, as described.

4.2   Classification of Distorted Images

In the first stage of the conducted experimental analysis we considered the impact of distortions with known intensity on the classification accuracy. To this end we compared two feature types: classification using either HOG feature descriptors or plain, vectorized images. Furthermore, we considered two settings: distortions applied either only to the test data, or to both training and test data. The former corresponds to the case, in which the image quality differs between the training and evaluation stages, respectively. This can occur when we either do not expect to observe the distortions during evaluation of the model, or we are forced to train the model on undistorted images, possibly due to unavailability of the distorted data or due to uncertainty about the underlying distortion model.

The results for this part of the experimental study were grouped by the distortion type and are presented in Fig. 4. As can be seen, in all cases, presence of distortions had a significant impact on the classification accuracy, even for distortions of seemingly small values. This can be compared with a sample of distorted images presented in Sect. 3, where especially the small distortion levels do not affect the possibility of image recognition by a human examiner.

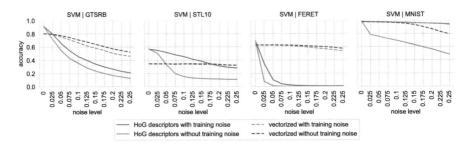

Fig. 4. Classification results after applying Gaussian noise.

In the case of undistorted images using HOG features led to a better performance than using vectorized images (i.e. a pure intensity signal) in all cases. However, in the presence of distortions, especially of higher intensity, using vectorized images led to a better performance in some of the cases. In isolated cases, presence of noise actually improved the final performance of the model, especially when noise was also applied in a training phase. This is likely due to the noise acting as an additional regularizer. Presence of training noise was also beneficial for the classification with HOG feature descriptors in almost all of the cases (Fig. 5).

4.3   Applying Denoising Prior to Classification

In the second stage of the conducted experiment we considered the possibility of training the classifier on clean, undistorted images, and removing the distortions

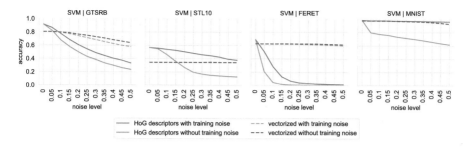

Fig. 5. Classification results after applying quantization noise.

present in the test data prior to classification. This corresponds to the case, in which we do expect to observe distortions during the model evaluation, but are still unable to obtain sufficient amount of distorted images for training. We considered all of the previously used noise models, that is Gaussian, quantization and salt-and-pepper noise, as well as three denoising algorithms: median filtering, bilateral filtering and BM3D algorithm [7]. Optimal denoising algorithm and its parameters were chosen by performing a grid-search on training set for each dataset, noise type and noise level. For the median filtering we tested the following kernels 3×3, 5×5, 7×7, 9×9, 11×11 and 13×13, respectively. For the bilateral filtering we tested $\sigma_s \in \{0.05, 0.1, 0.2, 0.3, 0.4, 0.5\}$ and $\sigma_r \in \{3, 5, 7\}$, respectively. Lastly, for the BM3D algorithm values of $\sigma \in \{0.05, 0.1, 0.2, 0.4, 0.5\}$ were used. Tested types of noise were discussed in Sect. 4.1 (Fig. 6).

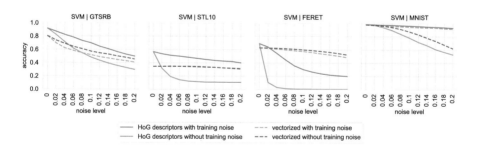

Fig. 6. Classification results after applying Salt-and-Pepper noise.

Results for this part of the experimental study were also grouped by the distortion type. For the reference we presented both, the case in which no method of dealing with noise was applied, as well as the strategy of applying the same distortions on the training data. We considered only the classification using HOG feature descriptors. The results were presented in Figs. 7, 8, and 9. As can be seen, in all of the cases applying either strategy of dealing with noise resulted in as-good-as or better performance than the baseline, in which no strategy was applied. For both Gaussian and quantization noise, applying training noise

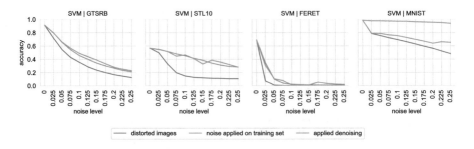

Fig. 7. Classification after applying Gaussian noise, with denoising algorithms.

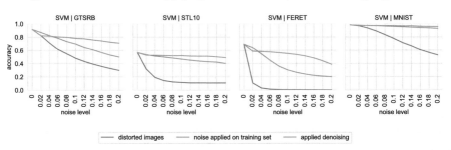

Fig. 8. Classification after applying Salt-and-Pepper noise, with denoising algorithms.

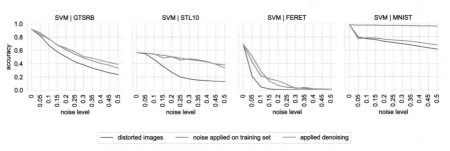

Fig. 9. Classification after applying quantization noise, and with denoising algorithms.

led to at least as-good-as or better classification accuracy than denoising for all datasets. On the other hand, in the case of salt-and-pepper noise, applying denoising led to a better performance for medium to high noise intensity for GTSRB, STL-10 and MNIST, and worse for FERET, respectively.

4.4 Handling Unknown Distortion Level

In the final stage of the conducted experimental study we considered the case in which the intensity of distortion was selected randomly for every image, as well as the case, in which not only the intensity, but also the type of noise was selected randomly. This setting is the closest to a real conditions, in which both the type and intensity of the distortions can change on a case-by-case basis. Once again,

Table 2. Classification results with random distortion intensity using HOG features combined with **SVM** classifier. Two strategies of dealing with distortions were presented: applying same distortions on training data (TD) and, when applicable, denoising (DN)

Distortion type	Dataset	Baseline	TD	DN
Gauss. noise	GTSRB	0.32	0.40	**0.41**
	STL-10	0.19	**0.38**	0.32
	MNIST	0.64	**0.96**	0.68
	FERET	0.01	0.02	**0.03**
S&P noise	GTSRB	0.50	0.65	**0.77**
	STL-10	0.14	0.45	**0.51**
	MNIST	0.75	0.95	**0.97**
	FERET	0.02	0.27	**0.53**
Quant. noise	GTSRB	0.45	**0.54**	0.54
	STL-10	0.25	**0.45**	0.37
	MNIST	0.70	**0.97**	0.73
	FERET	0.03	0.06	**0.16**
Random noise	GTSRB	0.43	0.51	**0.58**
	STL-10	0.19	**0.40**	0.39
	MNIST	0.70	**0.96**	0.70
	FERET	0.02	0.05	**0.23**
None	GTSRB	0.92		
	STL-10	0.57		
	MNIST	0.98		
	FERET	0.69		

the following strategies were examined - The baseline case, in which we apply no strategy of dealing with noise, and the strategy of applying the same, in this case random, distortions on the training data, as well as applying denoising. For the reference, we also present the classification accuracy in the case, in which no distortions were applied.

The results of these experiments were presented in Table 2. As can be seen, applying distortions with random intensity also leads to a significant performance drop in the observed classification accuracy. Furthermore, applying one of the strategies of dealing with distortions leads to an improved performance. Similar to the results with a known noise intensity, denoising was beneficial when applied on training data in all cases. For other types of noise no clear trends regarding the choice of strategy of dealing with distortions were observed, and the performance varied depending on the dataset and the used classifier.

5   Conclusions

In this paper we experimentally evaluated the impact of different types of distortions on image recognition with the HOG feature descriptors, evaluated with four different classification methods, and on four reference databases. For reference, we also considered the impact of the same distortions on classification with vectorized images with pure intensity. Furthermore, we evaluated two strategies of dealing with distortions: applying similar distortions on the training data, and in the case of noise, applying denoising. The main findings of this paper are as follows:

- Presence of distortions significantly affects classification performance in the image recognition task, whether using HOG feature descriptors or vectorized images. Classification with HOG feature descriptors is especially susceptible to high distortion intensity, more so than classification using vectorized images. As a result, despite better performance observed for HOG feature descriptors in case on undistorted images, classification with vectorized images can be a potentially useful alternative in case of significant image distortions. Such performance of HOG features is caused by significantly distorted process of computing image derivatives in presence of noise. Hence, consecutive computation of orientation histograms leads to poor results.
- Both of the considered strategies of dealing with distortions, that is applying a similar distortion on the training data and denoising, led to significant improvements in the classification accuracy. Particularly, in the case of the salt-and-pepper noise, denoising turned out to be the preferable approach. However, for other types of noise the choice of a strategy had to be made based on the dataset and used classification algorithm.
- Similar trends to these observed for known distortions were also observed in the case of unknown, random distortions, with both unknown noise type and its values. In particular, both of the considered strategies of dealing with distortions led to an improved performance compared to the baseline, in which presence of distortions in test data was not accounted for.

However, while both of the considered strategies of dealing with distortions led to a significant improvement in classification accuracy on distorted images, complete restoration of the baseline performance on undistorted data was rarely possible. This highlights the importance of further research in the area of image restoration techniques, which could potentially lead to reducing the negative impact of distortions on the image recognition, in particular with HOG feature descriptors.

Acknowledgment. This work was supported by the Polish National Science Center under the grant no. 2014/15/B/ST6/00609 and the PLGrid infrastructure.

References

1. Bai, Y., Guo, L., Jin, L., Huang, Q.: A novel feature extraction method using pyramid histogram of orientation gradients for smile recognition. In: 2009 16th IEEE International Conference on Image Processing (ICIP), pp. 3305–3308. IEEE (2009)
2. Bertozzi, M., Broggi, A., Del Rose, M., Felisa, M., Rakotomamonjy, A., Suard, F.: A pedestrian detector using histograms of oriented gradients and a support vector machine classifier. In: 2007 IEEE Intelligent Transportation Systems Conference, ITSC 2007, pp. 143–148. IEEE (2007)
3. Chuang, C.H., Huang, S.S., Fu, L.C., Hsiao, P.Y.: Monocular multi-human detection using augmented histograms of oriented gradients. In: 2008 19th International Conference on Pattern Recognition, ICPR 2008, pp. 1–4. IEEE (2008)
4. Coates, A., Ng, A., Lee, H.: An analysis of single-layer networks in unsupervised feature learning. In: Proceedings of the Fourteenth International Conference on Artificial Intelligence and Statistics, pp. 215–223 (2011)
5. Cyganek, B.: Recognition of solid objects in images invariant to conformal transformations. In: Conference on Computer Recognition Systems CORES 2009. Advances in Soft Computing, vol. 57, pp. 247–255 (2009)
6. Cyganek, B.: Object Detection and Recognition in Digital Images: Theory and Practice. Wiley, Hoboken (2013)
7. Dabov, K., Foi, A., Katkovnik, V., Egiazarian, K.: Image restoration by sparse 3D transform-domain collaborative filtering. In: Image Processing: Algorithms and Systems VI, vol. 6812, p. 681207. International Society for Optics and Photonics (2008)
8. Dalal, N., Triggs, B.: Histograms of oriented gradients for human detection. In: 2005 IEEE Computer Society Conference on Computer Vision and Pattern Recognition, CVPR 2005, vol. 1, pp. 886–893. IEEE (2005)
9. Déniz, O., Bueno, G., Salido, J., De la Torre, F.: Face recognition using histograms of oriented gradients. Pattern Recogn. Lett. **32**(12), 1598–1603 (2011)
10. Do, T.T., Kijak, E.: Face recognition using co-occurrence histograms of oriented gradients. In: 2012 IEEE International Conference on Acoustics, Speech and Signal Processing (ICASSP), pp. 1301–1304. IEEE (2012)
11. Dodge, S., Karam, L.: Understanding how image quality affects deep neural networks. In: 2016 Eighth International Conference on Quality of Multimedia Experience (QoMEX), pp. 1–6. IEEE (2016)
12. Dutta, A., Veldhuis, R.N., Spreeuwers, L.J.: The impact of image quality on the performance of face recognition. In: 33rd WIC Symposium on Information Theory in the Benelux. Centre for Telematics and Information Technology (CTIT) (2012)
13. Ebrahimzadeh, R., Jampour, M.: Efficient handwritten digit recognition based on histogram of oriented gradients and SVM. Int. J. Comput. Appl. **104**(9), 10–13 (2014)
14. Freeman, W.T., Roth, M.: Orientation histograms for hand gesture recognition. In: International Workshop on Automatic Face and Gesture Recognition, vol. 12, pp. 296–301 (1995)
15. Karahan, S., Yildirum, M.K., Kirtac, K., Rende, F.S., Butun, G., Ekenel, H.K.: How image degradations affect deep CNN-based face recognition? In: 2016 International Conference of the Biometrics Special Interest Group (BIOSIG), pp. 1–5. IEEE (2016)

16. Karami, E., Prasad, S., Shehata, M.: Image matching using SIFT, SURF, BRIEF and ORB: performance comparison for distorted images. arXiv preprint arXiv:1710.02726 (2017)
17. Khan, N.Y., McCane, B., Wyvill, G.: SIFT and SURF performance evaluation against various image deformations on benchmark dataset. In: 2011 International Conference on Digital Image Computing Techniques and Applications (DICTA), pp. 501–506. IEEE (2011)
18. Kobayashi, T., Hidaka, A., Kurita, T.: Selection of histograms of oriented gradients features for pedestrian detection. In: International Conference on Neural Information Processing, pp. 598–607. Springer (2007)
19. Koziarski, M., Cyganek, B.: Image recognition with deep neural networks in presence of noise-dealing with and taking advantage of distortions. Integr. Comput.-Aided Eng. **24**(4), 337–349 (2017)
20. LeCun, Y., Bottou, L., Bengio, Y., Haffner, P.: Gradient-based learning applied to document recognition. Proc. IEEE **86**(11), 2278–2324 (1998)
21. Phillips, P.J., Moon, H., Rizvi, S.A., Rauss, P.J.: The FERET evaluation methodology for face-recognition algorithms. IEEE Trans. Pattern Anal. Mach. Intell. **22**(10), 1090–1104 (2000)
22. Schmidhuber, J.: Deep learning in neural networks: an overview. Neural Netw. **61**, 85–117 (2015)
23. Stallkamp, J., Schlipsing, M., Salmen, J., Igel, C.: Man vs. computer: benchmarking machine learning algorithms for traffic sign recognition. Neural Netw. **32**, 323–332 (2012)
24. Suard, F., Rakotomamonjy, A., Bensrhair, A., Broggi, A.: Pedestrian detection using infrared images and histograms of oriented gradients. In: 2006 IEEE Intelligent Vehicles Symposium, pp. 206–212. IEEE (2006)
25. Tan, H., Yang, B., Ma, Z.: Face recognition based on the fusion of global and local HOG features of face images. IET Comput. Vis. **8**(3), 224–234 (2013)
26. Vasiljevic, I., Chakrabarti, A., Shakhnarovich, G.: Examining the impact of blur on recognition by convolutional networks. arXiv preprint arXiv:1611.05760 (2016)
27. Wang, C.C.R., Lien, J.J.J.: AdaBoost learning for human detection based on histograms of oriented gradients. In: Asian Conference on Computer Vision, pp. 885–895. Springer (2007)
28. Watanabe, T., Ito, S., Yokoi, K.: Co-occurrence histograms of oriented gradients for pedestrian detection. In: Pacific-Rim Symposium on Image and Video Technology, pp. 37–47. Springer (2009)
29. Zhu, Q., Yeh, M.C., Cheng, K.T., Avidan, S.: Fast human detection using a cascade of histograms of oriented gradients. In: 2006 IEEE Computer Society Conference on Computer Vision and Pattern Recognition, vol. 2, pp. 1491–1498. IEEE (2006)

Information and Communication
Technology Forum 2019

Traffic Feature-Based Botnet Detection Scheme Emphasizing the Importance of Long Patterns

Yichen An, Shuichiro Haruta, Sanghun Choi, and Iwao Sasase[✉]

Department of Information and Computer Science, Keio University,
3-14-1 Hiyoshi, Kohoku, Yokohama, Kanagawa 223-8522, Japan
anyichen@sasase.ics.keio.ac.jp, sasase@ics.keio.ac.jp

Abstract. The botnet detection is imperative. Among several detection schemes, the promising one uses the communication sequences. The main idea of that scheme is that the communication sequences represent special feature since they are controlled by programs. That sequence is tokenized to truncated sequences by n-gram and the numbers of each pattern's occurrence are used as a feature vector. However, although the features are normalized by the total number of all patterns' occurrences, the number of occurrences in larger n are less than those of smaller n. That is, regardless of the value of n, the previous scheme normalizes it by the total number of all patterns' occurrences. As a result, normalized long patterns' features become very small value and are hidden by others. In order to overcome this shortcoming, in this paper, we propose a traffic feature-based botnet detection scheme emphasizing the importance of long patterns. We realize the emphasizing by two ideas. The first idea is normalizing occurrences by the total number of occurrences in each n instead of the total number of all patterns' occurrences. By doing this, smaller occurrences in larger n are normalized by smaller values and the feature becomes more balanced with larger value. The second idea is giving weights to the normalized features by calculating ranks of the normalized feature. By weighting features according to the ranks, we can get more outstanding features of longer patterns. By the computer simulation with real dataset, we show the effectiveness of our scheme.

Keywords: Botnet detection · Feature emphasizing · Detection algorithms

1 Introduction

Recently, the computer networks are exposed to the crisis of the botnets. The botnet is composed of compromised computers and our computer can become its member through Trojan horses [1]. The attacks by botnet include spreading spams, conducting click-fraud scams, DDoS (Distributed Denial of Service) [2],

M. Choraś and R. S. Choraś (Eds.): IP&C 2019, AISC 1062, pp. 181–188, 2020.
https://doi.org/10.1007/978-3-030-31254-1_22

and so on [3]. The botnet consists of two components called bots and C&C (Command and Control) servers [11]. The C&C server sends instructions to bots and the bots follow them. Figure 1 shows the relationship between our system model and the botnet. As shown in Fig. 1, botnets are scattered all over the network and the attacker can control the botnets through multiple C&C servers. According to [4], about 40% of the 800 million computers connected to the Internet are botnets. For the secure use of the network, botnet detection is urgent demand.

In order to detect botnets, many approaches have been proposed and they are classified into the deep payload inspection and the shallow inspection. In the deep payload inspection, application layer's data is used. Goebel et al. propose an approach to use IRC (Internet Relay Chat) nicknames because the IRC based botnets tend to use same or unusual IRC nicknames in many cases [5]. However, recently, since more and more botnets begin to use P2P (peer to peer) or HTTP (HyperText Transfer Protocol), it has been difficult to apply that study.

Kapre et al. propose a technique by analyzing the behavior of TCP/HTTP in the host side [6]. While it can capture abnormal activities with high probability, it cannot be used in all over the network. As opposed to the deep payload inspection, the approaches of the shallow inspection rely on the packet header's information. This is because the recent botnets encrypt communications and the detailed features cannot be captured in the deep payload inspection.

In [7], Strayer et al. propose an approach which focuses on the fact that the botnets and ordinary users have different features such as size of packets, sending rate and so on. Those features are fed into machine learning classifier such as the SVM (Support Vector Machine) [8]. In addition to those features, Lee et al. focus on analyzing the connection failures [9]. However, it is probable that the connection failure also occurs in ordinary users' cases, that scheme has high false positive rate. Moreover, the features mentioned above can be manipulated by attackers who try to avoid detection. In order to deal with this, Su et al. propose a scheme which uses the communication sequence as a feature [10].

The main idea behind scheme [10] is that the communication sequences of bots are not easily changed and represent special feature since they are controlled by programs which are not frequently updated. In that scheme, the communication sequence is tokenized to truncated sequences by n-gram. The occurrences of patterns appeared in the truncated sequences are used as a feature vector. We focus on the scheme [10] as the previous scheme since it achieves high accuracy based on the feature which is not easily manipulated by attackers.

However, although the feature value of the previous scheme is normalized by the total number of all patterns' occurrences, the number of occurrences in larger n is less than those of smaller n. That is, regardless of the value of n, the previous scheme normalizes the feature value by the fixed number of all patterns' occurrences. As a result, the value of normalized longer patterns' features become smaller and are hidden by other features.

In this paper, first of all, we investigate the same patterns in the botnets communication contained in the dataset, and find most of botnets have the same

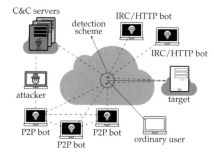

Fig. 1. System and attacker model

Table 1. The feature creation in previous scheme

direction	forward-backward string	XOR operation	CDLS
out	0		
in	1	>	1
in	1	>	0
out	0	>	1
out	0	>	0
out	0	>	0
in	1	>	1
in	1	>	0

CDLS = 1010010 ($L = 7$)	truncated sequences						
1-gram	1	0	1	0	0	1	0
2-gram	10	01	10	00	01	10	
3-gram	101	010	100	001	010		
...							
6-gram	1101001	010010					
7-gram	1010010						

n	1-gram		2-gram			...	6-gram	7-gram	total
occurred patterns	0	1	00	01	10	...	101001	010010	1010010	
x (occurrences)	4	3	1	2	3	...	1	1	28
f (feature)	4/28= 0.142	3/28= 0.107	1/28= 0.035	2/28= 0.071	3/28= 0.107	...	1/28= 0.035	1/28= 0.035	1

patterns whose length is longer than 5. Then, we consider the feature in the previous scheme can be improved by attaching importance to the long pattern. To realize this, we propose a traffic feature-based botnet detection scheme emphasizing the importance of long patterns. We realize the emphasizing by two ideas. The first idea is normalizing occurrences by the total number of occurrences in each n instead of the total number of all patterns' occurrences. By doing this, smaller occurrences in larger n are normalized by small sum and the feature becomes more balanced with larger value. The second idea is giving weights to the normalized features according to the importance. In the condition where longer pattern's occurrence is very large, it is a useful feature for detecting botnet. Thus, we calculate ranks of the normalized feature and create new feature according to the ranks. We evaluate our scheme by using two datasets and the simulation results demonstrate that maximum improvement in our scheme is 12% compared with the previous scheme (Table 1).

The rest of this paper is constructed as follows: we explain the previous scheme and its shortcoming in Sect. 2. The proposed scheme is described in Sect. 3. Simulation results are shown in Sect. 4. We conclude this paper and mention future work in Sect. 5.

2   Previous Scheme

The main idea of the previous scheme is that the communication sequences of bots are not easily changed and represent special feature since they are controlled by programs which are not frequently updated. Due to the control by the program, the directional information between bots and C&C servers can be used as a feature. The flow of creating the feature in the previous scheme is shown in Fig. 1. As shown in Fig. 1, the original source of feature is called "forward-backward string". In order to identify the communications between botnet-C&C server and C&C server-botnet, they are calculated by XOR (exclusive OR) and the result is called "corresponding direction less string (CDLS)". CDLS is tokenized to truncated sequences by n-gram and the occurrences of patterns appeared in the truncated sequences are counted. However, the numbers of occurrences of each pattern are highly dependent on the length of CDLS. In order to mitigate this, these occurrences are normalized by the total number of all patterns' occurrences and they are used as a feature vector.

2.1 Shortcomings of Previous Scheme

Although the feature value of the previous scheme is normalized by the total number of all patterns' occurrences, the number of occurrences in larger n is less than those of smaller n. As a result, normalized long patterns' features become very small value and are hidden by other features. The bottom table in Fig. 1 shows an example of the previous scheme's normalization in the case of CDLS $= 1010010(L = 7)$. As shown in this table, when $n = 7$, the number of occurrence is only one. Although this is much smaller value than the case $n = 1$ where the numbers of occurrences are four and three, these values are normalized by a same value in the previous scheme. This shortcoming is more specifically explained by the equation. Let x, L, N, and f denote the number of occurrence of the pattern, the length of the CDLS, the maximum n in the n-gram and the feature value, respectively. We can calculate f as

$$f = \frac{x}{\sum_{i=1}^{N}(L - i + 1)} = \frac{x}{-\frac{1}{2}N^2 + N(L + \frac{1}{2})} \tag{1}$$

on the condition $L \geq N$. From this equation, regardless of the value of n (Note that not N), the previous scheme normalizes it by the fixed total number of all patterns' occurrences. We consider this is the reason why the previous scheme gets similar detection results when $N > 3$. However, in fact, the bots communications have a strong regularity because it is controlled by the program. In order to demonstrate it, we investigated the same patterns in the botnets communication contained in the dataset [12]. The results are shown in the Table 2. From this table, we can see most of botnets have the same patterns whose length is longer than five. From the fact mentioned above, the previous scheme does not attach importance to the long patterns.

3   Proposed Scheme

We argue the feature in the previous scheme can be improved by attaching importance to the long patterns. From this consideration, we propose a traffic feature-based botnet detection scheme emphasizing the importance of long patterns. We realize that emphasizing by two ideas. The first idea is normalizing occurrences by total number of occurrences in each n instead of the number of all patterns' occurrences. By doing this, smaller occurrences in larger n are normalized by smaller value and become more balanced with larger values. The second idea is giving weights to the normalized features according to the importance. In the condition where longer pattern's occurrence is more frequent, it is useful feature for detecting botnet. Thus, we calculate ranks of the normalized features and create new feature according to the ranks. In the following sections, we explain our normalizing and ranking procedures in detail.

3.1 Normalizing Procedure

Let F denote the feature value calculated by normalizing in our scheme. Since we normalize by total number of occurrences in each n, we can describe F as

$$F = \frac{x}{L - n + 1} \tag{2}$$

on the condition $L \geq N$. Compared with Eq. (1), the value of F is more balanced and larger than f because the occurrence x is smaller when n is larger. By comparing it with Fig. 1, our scheme's features have larger value than the previous ones. In particular, we focus on the patterns where "1" and "10" have the same occurrence. In this case, while the previous scheme has the same values of f, F of 2-gram is larger in our scheme. This indicates that the longer patterns are emphasized in the proposed normalization.

3.2 Ranking Procedure

The normalized feature vector is weighted by ranking procedure. We first calculate ranks for each n-gram. The rank of the feature with smallest occurrence is one and that of largest is 2^n. Note that the ranks become the same values if the occurrences are the same and they become one if the occurrences are zero. We weight the feature F by multiplying rank r. Let F' denote the new feature weighted by ranking procedure. The new feature is represented as

$$F' = F * r = \frac{rx}{L - n + 1} \tag{3}$$

on the conditions $1 \leq r \leq 2^n$ and $L > N$. From the former condition, as n is larger, the rank of larger occurrences become larger so that the value of the feature is emphasized. Figure 4 shows the example of our weighting. As we can see from this figure, the pattern "10" can be emphasized compared with others. If n is much larger, more effective emphasizing can be expected (Tables 3 and 4).

Table 2. The length of same patterns in the dataset

l: length of same patterns	The number of same pairs
$1 < l < 5$	1426
$5 \leq l$	3233
Total	4659

Table 3. The number of traffic in the dataset

Dataset name	Botnet pairs	Ordinary user pairs
ISCX	4659	3240
ISOT	1072	6996

Table 4. The example of weighting in the proposed scheme

	n	1-gram		2-gram			...	6-gram	7-gram
								CDLS = 1010010 ($L = 7$)	
occurred patterns		0	1	00	01	10	...	1010010100101010010	1010010
x (occurrences)		4	3	1	2	3	...	1	1
F (feature)		4/7= 0.571	3/7= 0.428	1/6= 0.166	2/6= 0.333	3/6= 0.5	...	1/2= 0.5	1/2= 0.5	1/1= 1
rank(r)		2	1	2	3	4	...	63	63	128
F' (new feature)		1.142	0.428	0.332	0.999	2	...	31.5	31.5	128

4   Simulation Results

In order to show the effectiveness of our scheme, we evaluate the detection accuracy calculated as

$$accuracy = \frac{TP + TN}{TP + TN + FP + FN},\tag{4}$$

where TP, TN, FP, and FN denote the number of True Positive, True Negative, False Positive, and False Negative, respectively. Each result is yielded by 10-fold cross validation. We use SVM as the machine learning classifier with a parameter γ. That parameter is defined in SVM and its value indicates how far the influence of each training samples reaches. Although the larger value of γ brings better result for SVM, that includes the risk of overlearning. In order to prove our idea, we test the effect of this parameter. We use ISCX [12, 13] as a primary dataset and supplementally use ISOT dataset [14]. This is because we would like to show our result does not depend on datasets. Moreover, we also evaluate the effect of noise under the same condition of the previous scheme.

4.1   Detection Accuracy

Overall Tendency. Figure 2 shows N versus the detection accuracy in the dataset ISCX and ISOT. The parameter γ is fixed to 10. As we can see from Fig. 2, the accuracy of the previous scheme [10] and the proposed scheme increases by degrees in both of dataset. This is because the number of patterns increases and longer patterns are more valuable. However, in the previous scheme, the increase of accuracy is slowly. On the other hand, our schemes achieve rapid increase in the both of dataset. Especially in ISCX, when $N = 4$, a value the previous scheme recommend to use, our normalizing and ranking improve the previous scheme by 5% and 12%, respectively. In ISOT, those improve by 3% and 4%, respectively. We can say this is the effectiveness of our emphasizing. The reason why there is different performance between two datasets is that ISOT includes fewer types of botnets. Thus, it is originally easy

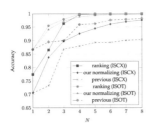

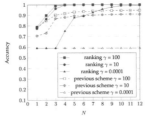

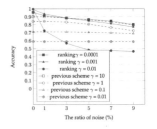

Fig. 2. N versus the detection accuracy with $\gamma = 10$

Fig. 3. N versus the detection accuracy in ISCX

Fig. 4. The ratio of noise versus detection accuracy

to classify bots and ordinary users and as a result, the difference of the accuracy between our schemes is small.

Effectiveness of Emphasizing Longer Pattern. Figure 3 shows N versus detection accuracy with multiple γ. ISCX is used as a dataset. We set the parameter γ to 100, 10, and 0.0001. As we can see from Fig. 3, our scheme improves the previous scheme in all γ pairs. This is similar situation in Fig. 2. Focusing on the lines whose $\gamma = 0.0001$, the accuracy of our scheme approaches 1.0. Especially, when $N \geq 4$, our scheme continues to improve the accuracy. This result shows that the long patterns' importance is emphasized in our ranking phase.

False Negative Analysis. Although the accuracy in our scheme approaches 1.0, there exist two types of botnets that are not detected. The type of botnets has same CDLS with some ordinary users. In the other type, some botnets do not have enough data to build a longer CDLS. In this case, the feature values are small and become same with others. As a result, we cannot rank properly and make a misjudgment. Thus, we can say that our scheme needs longer time to collect the communications data of the botnets in order to get higher accuracy.

4.2   Robustness to Noise

As is the case with the previous scheme, we define noise as packets the attacker inserts to communications between a bot and a C&C server in order to avoid detection. We add the noise packet behind each packet with α percent of the probability. If the noise is added, that direction is the same as preceding packet. We set α to $0, 1, 3, 5, 7, 9$. Figure 4 shows the ratio of noise versus the detection accuracy when $N = 8$. As we can see from Fig. 4, the accuracy decreases with increase of ratio of noise in all cases. However, by adjusting the parameter γ, the proposed scheme and the previous scheme can become robust to the noise to a certain extent. However, we can say that our scheme is more sensitive to the noise because we emphasize the feature. For example, when $\gamma = 0.01$, our scheme rapidly drops the accuracy. This is the same situation of other machine learning based approaches. If the ratio of noise increases a lot, all of the machine learning based approaches cannot detect botnet.

5   Conclusion

We have proposed a traffic feature-based botnet detection scheme emphasizing the importance of long patterns. We focus on the fact that long communication patterns of botnets are useful for detecting. The proposed scheme emphasizes the long pattern's importance by normalizing and ranking procedures. By the computer simulation with real dataset, we show the maximum improvement in our scheme is 12% compared with the previous scheme. As a future work, we should consider the case where more noise exists.

Acknowledgment. This work is partly supported by the Grant in Aid for Scientific Research (No. 17K06440) from Japan Society for Promotion of Science (JSPS).

References

1. Saha, B., Gairola, A.: Botnet: an overview. In: CERT-In White Paper, CIWP-2005-05 (2005)
2. Hoque, N., Bhattacharyya, D.K., Kalita, J.K.: Botnet in DDoS attacks: trends and challenges. IEEE Commun. Surv. Tutor. **17**(4), 2242–2270 (2015)
3. Sahi, A., Lai, D., Li, Y., Diykh, M.: An efficient DDoS TCP flood attack detection and prevention system in a cloud environment, pp. 6036–6048 (2017)
4. Li, C., Jiang, W., Zou, X.: Botnet: survey and case study, pp. 1184–1187 (2009)
5. Goebel, J., Holz, T.: Rishi: identify bot contaminated hosts by IRC nickname evaluation. HotBots **7**, 8 (2007)
6. Kapre, A., Padmavathi, B.: Behaviour based botnet detection with traffic analysis and flow interavals using PSO and SVM. In: ICICCS, pp. 718–722 (2017)
7. Livadas, C., Walsh, R., Lapsley, D., Strayer, W.T.: Using machine learning techniques to identify botnet traffic. In: IEEE 2006 Proceedings of the 31st IEEE Conference, pp. 967–974 (2006)
8. Vapnik, V.: Pattern recognition using generalized portrait method. Autom. Remote Control **24**, 774–780 (1963)
9. Lee, Y.C., Tseng, C.M., Liu, T.J.: A HTTP botnet detection system based on ranking mechanism. In: 2017 Twelfth International Conference, pp. 115–120. IEEE (2017)
10. Su, Y.H., Rezapour, A., Tzeng, W.G.: The forward-backward string: a new robust feature for botnet detection. In: 2017 IEEE Conference on Dependable and Secure Computing, pp. 485–492 (2017)
11. Dietrich, C.J., Rossow, C., Freiling, F.C., Bos, H., Van Steen, M., Pohlmann, N.: On botnets that use DNS for command and control. In: Seventh European Conference on Computer Network Defense, pp. 9–16 (2011)
12. Leskovec, J., Mcauley, J.J.: Towards effective feature selection in machine learning-based botnet detection approaches. In: Communications and Network Security (CNS). IEEE (2014)
13. ISCX botnet dataset university of new Brunswick. http://www.unb.ca/cic/research/datasets/botnet.html
14. Saad, S., Traore, I., Ghorbani, A., Sayed, B., Zhao, D., Lu, W., Felix, J., Hakimian, P.: Detecting P2P botnets through network behavior analysis and machine learning. In: Proceedings of 9th Annual Conference on Privacy, pp. 174–180. IEEE (2011)

Performance Evaluation of the WSW1 Switching Fabric Architecture with Limited Resources

Mustafa Abdulsahib, Wojciech Kabaciński, and Marek Michalski[(⊠)]

Faculty of Electronics and Telecommunications, Poznan University of Technology,
ul. Polanka 3, 60-965 Poznań, Poland
mustafa.abdulsahib@doctorate.put.poznan.pl,
{wojciech.kabacinski,marek.michalski}@put.poznan.pl

Abstract. The evaluation of the Wavelength-Space-Wavelength (W-S-W) switching fabrics is considered in this paper. The combinatorial properties of such switching fabrics have been discussed in several papers. The strict-sense nonblocking (SSNB) conditions require tunable spectrum converters (TSCs) of wide conversion range in order to utilize all the interstage frequency slot units (FSUs). In this paper, we simulate the mentioned switching fabric and estimate the required numbers of both frequency slot units and tunable spectrum converters so the internal blocking probability is no more than 10^{-10}.

1 Introduction

By emerging new services such as high-definition video distribution, real-time video communication or online storage, the traffic volume has been multiplied to date. However, the traffic growth rate will not stop here due to the day by day technology advances, and its fast increase will be expected for the years to come. These applications generate data flows of 10 Gb/s up to terabit level. The predictable consequence in the near future is that the network operators will require a new generation of optical transport networks to serve this huge and heterogeneous volume of traffic in a cost-effective and scalable manner. In response to these large capacity and diverse traffic needed in the near future, the elastic optical network (EON) architecture has been proposed. The EON is a new networking paradigm that is capable of assigning bandwidth to optical paths flexibly, this is why they are called "flexible optical networks" or "elastic optical networks" [5].

In [4], the International Telecommunication Union (ITU) proposed the flexible spectrum grid, where the minimum portion of a spectrum, which can be occupied by a connection, is called the slot width granularity. This slot width granularity is of 12.5 GHz and it is also known as the FSU [3]. A connection may occupy a spectrum of width equal to $m \times 12.5$ GHz, and this connection is named an m-slot connection. In EON, when an m-slot connection requires bandwidth higher than 12.5 GHz, it can spread over multiple FSUs, however, these FSUs must be adjacent to each other as presented in Fig. 1 [5].

© Springer Nature Switzerland AG 2020
M. Choraś and R. S. Choraś (Eds.): IP&C 2019, AISC 1062, pp. 189–196, 2020.
https://doi.org/10.1007/978-3-030-31254-1_23

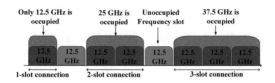

Fig. 1. The spectrum assignment in the flexible grid.

Flexible optical connections must be served by flexible optical switching nodes [11]. The simplest elastic optical switch can be implemented using an element called the bandwidth-variable wavelength selective switch (BV-WSS). The BV-WSS can forward a connection from an input set of FSUs to any set of FSUs in the output links. In [2], a flexible optical switch is constructed using q BV-WSSs, each of capacity $1 \times q$. The main drawback of such a switch is its blocking characteristics. Since it lacks TSCs, each connection must occupy the same set of adjacent FSUs in input and output links. Networks that operate statically or incrementally can be designed using such switches. However, when the network is operating dynamically, the blocking probability parameter will be unsatisfactory. By adding TSCs, we can improve the switch performance by significantly reducing this parameter. It is not economical and sometime not necessary to implement a switching node with full spectrum conversion capability, in which any input spectrum can be interchanged to any output spectrum, and dedicated TSCs have to be equipped for all the FSUs, alternatively, partially equipped TSCs can be shared in an optical node [13].

In [14], four types of switching architectures were proposed, which are called NA-I, NA-II, NA-III and NA-IV, respectively. Architectures proposed in [14] are considered as one-stage and two-stage switching fabrics. All of these architectures are blocking, in other words, there are some cases where it is not possible to establish a connection between a certain set of input FSUs and another set in the output.

The SSNB conditions for architecture Space-Wavelength-Space (S-W-S) were considered in [1] while the nonblocking conditions for the W-S-W switching fabric architectures were discussed in [8–10]. The W-S-W architectures are considered as 3-stage switching fabrics, and are called W-S-W architectures, since they utilize spectrum conversion in the first and third stages, and space switching in the center stage. The SSNB, wide-sense nonblocking (WSNB), and rearrangeable nonblocking (RNB) conditions for this architecture were derived and proved in [7,8] and [10], respectively.

In this paper, we investigate a version of the W-S-W switching fabric architectures, which is named the WSW1 switching fabric. At first, we calculate the number of FSUs required by the switching fabric to be strictly-nonblocking, then the switching fabric is provided with the less number of FSUs in the interstage links. We show that a reasonable blocking probability can be reached when the number of FSUs in the interstage links is much lower than the SSNB requirement. Furthermore, the number of TSCs provided in first and last stages is also reduced, however, a certain level of blocking probability must always be maintained.

This paper consists of four sections. Section 2 presents the investigated architecture along with the definition of connections served by the switch. The simulation methodology is presented in Sect. 3. Numerical analysis and experimental observations are presented in Sect. 4. The paper ends with conclusions.

2   The Switching Fabric

The investigated architecture is composed of three stages, the first and last stages contain r of 1×1 bandwidth-variable waveband converting switches (BV-WBCS), while the central stage (stage two) consists of a single $r \times r$ bandwidth-variable waveband selective space switch (BV-WBSSS). Each input and output fiber has n FSUs while each interstage link contains k of FSUs [7,8,10]. In the rest of the paper, the WSW1 switching fabric is refereed to as the WSW1(r, n, k) switching fabric, when a certain configuration is discussed. The mentioned switching fabric is presented in Fig. 2.

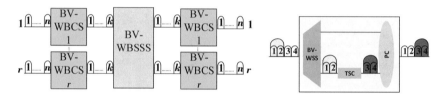

Fig. 2. The WSW1(r, n, k) switching fabric architecture and BV-WBCS that contains only one TSC.

In flexible optical nodes, a connection is set up between sets of FSUs in the input and output fibers, and might occupy a bandwidth equal to m FSUs. Such a connection will be called an m-slot connection. The number of FSUs used by one m-slot connection may be limited by $m_{\max}$, i.e., $1 \leqslant m \leqslant m_{\max} \leqslant n$. To establish an m-slot connection, a set of free m adjacent FSUs must be available in the interstage links between the input and the output. When there is no such set, the new connection is blocked.

3   The Simulation Environment

The simulation environment is designed using C++ programming language. All the elements of the simulator are implemented as objects, i.e., the traffic generators, converters, connections, and the switching fabric is a combination of objects. Generally, the evaluation method is similar to what was done in [6], however, some improvements are introduced to reduce the effect of externally blocked connections. An externally blocked connection can be defined as an

m-slot connection which is rejected due to lack of free adjacent FSUs in the destination link. The external blocking is independent of the switching architecture and can distort the internal blocking results. This issue was considered in [9], where the authors analyzed multiple methods, as an effort to reduce the external blocking.

When a new m-slot connection arrives to the switching fabric, the simulator looks for a free destination to forward the request, and if found, the predefined source link will be checked for m free adjacent FSUs. When both links have at least m free adjacent FSUs, the connection is considered as valid and simulator begins internal links checking, if not, the connection is considered as externally blocked and is not considered or counted as a valid connection. If the connection is not externally blocked, the simulator checks the status of the interstage links which are responsible for establishing the connection. The new connection can be established either with using or without using TSCs, which gives us four cases to establish the connection. If any of these cases is met, the connection is established. When all cases are checked and none of them is met, the connection is reported as internally blocked. The algorithm of setting up one connection is presented in Algorithm 1.

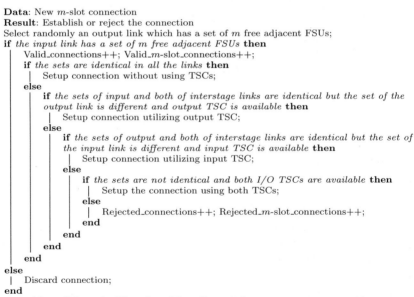

Data: New m-slot connection
Result: Establish or reject the connection
Select randomly an output link which has a set of m free adjacent FSUs;
if *the input link has a set of m free adjacent FSUs* **then**
  Valid_connections++; Valid_m-slot_connections++;
  if *the sets are identical in all the links* **then**
    Setup connection without using TSCs;
  else
    if *the sets of input and both of interstage links are identical but the set of the output link is different and output TSC is available* **then**
      Setup connection utilizing output TSC;
    else
      if *the sets of output and both of interstage links are identical but the set of the input link is different and input TSC is available* **then**
        Setup connection utilizing input TSC;
      else
        if *the sets are not identical and both I/O TSCs are available* **then**
          Setup the connection using both TSCs;
        else
          Rejected_connections++; Rejected_m-slot_connections++;
        end
      end
    end
  end
end
else
  Discard connection;
end

Algorithm 1. The algorithm for setting up a new connection

As presented in Algorithm 1, the internal blocking probability (Bp) is calculated by dividing the number of rejected connections by the total number of valid connections. Sometimes, it is better to calculate the blocking probability of each connection rate (Bpm), to have a better understanding of the switch performance. Equations below define how mentioned parameters are calculated.

$$B_p = \frac{All\ Rejected\ Connections}{Total\ of\ Valid\ Connections} \qquad (1)$$

$$Bpm = \frac{Rejected\ m\text{-}slot\ Connections}{Sum\ of\ Valid\ m\text{-}slot\ Connections} \qquad (2)$$

The evaluation procedures in this paper are divided into several tests. For all tests, common parameters are $m_{max} = 4$, number of events (connections and disconnections requests) is set to 10^8, number of samples for each simulation is 10, other effective parameters are specific for each test. The number of TSCs in each BV-WBCS varies from 0 to n. To clarify this variation, in Fig. 2, we assumed that there is only one TSC that can be used. As presented in the figure, only one connection that require spectrum conversion can established at a time, other connections must pass through without conversion. It should be noted that the conversion capability of the BV-WBCS could be implemented either as presented in Fig. 2 or it could be of other types, such as LCoS based full wavelength converter BV-WSS [12].

4 Simulation Results

At first, we evaluate the effect of reducing the number of FSUs in the interstage links. A WSW1$(2, 20, k)$ switching fabric is evaluated. According to SSNB conditions presented in [8], mentioned switching fabric needs 132 of FSUs in the interstage links to be nonblocking. The switching fabric was tested with number of FSUs $k = 20$, with an increase of 1 FSU in each execution, until we find the number of FSUs needed so the blocking probability will be no more than 10^{-10}.

In Fig. 3, we can notice that, for ascending traffic intensity, 34 FSUs are required for intensity of 0.4 Erlang while 37 FSUs are needed when the intensity of traffic is increased to 0.9 Erlang. Usually, when traffic intensity increases, the demand for more FSUs also increases, due to higher number of connections and less time between two successive calls. However, we can see in the mentioned chart, the demand for more FSUs reaches a stability when the intensity exceeds 0.8 Erlang, this probably happens due to higher external blocking which is reflected as almost the same number of connections processed by the switching fabric. When we analyze the results in more details, we must consider the required number of FSUs for each m-slot connection to reach the desired blocking probability. Since m_{max} was set to 4, we have 4 rates of m, which are $m = 1, 2, 3,$ and 4. The FSU requirements for m-slot connections with intensities $0.5, 0.7,$ and 0.9 Erlang are presented in Table 1.

Next, we will consider the number of I/O FSUs effect. Three configurations of the WSW1$(2, n, k)$ switching fabric are evaluated under 0.7 Erlang traffic intensity. The difference between these configurations is the number of FSUs in the I/O links. The first configuration contains only 20 FSUs in each I/O links while others contain 40 and 80, respectively. Similar to what was done earlier, we will evaluate the mentioned configurations with a number of interstage FSUs less than the SSNB requirement. To clarify this principle, each configuration was

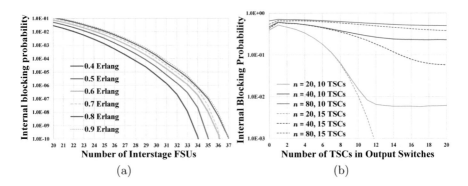

Fig. 3. Internal Bp vs k under different intensities of traffic (a) and Bp vs number of TSCs in the output BV-WBCSs for different n when 10 and 15 TSCs are provided in the input BV-WBCSs.

tested with number of interstage FSUs k_e, where k_e is the minimum number of FSUs required in the interstage links to maintain a blocking probability of 10^{-10}. For each of the configurations, k_e and number of FSUs required in SSNB (k_{ssnb}) are presented in Table 2.

Table 1. The required number of FSUs in the interstage links for each m-slot connection in order to reach 10^{-10} blocking probability.

Intensity	$m = 1$	$m = 2$	$m = 3$	$m = 4$
0.5	29	31	34	35
0.7	29	32	35	36
0.9	30	33	35	37

Table 2. The number of interstage FSUs required in each configuration so the blocking probability is no more than 10^{-10}.

n	k_{ssnb}	k_e	$m_{\max}$
20	132	37	4
40	292	63	4
80	612	110	4

We can conclude from Fig. 3b, that as n increases, more TSCs are needed for the same traffic intensity. In Fig. 3b, all of these switching fabric configurations are provided with 10 TSCs in each input BV-WBCS. For all the configurations, the desired blocking probability could not be reached with this number of input TSCs no matter the number of provided output TSCs. However, the blocking probability of WSW1$(2, 20, 37)$ is less than the other two, and its value was around 10^{-2}. When the input TSCs are raised to 15 in Fig. 3b, the blocking probability of WSW1$(2, 20, 37)$ reached almost the desired blocking probability, however, the blocking probability of both WSW1$(2, 40, 63)$ and WSW1$(2, 80, 110)$ configurations is still high and more input TSCs are needed.

Now let us move to the next step, where we consider the effect of changing the numbers of TSCs in input and output BV-WBCSs of WSW1$(2, 20, 37)$.

When the mentioned switching fabric has only one TSC in the input BV-WBCS, the blocking probability for all the intensities is high. Increasing the number of TSCs in the output stage BV-WBCS didn't lead to a significant reduction in the blocking probability. Furthermore, the dropping stabilizes when 5 output TSCs are utilized for 0.4 Erlang while for 0.9 Erlang, the dropping stabilizes at 7. Providing more TSCs in the output doesn't lead to any reduction in the blocking probability. The situation start to change when more input TSCs are provided. With 8 TSCs in each input BV-WBCS, lower intensities began to reach a blocking probability between the values of 10^{-2} and 10^{-3}, as presented in Fig. 4a. As presented in Fig. 4b, when we equip the input BV-WBCSs with 16 TSCs each, a blocking probability of 10^{-10} is reached for intensities $0.4, 0.5$ and 0.6 Erlang. For higher intensities of traffic, more input TSCs are needed. Traffic intensities of 0.7 and 0.8 Erlang required a combination of 17 input and 18 output TSCs while for 0.9 Erlang, 18 TSCs are required in each BV-WBCS to reach a blocking probability of 10^{-10}.

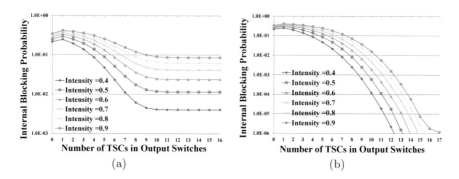

Fig. 4. Internal blocking probability vs number of TSCs in the output BV-WBCS when: (a) 8 and (b) 16 TSCs are provided in each input BV-WBCSs.

5   Conclusion

Simulation of the $WSW1(r, n, k)$ elastic optical switch is considered in this paper. Results showed that there is a certain number of TSCs needed to reach a very low blocking probability, and these numbers are much less than numbers required for SSNB. As expected, having more FSUs in I/O links requires more TSCs to maintain a very low blocking probability. What is important is, for small values of n, i.e. 20, utilizing less than 28% of FSUs required in the SSNB [8] is satisfactory to maintain an internal blocking probability around 10^{-10}, and this percentage is further reduced when n grows to reach 18% of SSNB FSUs when $n = 80$.

Acknowledgements. The work of Mustafa Abdulsahib was supported by the National Science Centre, Poland (NCN) under Grant UMO-2016/21/B/ST7/02257

(ERP: 08/ 84/PNCN/2257), Wojciech Kabaciński and Marek Michalski were supported by funding by the Ministry of Science and Higher Education, Poland under Grant 08/82/SBAD/8230.

References

1. Danilewicz, G., Kabaciński, W., Rajewski, R.: Strict-sense nonblocking space-wavelength-space switching fabrics for elastic optical network nodes. IEEE/OSA J. Opt. Commun. Netw. **8**(10), 745–756 (2016)
2. FINISAR: 1x9/1x20 Flexgrid Wavelength Selective Switch (WSS) (2015). https://www.finisar.com/sites/default/files/downloads/1x9_1x20_flexgrid_wss_pb_v3.pdf
3. Gerstel, O., Jinno, M., Lord, A., Yoo, S.J.B.: Elastic optical networking: a new dawn for the optical layer? IEEE Commun. Mag. **50**(2), S12–S20 (2012)
4. ITU-T: Recommendation G.694.1. Spectral Grids for WDM Applications: DWDM Frequency Grid. International Telecommunication Union - Telecommunication Standardization Sector (ITU-T) (2012)
5. Jinno, M., Takaraa, H., Kozicki, B., Tsukishima, Y., Sone, Y., Matsuoka, S.: Spectrum-efficient and scalable elastic optical path network: architecture, benefits, and enabling technologies. IEEE Commun. Mag. **47**(11), 66–73 (2009)
6. Kabaciński, W., Abdulsahib, M., Michalski, M.: Performance evaluation of WSW2 switching fabric architecture with limited number of spectrum converters. In: The International Scientific Conference Advances in Wireless and Optical Communications (RTUWO), Riga (2018)
7. Kabaciński, W., Abulsahib, M., Michalski, M.: Wide-sense nonblocking W-S-W node architectures for elastic optical networks. IEICE Trans. Commun. **E102-B**(5) (2019, accepted for publication)
8. Kabaciński, W., Michalski, M., Abdulsahib, M.: The strict-sense nonblocking elastic optical switch. In: IEEE 15th International Conference on High Performance Switching and Routing (HSPR), Budapest, Hungary (2015)
9. Liew, S.C., Ng, M.H., Chan, C.W.: Blocking and nonblocking multirate Clos switching networks. IEEE/ACM Trans. Netw. **6**(3), 307–318 (1998)
10. Lin, B.: Rearrangeable W-S-W elastic optical networks generated by graph approaches. IEEE/OSA J. Opt. Commun. Netw. **10**(8), 675–685 (2018)
11. Tomkos, I., Azodolmolky, S., Solé-Pareta, J., Palkopoulou, E.: A tutorial on the flexible optical networking paradigm: state of the art, trends, and research challenges. Proc. IEEE **102**(9), 1317–1337 (2014)
12. Xie, D., Wang, D., Zhang, M., Liu, Z., You, Q., Yang, Q., Yu, S.: LCoS-based wavelength-selective switch for future finer-grid elastic optical networks capable of all-optical wavelength conversion. IEEE Photonics J. **2**(2) (2017)
13. Yan, F., Hu, W., Sun, W., Gue, W., Jin, Y., He, H., Dong, Y.: Placements of shared wavelength converter groups inside a cost-effective permuted Clos network. IEEE Photonics Technol. Lett. **19**(13), 981–983 (2007)
14. Zhang, P., Li, J., Guo, B., He, Y., Chen, Z., Wu, H.: Comparison of node architectures for elastic optical networks with waveband conversion. China Commun. **10**(8), 77–87 (2013)

AI-Based Analysis of Selected Gait Parameters in Post-stroke Patients

Prokopowicz Piotr[1]([envelope]) [ORCID], Mikołajewski Dariusz[1] [ORCID], Tyburek Krzysztof[1] [ORCID], Mikołajewska Emilia[2] [ORCID], and Kotlarz Piotr[1] [ORCID]

[1] Institute of Mechanics and Applied Computer Science, Kazimierz Wielki University, Kopernika 1, 85-074 Bydgoszcz, Poland
{piotrekp,dmikolaj,krzysiekkt,piotrk}@ukw.edu.pl
[2] Department of Physiotherapy, Ludwik Rydygier Collegium Medicum in Bydgoszcz, Nicolaus Copernicus University in Toruń, Torun, Poland
emiliam@cm.umk.pl

Abstract. In this paper we propose solution of the problem of clinical gait analysis in post-stroke patients using advanced artificial intelligence approaches: fuzzy logic, neural networks, and fractal dimension. We focus on the stroke influence on gait pattern and features due to stroke is regarded one of the major causes of disability, including gait disorders. No doubt gait may be described by many parameters but it still needs advanced computational approach. Statistical analysis and simulation of gait features allow for relatively early detection of many limitations, selection of the proper therapeutic method, and assessment of the therapy progress. Results presented here are promising despite our approach needs for further studies toward clinical application.

Keywords: Fuzzy logic · Neural networks · Fractal dimension · Gait analysis · Walking function

1 Introduction

This paper focuses on the novel solution of the problem of clinical gait analysis in post-stroke patients using advanced artificial intelligence approaches: fuzzy logic, neural networks, and fractal dimension. We focus on the stroke influence on gait pattern and features due to stroke is regarded one of the major causes of disability, including gait disorders. Gait may be described by many various parameters. What more gait pattern significantly changes with age and health of status, both in positive and negative (e.g. due to neurodegenerative disorders) manner. Their statistical analysis and simulation allow for relatively early detection of many limitations and disorders, selection of the proper therapeutic method, and assessment of the running therapy progress for reassessment purposes. Computational objectivization of clinical gait analysis is significant from clinical, scientific and societal points of view, but still needs for novel, more advanced, approaches.

© Springer Nature Switzerland AG 2020
M. Choraś and R. S. Choraś (Eds.): IP&C 2019, AISC 1062, pp. 197–205, 2020.
https://doi.org/10.1007/978-3-030-31254-1_24

Patients after stroke constitute a particular group of patients with gait limitations. Main causes of the aforementioned situation are wide spread of stroke, associated disorders both in the motor area (including motor control) and sensoric area, wide diversity of areas and levels of damages and deficits, importance of lower limb function recovery for patients and their families, necessity of individual approach in each case, and necessity of the prevention of secondary changes as far as subsequent stroke. Stroke is perceived a major cause of disability, including gait disorders. Mortality, case-fatality, and health-related quality of life (HRQoL) for stroke data reported by different countries, even European, are inconsistent. Incidence of stroke, adjusted to the World Health Organization (WHO) world standard population, ranged from 76 per 100,000 population per year in Australia up to 119 per 100,000 population per year in New Zealand. Crude mortality and crude incidence of stroke were positively correlated with the proportion of the population aged $\geq$ 65 years, but not with time [18,19].

The risk of stroke among European populations varies more than 2-fold in men and women. Higher rates of stroke were observed in eastern European countries, and lower rates in southern European countries. What more there is considerable geographic variation in stroke mortality even around one country, e.g. the United Kingdom (UK) [3]. More than 40% of patients after first-ever-lifetime stroke had a poor outcome, i.e. 3 months after stroke they were dead, dependent, or institutionalized [5]. The study by Ayis et al. [2] demonstrated significant variations in survival, HRQoL and utilities across populations at 3 and 12 months after stroke, that could not be explained by stroke severity and sociodemographic factors. HRQoL was assessed by the physical component summary (PCS) and mental component summary (MCS) of the Short-Form Health Survey (SF-12), mapped into the EuroQoL-5D (EQ-5D). Strong associations between HRQoL at 3 months and survival to 1 year after stroke were identified [2]. Variation is observed in the content of health state descriptions for all levels of stroke severity, e.g. between the content of descriptions and how stroke is experienced by patients for domains related to HRQoL. There is no systematic/standardized method for content/scope of health state descriptions for stroke, and the patient perspective is not incorporated [4]. Although walking speed is the most common measure of gait performance post-stroke, improved walking speed following rehabilitation does not always indicate the recovery of paretic limb function.

Looking for more effective methods of gait clinical analysis and reeducation in post-stroke survivors is one of the most important issues in contemporary neurorehabilitation. Walking dysfunctions persist following post-stroke rehabilitation. A major limitation of current rehabilitation efforts is the inability to identify modifiable deficits that, when improved, will result in the recovery of walking function [8–10]. Previous studies have relied on cross-sectional analyses to identify deficits to target during walking rehabilitation; however, these studies did not account for the influence of a key covariate - maximum walking speed.

Fortunately previous studies by Prokopowicz et al. showed significance of studies incorporating computational intelligence, especially directed fuzzy numbers, into clinical gait analysis [12–14]. Thus this paper constitutes development of the aforementioned studies joining various methods: fuzzy logic, neural networks, and fractal dimension. Such hybrid approach may provide increase of exactness and variety of described gait features, e.g. irregularity of gait reflecting hemiparetic gait.

Scientists and clinicians still look for more effective rehabilitation methods in post-stroke patients, including rehabilitation of gait disorders, paying particular attention to efficient methods according to the evidence-based medicine paradigm. This situation also concerns the most widely applied so far neurorehabilitation methods: NeuroDevelopmental Treatment-Bobath (NDT-Bobath) approach for adults. There is still a lack of randomized trials showing its efficiency unambiguously [6,15]. There was a need also for simple, cheap, quick, exact, test-retest reliable, easy in application and interpretation, not requiring high computational cost markers of physiological and pathological gait used in everyday clinical pracThe aim is threefold: (1) to investigate and describe a number of basic computational tools reflecting human walking, (2) to present the outcome of the gait reeducation using the NDT-Bobath method, (3) to determine the relationships between commonly studied spatio-temporal parameters reflecting post-stroke walking function (including home-based rehabilitation) [1,7,11,16,17]. The aim is threefold: (1) to investigate and describe a number of basic computational tools reflecting human walking, (2) to present the outcome of gait reeducation using the NeuroDevelopmental Treatment-Bobath (NDT-Bobath) method, (3) to determine the relationships between commonly studied spatio-temporal parameters reflecting post-stroke walking function.

2   Materials and Methods

The current study included 40 adult patients after stroke. Patients were randomly assigned to one of the treatment groups: study group (n = 20, treated with NDT-Bobath, 10 sessions during 2 weeks), and the reference group (n = 20, treated with traditional method only, 10 sessions). Patients' overall profile shows Table 1. Patients' flow shows Fig. 1. Inclusion criteria were following: diagnosis: stroke (type haemmorhagic/ischemic not specified), age at least 18 years, time after cerebrovascular accident (CVA): since 1 month to 3 years, gait function presented.

2.1   Methods

Study organization: prospective study, before-after study with randomization. Ten sessions of the therapy according to the current rules of NDT-Bobath for adults method was provided by recognized NDT-Bobath therapists with 15 years of experience in neurorehabilitation. The measurements (spatio-temporal gait parameters based on 10 m walking test: gait velocity, normalized gait velocity, cadence, normalized cadence, stride length, and normalized stride length) were administered twice: on admission (before the therapy) and after the last session of the therapy. The assessment was made also using novel markers of gait:

Table 1. Patients overall profile.

Feature	Study group n = 20 (100%)	Reference group n = 20 (100%)
Age [years]:		
Min	44	46
Max	81	85
Mean	63.5	65.67
SD	5.77	7.76
First quartile (Q1)	57	58
Median (second quartile)	68	66
Third quartile (Q3)	72	74
Gender: female (F)	10 (50%)	10 (50%)
Gender: male (M)	10 (50%)	10 (50%)
Height [m]:		
Min	1.55	1.62
Max	1.79	1.76
Mean	1.65	1.66
SD	0.05	0.05
First quartile (Q1)	1.63	1.67
Median (second quartile)	1.66	1.65
Third quartile (Q3)	1.73	1.72
Side affected: left (L)	10 (50%)	10 (50%)
Side affected: right (P)	10 (50%)	10 (50%)
Time after CVA [weeks]:		
Min	6	6
Max	155	154
Mean	63.50	64.20
SD	15.63	15.82
First quartile (Q1)	22	20
Median (second quartile)	58	53
Third quartile (Q1)	104	103

fractal dimension, fuzzy parameter, and classification based on artificial neural networks. Such selection was made because novel markers may measure better e.g. gait symmetry, and their values for healthy people were assessed before according to the norms established by Mikołajewska et al. [10].

Statistical analysis was made using software Statistica 12. Where available, measured data was described as mean with standard deviation (SD) or median with minimal and maximal values. Normality of distribution was checked using Shapiro-Wilk test. According to the needs, for data sets with normal distribu-

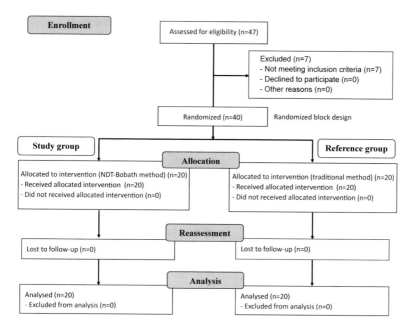

Fig. 1. Patiens flow diagram (CONSORT 2010 i.e. CONsolidated Standards of Reporting Trials 2010, http://www.consort-statement.org/consort-2010.)

tion t-test was applied, and for the other data sets U-Mann Whitney test was applied. Spearmans' rho correlation coefficient was used to measure relationships among gait parameters and age, gender, time from CVA, and side of paresis. The statistical significance represented by p-value (probability value) was set at 0.05.

Fuzzy-based analysis, fractal analysis, a artificial neural networks (ANN) simulation was made using Matlab software (Neural Networks toolbox). Where available computational model of data set was created. Tendencies within data were acquired and described. Fuzzy parameter (and its relative changes) reflected distance from so called "normal values" (i.e. values of gait parameters for healthy people). Fractal parameter reflected irregularity of the gait pattern, reflecting not only hemiplegic gait, but also diverse step length between left and right leg as far as shortened step during second part of the walk reflecting increasing fatigue during the test. Thus fractal parameter was also assessed graphically. ANN-based analysis showed possibility of semi-automated deficit detection, even in moderate and mild states of gait disorders. Construction of the ANN-based gait models may increase possibility to semi-automatically gather e.g. gait disorders as a preclinical study, without special equipment required in gait laboratories.

Ethical Issues. This study was conducted in accordance with the Declaration of Helsinki and the guidelines for Good Clinical Practice (GCP). Freely given written informed consent was obtained from every patient before the study.

Table 2. Changes of spatio-temporal gait parameters.

Value		Parameter					
		Velocity	Cadence	Stride length	Normalized velocity	Normalized cadence	Normalized stride length
Change in study group	Mean	0.31	19.17	0.53	0.11	0.09	0.63
	SD	0.22	14.04	0.29	0.08	0.07	0.32
	Min	−0.1	−5	0.15	−0.02	0	0.26
	Q1	0.125	10.75	0.33	0.06	0.05	0.385
	Median	0.3	18	0.49	0.09	0.08	0.55
	Q3	0.4	23	0.68	0.13	0.11	0.88
	Max	0.8	61	1.32	0.25	0.31	1.36
Change in reference group	Mean	0.11	11.87	0.14	0.04	0.05	0.12
	SD	0.18	16.46	0.14	0.06	0.08	0.19
	Min	−0.1	−8	−0.13	−0.03	−0.11	−0.6
	Q1	0	0	0.05	0	0.01	0.01
	Median	0.05	9.5	0.15	0.015	0.04	0.175
	Q3	0.2	22.25	0.22	0.085	0.1	0.24
	Max	0.7	67	0.55	0.23	0.34	0.33
p-value		0.017	0.023	0.005	0.017	0.019	0.003

3   Results

Among the 40 patients involved in the study, the results were as follows: in terms of gait velocity and normalized gait velocity recovery was observed in 32 cases (80%), in terms of cadence and normalized cadence recovery was observed in 30 cases (75%), in terms of stride length and normalized stride length recovery was observed in 34 cases (85%), in terms of fuzzy parameter recovery was observed in 35 cases (87.5%), and in terms of fractal parameter recovery was observed in 32 cases (80%) (Tables 2, 3 and 4). Benefits were observed after short-term therapy, reflected by measurable statistically significant changes in the patients' gait parameters.

As a result of artificial neural network (ANN) analysis there were observed very good matches of the network to the proposed models. Errors in classification weren't observed. Usefulness of fractal dimension as a measure of gait irregularity allows for recognition of hemiplegic gait. Data set was divided into teaching set (70%), validation set (20%) and test set (10%). There were observed following errors values: learning error was 0.01, validation error: 0.03, test error: 0.05.

There were observed statistically relevant moderate relationships among aforementioned parameters and age, gender, time from CVA, and side of paresis reflected in the value of Spearmans' rho correlation coefficient. They may suggest prognostic signs, but computational tendency analysis in gait re-education still requires further studies.

Table 3. Changes of fuzzy parameter.

Change in study group	Mean	0.04
	SD	0.01
	Min	−0.10
	Q1	0
	Median	0.03
	Q3	0.11
	Max	0.25
Change in reference group	Mean	0.03
	SD	0.01
	Min	−0,20
	Q1	0.01
	Median	0.02
	Q3	0.11
	Max	0.34
p-value		0.039

Table 4. Changes of fractal parameter.

Change in study group	Mean	0.07
	SD	0.03
	Min	0.01
	Q1	0.03
	Median	0.04
	Q3	0.05
	Max	0.06
Change in reference group	Mean	0.04
	SD	0.01
	Min	0.01
	Q1	0.03
	Median	0.04
	Q3	0.05
	Max	0.06
p-value		0.012

4   Discussion and Summary

Main idea of this paper was solving the problem of relatively efficient, cheap, and quick clinical gait analysis in post-stroke patients using advanced artificial intelligence approaches: fuzzy logic, neural networks, and fractal dimension. Obtained results of the study confirmed better exactness of the proposed computational methodology of the gait analysis. New gait markers (fractal dimension, fuzzy parameter, results of artificial neural networks classification) showed sensitive, quick, and cheap gait diagnostic tool, supplementing currently used methods of clinical gait analysis. Moreover, fractal dimension and the fuzzy parameter may constitute standalone tool. Thus our results constitute novelty, and their application may provide another progress in everyday clinical gait analysis, including remote for eHealth purposes.

As a result of rehabilitation, statistically significant changes in the spatio-temporal gait parameters as far as novel gait markers have been observed. They were more favorable in the group of patients treated with NDT-Bobath method compared to the group treated with the traditional approach. The greatest recovery was observed in fuzzy parameter and gait velocity. Prognostic signs of the greatest recovery were younger age, longer time after CVA, males, and right hemiplegia.

There is lack of studies to compare results of our study. Main limitation constitutes limited sample of patients - there is need for further randomized studies on bigger sample. Knowledge gathered as a result of this study allows for enhancement of the current methodology of clinical gait analysis incorporating new markers, including new software for mobile devices developed by our interdisciplinary team. Described study gives new tools increasing the objectivity of the rehabilitation process assessment in inpatient conditions, outpatient conditions, as far as home-based rehabilitation. Moreover, the study has presented the prevalence of the NDT-Bobath concept over traditional method in post-stroke gait rehabilitation in the most objective way, and relatively simple methodology allows for the immediate transfer of the described experiences to everyday clinical practice in neurorehabilitation, as far as to replication of this study or as a starting point for further development of research. The aforementioned approach may provide further progress in gait rehabilitation in described group of patients. In patients in the chronic phase of stroke recovery, also improving maximum walking speed may be necessary to improve long-distance walking function - it may constitute another parameter need to be taken into consideration in calculations.

References

1. Awad, L.N., Reisman, D.S., Wright, T.R., Roos, M.A., Binder-Macleod, S.A.: Maximum walking speed is a key determinant of long distance walking function after stroke. Top. Stroke Rehabil. **21**(6), 502–509 (2014). https://doi.org/10.1310/tsr2106-502
2. Ayis, S., Wellwood, I., Rudd, A.G., McKevitt, C., Parkin, D., Wolfe, C.D.A.: Variations in health-related quality of life (HRQoL) and survival 1 year after stroke: five European population-based registers. BMJ Open **5**(6) (2015). https://doi.org/10.1136/bmjopen-2014-007101
3. Bhatnagar, P., Scarborough, P., Smeeton, N.C., Allender, S.: The incidence of all stroke and stroke subtype in the United Kingdom, 1985 to 2008: a systematic review. BMC Public Health **10**(1), 539 (2010). https://doi.org/10.1186/1471-2458-10-539
4. Gray, J., Lie, M.L.S., Murtagh, M.J., Ford, G.A., McMeekin, P., Thomson, R.G.: Health state descriptions to elicit stroke values: do they reflect patient experience of stroke? BMC Health Serv. Res. **14**(1), 573 (2014). https://doi.org/10.1186/s12913-014-0573-6
5. Heuschmann, P., Wiedmann, S., Wellwood, I., Rudd, A., Di Carlo, A., Bejot, Y., Ryglewicz, D., Rastenyte, D., Wolfe, C.: Three-month stroke outcome. Neurology **76**(2), 159–165 (2011). https://doi.org/10.1212/WNL.0b013e318206ca1e
6. Klimkiewicz, P., Kubsik, A., Woldańska-Okońska, M.: NDT-Bobath method used in the rehabilitation of patients with a history of ischemic stroke. Wiad. Lek. **65**(2), 102–107 (2012)
7. Lozano-Ortiz, C.A., Muniz, A.M.S., Nadal, J.: Human gait classification after lower limb fracture using artificial neural networks and principal component analysis. In: 2010 Annual International Conference of the IEEE Engineering in Medicine and Biology, pp. 1413–1416, August 2010
8. Mikołajewska, E.: Associations between results of post-stroke NDT-Bobath rehabilitation in gait parameters, ADL and hand functions. Adv. Clin. Exp. Med. **22**(5), 731–738 (2013)

9. Mikołajewska, E.: The value of the NDT-Bobath method in post-stroke gait training. Adv. Clin. Exp. Med. **22**(2), 261–272 (2013)
10. Mikołajewska, E., Prokopowicz, P., Mikolajewski, D.: Computational gait analysis using fuzzy logic for everyday clinical purposes – preliminary findings. Bio-Algorithms Med-Syst. **13**(1) (2017). https://doi.org/10.1515/bams-2016-0023
11. Muniz, A., Nadal, J.: Application of principal component analysis in vertical ground reaction force to discriminate normal and abnormal gait. Gait Posture **29**(1), 31–35 (2009). https://doi.org/10.1016/j.gaitpost.2008.05.015
12. Prokopowicz, P., Mikołajewska, E., Mikołajewski, D., Kotlarz, P.: Analysis of temporospatial gait parameters, pp. 289–302. Springer, Cham (2017). https://doi.org/10.1007/978-3-319-59614-3_17
13. Prokopowicz, P., Mikołajewski, D., Mikołajewska, E., Kotlarz, P.: Fuzzy system as an assessment tool for analysis of the health-related quality of life for the people after stroke. In: Rutkowski, L., et al. (ed.) Artificial Intelligence and Soft Computing, pp. 710–721. Springer, Cham (2017)
14. Prokopowicz, P., Mikołajewski, D., Mikołajewska, E., Tyburek, K.: Modeling trends in the hierarchical fuzzy system for multi-criteria evaluation of medical data. In: Kacprzyk, L., et al. (ed.) Advances in Fuzzy Logic and Technology 2017, pp. 207–219. Springer, Cham (2018)
15. Richards, C.L., Malouin, F., Dean, C.: Gait in stroke: assessment and rehabilitation. Clin. Geriatr. Med. **15**(4), 833–855 (1999)
16. Roelker, S.A., Bowden, M.G., Kautz, S.A., Neptune, R.R.: Paretic propulsion as a measure of walking performance and functional motor recovery post-stroke: a review. Gait Posture **68**, 6–14 (2019). https://doi.org/10.1016/j.gaitpost.2018.10.027
17. Sheffler, L.R., Chae, J.: Hemiparetic gait. Phys. Med. Rehabil. Clin. N. Am. **26**(4), 611–623 (2015). Stroke Rehabilitation
18. Thrift, A.G., Howard, G., Cadilhac, D.A., Howard, V.J., Rothwell, P.M., Thayabaranathan, T., Feigin, V.L., Norrving, B., Donnan, G.A.: Global stroke statistics: an update of mortality data from countries using a broad code of cerebrovascular diseases. Int. J. Stroke **12**(8), 796–801 (2017)
19. Thrift, A.G., Thayabaranathan, T., Howard, G., Howard, V.J., Rothwell, P.M., Feigin, V.L., Norrving, B., Donnan, G.A., Cadilhac, D.A.: Global stroke statistics. Int. J. Stroke **12**(1), 13–32 (2017). https://doi.org/10.1177/1747493016676285

Classification of Multibeam Sonar Image Using the Weyl Transform

Ting Zhao[1,2(✉)], Srđan Lazendić[2,3], Yuxin Zhao[1],
Giacomo Montereale-Gavazzi[4,5], and Aleksandra Pižurica[2]

[1] College of Automation, Harbin Engineering University, Harbin, China
{zhaoting,zhaoyuxin}@hrbeu.edu.cn
[2] Department of Telecommunications and Information Processing, TELIN-GAIM,
Ghent University, Ghent, Belgium
{Srdan.Lazendic,Aleksandra.Pizurica}@UGent.be
[3] Department of Mathematical Analysis, Ghent University, Ghent, Belgium
[4] Operational Directorate Natural Environment,
Royal Belgian Institute of Natural Sciences, Brussels, Belgium
gmonterealegavazzi@naturalsciences.be
[5] Renard Centre of Marine Geology Department of Geology,
Ghent University, Ghent, Belgium

Abstract. In this paper we develop a novel classification method for multibeam sonar images based on the Weyl transform. The texture descriptor based on Weyl coefficients describes effectively the multiscale correlation features appearing in the sonar images. Our classification approach combines the Weyl coefficients with statistical features that are commonly used in the analysis of seabed sonar images and captures the morphological variation and geoacoustic characteristics of the seafloor. We employ a neural network as a classifier. The proposed combined feature extraction method demonstrates better performance than the commonly used statistical methods in this application.

Keywords: Multibeam data processing · Multibeam sonar image ·
Feature extraction · Weyl transform · Acoustic sediment classification

1 Introduction

Backscattering from the seafloor is the result of an intricate interaction of the sound pulse with the water-sediment interface and relates to three basic quantities: the acoustic impedance contrasts between the propagation and sediment media, the volume inhomogeneity and the roughness. Due of this, backscatter directly relates to the seafloor nature and such hydroacoustic measurements can

This work was supported in part by Major Project of Chinese National Programs for Fundamental Research and Development (No. 613317) and in part by the China Scholarship Council.

M. Choraś and R. S. Choraś (Eds.): IP&C 2019, AISC 1062, pp. 206–213, 2020.
https://doi.org/10.1007/978-3-030-31254-1_25

be used to characterise it in the interest of geology, sedimentology and biology [1,2]. A conventional multibeam echosounder system is capable of collecting backscatter data and bathymetry data, from which we are able to obtain a variety of features of the seabed to distinguish the sediment type [3,4]. Texture-based techniques rely on the extraction and characterization of the textural information of each seabed type. All state-of-the-art methods, in order to have a reliable texture analysis, remove the angular dependency on each analysis zone which shares the same backscatter profile [5]. A well-established approach currently is to use the first-order [6] and the second-order statistical features [7]. The objective of this study is to develop more effective feature extraction methods to improve the reliability of acoustic sediment classification.

Recent studies have demonstrated that the Weyl transform [8,9] offers an excellent framework for data representation and texture analysis in general. The main contribution of this paper is to explore the potential of seabed sediment classification based on the Weyl transform. Furthermore, we develop an effective classification method for sonar images that combines the Weyl features and complementary statistical features, which are capturing the morphological variation and geoacoustic property. The experimental results show clearly that the proposed combined textural descriptor can effectively discriminate between the different classes of sediment. The paper is organized as follows: Sect. 2 reviews briefly the Weyl transform theory. Next, in Sect. 3 we present our proposed method for texture characterization of multibeam sonar images. The experiment results are presented in Sects. 4 and 5 concludes the paper.

2   Weyl Transform

The Weyl transform has recently shown remarkable results in the context of texture classification with standard texture images [8], outperforming some common textural descriptors including HOG [10] and LBP [11]. The transform has a desirable property of being invariant to a large class of multiscale signed permutations. In particular, different ways of orienting and translating the same texture will produce the same Weyl descriptor and patches sampled from the same texture should share similar Weyl transforms [8,12].

2.1   The Binary Heisenberg-Weyl Group

The binary Heisenberg-Weyl group HW_{2^m} is a group of permutation matrices and matrices that resemble permutation matrices with sign changes in some of the rows. Those square matrices of size 2^m exist for each power of 2 and are defined as tensor products

$$D(a,b) = D(a,0)\,D(0,b) = x^{a_{m-1}} z^{b_{m-1}} \otimes \cdots \otimes x^{a_0} z^{b_0}. \tag{1}$$

where

$$x = \begin{bmatrix} 0 & 1 \\ 1 & 0 \end{bmatrix}, \quad z = \begin{bmatrix} 1 & 0 \\ 0 & -1 \end{bmatrix}$$

and $a = (a_0, \ldots, a_{m-1})$, $b = (b_0, \ldots, b_{m-1}) \in \mathbb{Z}_2^m$ are two binary m-tuples.

Formally, the binary Heisenberg-Weyl group HW_{2^m} of order 2^{2m+2} is defined as $HW_{2^m} = \{i^\lambda D(a,b) \mid \lambda \in \{0,1,2,3\} \text{ and } a,b \in \mathbb{Z}_2^m\}$.

2.2   The Weyl Representation

As shown in [8], the signed permutation matrices $D(a,b)$ with $a^T b = 0$ form an orthonormal basis of the vector space of real square symmetric matrices with respect to the inner product given by $\langle R, S \rangle := \mathrm{tr}(R^T S)$. In particular, each real symmetric matrix R can be represented as a linear combination of the basis elements as

$$R = \sum_{\substack{a,b \in \mathbb{Z}_2^m \\ ab^T = 0}} \left\{ \frac{1}{2^{m/2}} \mathrm{tr}\left[R \cdot D(a,b)\right] \right\} \frac{1}{2^{m/2}} D(a,b). \tag{2}$$

Given a vectorized signal $y \in \mathbb{R}^{2^m}$, it's covariance matrix $yy^T \in \mathbb{R}^{2^m \times 2^m}$ is real, symmetric matrix and as such can be represented as

$$\begin{aligned}
yy^T &= \sum_{\substack{a,b \in \mathbb{Z}_2^m \\ ab^T = 0}} \left\{ \frac{1}{2^{m/2}} \mathrm{tr}\left[yy^T \cdot D(a,b)\right] \right\} \frac{1}{2^{m/2}} D(a,b) \\
&= \sum_{\substack{a,b \in \mathbb{Z}_2^m \\ ab^T = 0}} \omega_{a,b}(y) \frac{1}{2^{m/2}} D(a,b).
\end{aligned} \tag{3}$$

Coefficients $\omega_{a,b}(y)$ are the *Weyl coefficients* of the signal y and the corresponding isometric mapping $yy^T \mapsto \omega_{a,b}(y)$ is the *Weyl transform* [8].

3   Methodology

3.1   Texture Descriptor Based on Weyl Transform

The Weyl transform distinguishes the different textural structures by quantifying multiscale symmetry features [9]. Moreover, invariance to multiscale transformations ensures that the Weyl representation of image patches with the same textural structures exhibit similarity.

We divide the whole multibeam sonar image into a number of small patches using a moving window of size $S_w \times S_w$ ($S_w = 2^r, r \in \mathbb{Z}^+$). Each patch can be vectorized in a raster-scanning fashion which results in $S = 2^{2r}$ dimensional vector. Let $m = 2r$, $a = (a_{m-1} \ldots a_0)^T$ and $b = (b_{m-1} \ldots b_0)^T$. Then the Weyl coefficients of patch Y are computed by using (3). Figure 1 shows how we obtain the Weyl representation of a selected patch from a multibeam backscatter image.

An ideal texture descriptor should represent the samples of the same class with a compact and isolated cluster. We randomly select 800 patches of size 8×8 from 4 classes of multibeam backscatter images and compute the Weyl coefficients of

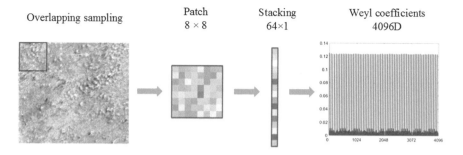

Fig. 1. Computation of the Weyl coefficients for a sonar image.

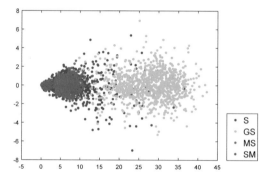

Fig. 2. Weyl coefficients with the dimension reduced to 2 using PCA.

all samples. For visualization purpose, we use PCA to reduce the dimensionality of the 4096-dimensional feature vector to two-dimensional one. Figure 2 shows the backscatter patches represented in Weyl coefficients after the dimensionality reduction. Different colors correspond to different seabed sedimentary classes: *Sand* (S, blue), *Gravelly Sand* (GS, green), *Muddy Sand* (MS, magenta), *Sandy Mud* (SM, red). This example shows that the proposed texture descriptor based on the Weyl transform discriminates well between GS/MS/SM and GS/S/SM classes, but not between S and MS classes. Due to the fact that sand and muddy sand show similar textures and similar distributions of pixel values (Fig. 3), the Weyl descriptor is not able to discriminate well between those two classes.

3.2 Combined Features for Multibeam Sonar Image

The Weyl transform captures textural characteristics more related to the local correlation. A complementary approach is to extract features, which mainly reveal the global zonal characteristics. Hence, we adopt statistical methods to extract characteristic features of the sonar image, that we refer to as Classical Statistical Features. In particular, we include: first-order statistics (backscatter-based), second-order statistics (backscatter-based) and terrain characterization (bathymetry-based).

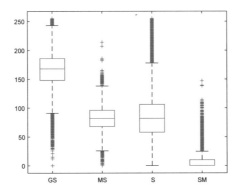

Fig. 3. Boxplot of the distribution of greyscale values for different sediments.

We calculate the first-order statistics from local patches using zonal statistics, including mean, maximum, minimum, quartile, standard deviation, kurtosis and skewness [6]. The second-order statistics are calculated from the Grey Level Co-occurrence Matrices (GLCM) [7]. We derive the entropy and homogeneity from the GLCM. Terrain modeling based on multibeam bathymetry data can make a significant contribution to the prediction of benthic habitat. The adopted terrain features include slope, rugosity and benthic position index [13,14]. We apply the feature selection algorithm of Boruta [15] to reduce the feature set to the more discriminative ones. Then the resulting most relevant statistical features are combined with the Weyl coefficients to generate a feature vector by stacking all the components. We normalize the features such that they are in the same range and thus contribute appropriately to the classification result.

4   Experimental Results

4.1   Dataset

We use the data set from a hydroacoustic survey conducted by Royal Belgian Institute of Natural Sciences in Oostende Harbour, Belgium, in November 2017. The multibeam data originates from the Kongsberg Maritime EM2040 dual system installed on RV Simon Stevin and were acquired at 300 KHz in normal mode, CW pulse form and 101 μs pulse length [16]. Backscatter and bathymetry data are both with a 1 m horizontal resolution (Fig. 4). The ground-truth data are collected from a number of grab samples, including Sand, Sandy Mud, Muddy Sand and Gravelly Sand. We demarcate 12 subblocks on the surveyed area, where the sediment type is already known by grab sampling analysis. Then we take 8×8 patches by overlapping sampling with a sliding step of 4 pixels from each of subblock and 17622 samples are available in total. In the experiments, we randomly take 1000 samples for every class of the sediment, including backscatter data, bathymetry data and their labels. The training set contains 200×4 samples and testing set contains another 800×4 samples, which are both randomly taken from the whole dataset.

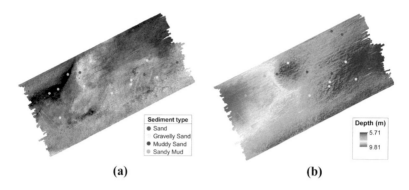

Fig. 4. (a) backscatter data and grab samples; (b) bathymetry data.

4.2 Results

To validate the performance of the proposed texture descriptor, we perform sediment classification on multibeam sonar images by feeding the combined Weyl-Statistical features to a 2-layer neural network. Each test patch is assigned to a sediment type. We compare the performance of Classical Statistical Features alone, Weyl coefficients alone and the Combined Features. From Sect. 3 we know that Classical Statistical Features are extracted both from backscatter data and bathymetry data, while the Weyl coefficients are computed only using backscatter data. Even though the bathymetry data is not used, Tables 1 and 2 indicate that the Weyl coefficients can isolate distinct sediment types with comparable accuracy as the Classical Statistical Features. Table 3 shows the classification accuracies using the combined Classical Statistical Features and Weyl Transform Features. The results in Fig. 5 show that the combined features significantly improve the classification accuracy for the sand class, compared to the first two methods. The overall accuracy of the combined method is also better than any single method.

Table 1. Classification results using Classical Statistical Features.

Ground truth	Prediction					
	S	GS	MS	SM	Total	Accuracy
S	384	71	324	21	800	48%
GS	13	774	13	0	800	97%
MS	73	3	724	0	800	91%
SM	62	0	12	726	800	91%
Total	532	848	1073	747	3200	82%

Table 2. Classification results using Weyl Transform Features.

Ground truth	Prediction					
	S	GS	MS	SM	Total	Accuracy
S	416	45	300	39	800	52%
GS	62	732	6	0	800	92%
MS	112	0	688	0	800	86%
SM	48	0	0	752	800	94%
Total	638	777	994	791	3200	81%

Table 3. Classification results using Combined Features.

Ground truth	Prediction					
	S	GS	MS	SM	Total	Accuracy
S	579	25	179	17	800	72%
GS	85	713	2	0	800	89%
MS	131	0	669	0	800	84%
SM	13	0	10	777	800	97%
Total	808	738	860	794	3200	86%

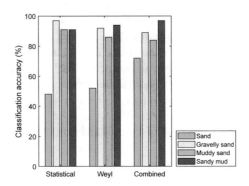

Fig. 5. Classification accuracy of the three feature extraction methods.

5   Conclusion

We designed a novel feature extraction method for seabed sediment classification based on the Weyl transform. We showed that the Weyl coefficients of multibeam sonar images can discriminate between different classes of sediment. We also proposed a combined feature extraction method based on the Weyl transform and Classical Statistical Features to capture better the characteristics of the seafloor both locally and globally. The combined feature vector proves to be more powerful in the classification of sediments than the Weyl transform alone or statistical features alone. Examples on Oostende Harbour dataset demonstrate the efficiency of the proposed feature extraction method for seabed sediment classification using multibeam sonar images.

References

1. Gaida, T., Tengku, A.T., Snellen, M., Amiri-Simkooei, A., Van Dijk, T., Simons, D.: Geosciences **8**, 455 (2018)
2. Montereale-Gavazzi, G., Roche, M., Lurton, X., Degrendele, K., Terseleer, N., Van Lancker, V.: Mar. Geophys. Res. **39**, 229–247 (2018)
3. Diesing, M., Mitchell, P., Stephens, D.: ICES J. Mar. Sci. **73**, 2425–2441 (2016)
4. Brown, C.J., Smith, S.J., Lawton, P., Anderson, J.T.: Estuarine, coastal and shelf. Science **92**, 502–520 (2011)
5. Nguyen, T.K.: Seafloor classification with a multi-swath multi-beam echo sounder. Ph.D. thesis, Ecole nationale supérieure Mines-Télécom Atlantique, Nantes (2017)
6. Janowski, Ł., Tęgowski, J., Nowak, J.: Ocean. Hydrobiol. Stud. **47**, 248–259 (2018)
7. Haralick, R.M., Shanmugam, K., Dinstein, I.H.: IEEE Trans. Syst. Man Cybern. **6**, 610–621 (1973)
8. Qiu, Q., Thompson, A., Calderbank, R., Sapiro, G.: IEEE Trans. Signal Process. **64**, 1844–1853 (2016)
9. Ahn, H.K., Qiu, Q., Bosch, E., Thompson, A., Robles, F.E., Sapiro, G., Warren, W.S., Calderbank, R.: Classifying pump-probe images of melanocytic lesions using the Weyl transform. In: 2018 IEEE International Conference on Acoustics, Speech and Signal Processing (ICASSP), Calgary, Alberta, Canada. IEEE (2018)
10. Dalal, N., Triggs, B.: Histograms of oriented gradients for human detection. In: International Conference on Computer Vision and Pattern Recognition (CVPR 2005). IEEE Computer Society, San Diego (2005)
11. Marko, H., Matti, P.: IEEE Trans. Pattern Anal. Mach. Intell. **28**, 657–662 (2006)
12. Howard, S.D., Calderbank, A.R., Moran, W.: J. Appl. Signal Process. **2006**, 111 (2006)
13. Wilson, M.F., O'Connell, B., Brown, C., Guinan, J.C., Grehan, A.J.: Mar. Geodesy **30**, 3–35 (2007)
14. Walbridge, S., Slocum, N., Pobuda, M., Wright, D.: Geosciences **8**, 94 (2018)
15. Kursa, M.B., Rudnicki, W.R.: J. Stat. Softw. **36**, 1–13 (2010)
16. Kongsberg Maritime. EM 2040 data sheet (2012)

Learning Local Image Descriptors
with Autoencoders

Nina Žižakić[1]([✉]), Izumi Ito[2], and Aleksandra Pižurica[1]

[1] Department of Telecommunications and Information Processing, TELIN – GAIM,
Ghent University – imec, Ghent, Belgium
{nina.zizakic,aleksandra.pizurica}@ugent.be
[2] Department of Human System Science, Tokyo Institute of Technology,
Tokyo, Japan
ito@ict.e.titech.ac.jp

Abstract. In this paper, we propose an efficient method for learning local image descriptors with convolutional autoencoders. We design an autoencoder architecture that yields computationally efficient extraction of patch descriptors through an intermediate image representation. The proposed approach yields significant savings in memory and processing time compared to a reference autoencoder-based patch descriptor. The results demonstrate improved robustness to noise and missing data.

Keywords: Local image descriptors · Autoencoders ·
Unsupervised deep learning

1 Introduction

Local feature descriptors are fundamental to image processing tasks such as image denoising, inpainting, object tracking, and saliency detection. These descriptors can be classified into two categories – the hand-crafted feature descriptors and the learned ones. Two common types of hand-crafted descriptors are distribution-based [1,3,10], and binary descriptors [4,11].

Learned descriptors have recently gained a lot of attention. The success of deep Convolutional Neural Networks (CNNs) in various image processing tasks has encouraged their usage for patch descriptors [2,7,12,15], showing excellent results in patch-matching. The CNN-based learning methods are supervised, i.e. trained with pairs of patches that are labeled as similar or dissimilar. The learned similarities and dissimilarities between image patches may not hold when they are affected by some degradation type that was not included in the training set.

An alternative is unsupervised learning based on autoencoders [8,9]. Chen et al. [5] applied autoencoders to the general problem of patch descriptors. Their autoencoder-learned descriptor shows promising results, however, its calculation time and memory is infeasible for higher resolution images. Their fully-connected network does not take advantage of the local similarity property of natural images, has more parameters to be trained, and requires longer training times

© Springer Nature Switzerland AG 2020
M. Choraś and R. S. Choraś (Eds.): IP&C 2019, AISC 1062, pp. 214–221, 2020.
https://doi.org/10.1007/978-3-030-31254-1_26

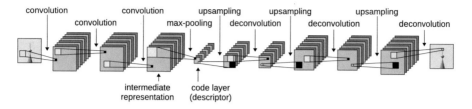

Fig. 1. The proposed autoencoder architecture. There are no max-pooling layers after the first two convolutions in order to obtain an intermediate representation (IR) of image that preserves the spatial information in the height-width plane.

than the convolutional autoencoder designs. Moreover, the descriptor from [5] does not allow different input sizes and therefore a separate autoencoder needs to be trained for every patch size, which renders the framework impractical for many applications.

We propose a novel autoencoder-based patch descriptor designed for applications with many patch comparisons within a single image. We design a specific network architecture that yields a special image representation that we call the *intermediate representation* (IR). The benefits of having a direct access to IR are twofold: (i) patch descriptors can be obtained from IR with a simple operation, and (ii) IR is structured such that overlapping patches' representations are overlapping themselves, resulting in a unique memory-saving property that, to our knowledge, does not hold for any other patch descriptor.

Besides, the introduction of IR enables incorporating contextual information beyond the patch borders into its descriptor, making the descriptors more robust to erroneous and missing parts of image patches. We employ convolutional layers in our method such that our descriptor can work with patches of different sizes without the need to retrain the autoencoder (which is an important practical advantage over [5]). This flexibility results in faster learning and wider applicability in the processing of natural images.

In the following section, we give a brief introduction to the autoencoders. Section 3 contains the description of our method, and in Sect. 4 we present and discuss the experimental results. We conclude the work in Sect. 5.

2   Preliminaries

Autoencoders are unsupervised neural networks used for learning efficient representations of data. An autoencoder consists of two parts, an encoder and a decoder, and is trained by minimising the reconstruction error between the input and output, while imposing some constraints on the middle layer. Formally, an autoencoder with encoder $\mathcal{E}$ and decoder $\mathcal{D}$ is trained to minimise the loss function $J(X, \mathcal{E}, \mathcal{D}) = \sum_{x \in X} \mathcal{L}(x, \mathcal{D}(\mathcal{E}(x))) + \Omega(\mathcal{E}(x))$, where $x \in X$ is a data sample, $\mathcal{L}$ is some metric and $\Omega(\mathcal{E}(x))$ is an optional sparsity regularisation term imposed on the hidden (code) layer. Autoencoders working with image data usually consist of alternating convolutional and max-pooling layers. The output neuron at location (i, j) in the k-th channel of the l-th convolutional layer is calculated as follows:

$$x_{ij}^{(l,k)} = \sum_{c \in C} \sum_{u=0}^{f^{(l)}-1} \sum_{v=0}^{f^{(l)}-1} w_{uv}^{(l,k)} x_{(i+u)(j+v)}^{(l-1,c)} + b^{(l)}, \tag{1}$$

where C is the set of channel indices, $w^{(l,k)}$ is the convolutional kernel for the l-th layer and k-th channel, $b^{(l)}$ the bias for the l-th layer, and $f^{(l)}$ is the size of the convolutional kernel (filter) for the l-th layer.

Chen et al. [5] proposed learning descriptors with an autoencoder that consists of a single hidden layer, which was fully connected with the input and output layers of the network. The encoder part of the network, i.e. the descriptor of the image patch x, was calculated as $\mathcal{E}(x) = S(W_\mathcal{E} x + b_\mathcal{E})$, where $W_\mathcal{E}$ and $b_\mathcal{E}$ are the weights and the bias in the encoding layer respectively, and $S(\cdot)$ is the sigmoid function. The authors used the loss function $J(X, \mathcal{E}, \mathcal{D}) = \frac{1}{2|X|} \sum_{x \in X} (x - \mathcal{D}(\mathcal{E}(x))) + \Omega_1(\mathcal{E}(x)) + \Omega_2(\mathcal{E}, \mathcal{D})$, with $\Omega_1(\cdot)$ being the dimensionality constraint on the middle (code) layer, and $\Omega_2(\cdot)$ the sparsity constraint on the weights of the network. The layers in this network can be represented as convolutions with a kernel of the same size as the input. Hence, they are fully-connected layers and not convolutional in the common sense of the term. The use of these fully-connected layers fails to exploit the local self-similarity property of natural images, requires longer training time compared to convolutional network designs, and requires training a new network for each patch size.

3   Proposed Method

Our primary contribution is a novel autoencoder architecture that provides an intermediate representation (IR) of an image. The use of IR has two main advantages: it is less memory-intensive than storing the descriptors of all patches within an image, and it allows a descriptor of a single patch to be extracted from IR with minimal computation. To accommodate this, we take advantage of the properties of the convolutional layers described in Sect. 2, but discard all the max-pooling layers in the encoder except for the last one. The convolutional layers exploit the self-similarity property of natural images to reduce the computational time while also obtaining better results than fully-connected layers. Moreover, the use of convolutional layers is critical for the ability to extract patch descriptors from the IR, since convolutional layers preserve spatial information in the height-with plane.

Max-pooling is typically applied since it adds extra non-linearity, plays the role of dimensionality reduction, and reduces the number of training parameters and hence the training time. However, successful neural network architectures have been recently reported also without max-pooling [13]. We omit max-pooling after the first two convolutions and employ non-linear activation functions. We leave only one max-pooling layer with large spatial extent at the end of the

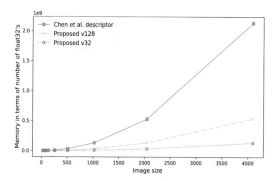

Fig. 2. A comparison of the memory requirements (expressed in the number of float 32's) as a function of image size in pixels, for the two versions of the proposed descriptor (v32 and v128) compared to Chen et al. [5].

encoder to reduce the dimension of the code layer. This architecture requires longer training time compared to standard use of max-pooling, but reduces the computational time and memory while using the descriptor. This is beneficial since the training needs to be done only once.

The reduction of computational time and the memory usage follows from the IR in our network. An IR is obtained by propagating the complete image (containing patches of interest) through the convolutional layers in the encoder, but not the max-pooling layer. Figure 1 shows the architecture of our network and the IR.

Let $x := x^{(0,:)}$ be the input image. We define the intermediate representation as $\mathcal{IR}(x) = x^{(L,:)}$, with

$$x^{(L,c)} = \mathcal{A}(\mathcal{C}_{l_L}(\mathcal{A}\dots(\mathcal{C}_{l_1}(x^{(0,:)})))),\tag{2}$$

where L is the number of convolutional layers in the encoder $\mathcal{E}$, $x^{(l_i,c)}$ is the c-th channel of the output of the l_i-th layer, $x^{(l_i,:)}$ denotes all channels of the output of the l_i-th layer, $\mathcal{A}$ is some activation function, and $\mathcal{C}_{l_i}$ is the l_i-th convolutional layer. From the intermediate representation of an image $\mathcal{IR}(x)$, we obtain the descriptor for a patch $x_{(i,j)}$, whose upper left corner is positioned at (i,j), as follows

$$\mathcal{E}(x_{(i,j)}) = \mathcal{MP}(\mathcal{IR}(x)_{(i,j)}),\tag{3}$$

i.e. by performing the max-pooling on the patch of the IR.

For the activation function $\mathcal{A}$ from (2), we have chosen the sigmoid function, $S(x) = \frac{1}{1+e^{-x}}$. This choice was made in order to avoid the case of having many zeros as an output, which was the case with some other activation functions, such as a rectifier. We trained the network with Adadelta optimizer [16], and we used binary cross entropy as a loss function (we scale the pixel values to be in $[0,1]$):

$$J(X, \mathcal{E}, \mathcal{D}) = \sum_{x \in X} \mathcal{L}(x, \hat{x}) = \sum_{x \in X} (x \log(\hat{x}) + (1 - x) \log(1 - \hat{x})) \qquad (4)$$

where $\hat{x} := \mathcal{D}(\mathcal{E}(x))$ is the output of the autoencoder. We trained the autoencoder in two different ways, creating two versions of the descriptor, v32 and v128 (named after the dimensionality of the descriptor for 16×16 patches). For the implementation, we use Keras library for neural networks. The autoencoder has been trained on a total of 150k 16×16 patches cropped from images from the ImageNet dataset [6]. The ratio between training, validation, and test set is $8 : 1 : 1$.

In Fig. 2 we compare the effective memory usage required by different descriptors depending on the image size. These results indicate potential for a tremendous decrease in memory usage for applications on a single image. This decrease could make some algorithms that use many patch comparisons feasible for use on larger images.

4   Experimental Evaluation

4.1   Robustness to Noise

We test the noise robustness of both versions of our descriptor, comparing them to the exhaustive search on pixel intensity values, and the existing descriptor trained with the autoencoder of [5]. We trained all the descriptors on the same set of colour patches.

The evaluation is performed as follows. We select a set of query patches within an image with added Gaussian noise. For each query patch, we retrieve the k most similar patches either by comparing their descriptors or by using exhaustive search over the pixel values. The quality of patch retrieval is evaluated based on the sum of square differences (SSD) between the pixels in query and retrieved patches before the noise was added. The standard deviation of the Gaussian noise was varied between 0 and 50.

Figure 3 summarises the results of our patch retrieval experiments and some visual examples are shown in Fig. 4 (left). When no noise is present, the exhaustive search retrieves the patches that are the most similar to the query patch. However, noise deteriorates the performance of the exhaustive search, whereas our descriptor v128 shows little decrease in performance.

Our descriptors also show superior performance compared to the existing descriptor learned with autoencoders. Our method v128 shows significantly better results than Chen [5], while having the same patch descriptor dimensionality. Furthermore, our method v32 that shows similar results to [5] has an order of magnitude lower dimensionality of the descriptor when encoding a single patch. The dimensionality comparison changes even more in our favour when encoding the whole image due to the usage of the IR (Fig. 2).

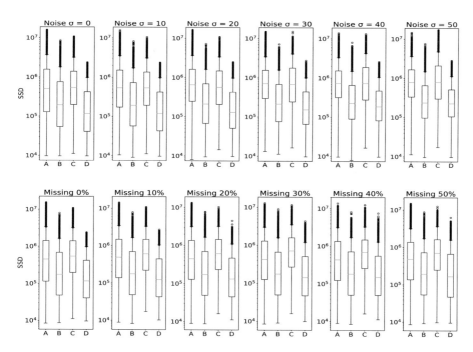

Fig. 3. Comparison of descriptors' robustness to noise (top) and missing data (bottom). A – proposed descriptor v32, B – proposed descriptor v128, C – Chen et al. [5], D – exhaustive search on pixel intensity values. The plots are showing SSDs of ground truth pixel values of patches found by the descriptors (in A-C) and exhaustive search (D), based on the noise (top) and percentage of missing area in a patch (bottom).

4.2   Robustness to Missing Data

We set up an experiment to determine the capability of the proposed descriptor when working with patches that contain missing regions. This type of operation is of interest for applications such as inpainting and scene reconstruction from multiview data. The setup is similar to the noise robustness evaluation, but here parts of the query patches have been randomly removed.

For the query patches with missing parts we want to find the best matching undamaged patches. We are searching for the matching patches by comparing the descriptors of the non-missing parts. The numerical evaluation is done based on the SSD values of the complete (undamaged) query and the found match. The results are shown in Fig. 3 (bottom), again comparing our two descriptors, descriptor from [5], and the exhaustive search over pixel intensity values. Visual comparison is shown in Fig. 4 (right).

The conclusions from these experiments are similar to those with the noisy patches. When the missing area in a patch is small, exhaustive search retrieves the best results. However, as the missing area is increasing, our descriptor v128 starts performing better and overtakes the exhaustive search, showing more robustness to missing data than the exhaustive search. Both of our proposed methods outperform the existing method [5], with a slightly larger margin than in the case of noisy patches. More elaborate analysis of the experiments is in our extended journal submission [14].

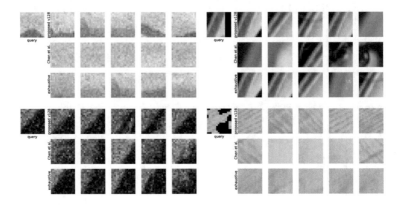

Fig. 4. Noisy patch retrieval (left) and patch retrieval where the query has missing parts (right). For each query, the first row corresponds to the proposed descriptor v128; the second row: the descriptor from [5], and the third row: exhaustive search. The missing parts of the query patches on the right are shown in black.

5   Conclusion

We propose a new method for learning local image descriptors using autoencoders. The proposed approach saves memory and computational time in comparison to existing methods when used for patch search and matching within a single image. We have evaluated the proposed descriptors' robustness to noise and missing data against an existing descriptor learned with autoencoders from [5] and exhaustive search over pixel intensity values. The proposed descriptors show improved results, and superior robustness to both noise and missing data in comparison with exhaustive search.

References

1. Arandjelović, R., Zisserman, A.: Three things everyone should know to improve object retrieval. In: 2012 IEEE Conference on Computer Vision and Pattern Recognition, pp. 2911–2918. IEEE (2012)
2. Balntas, V., Riba, E., Ponsa, D., Mikolajczyk, K.: Learning local feature descriptors with triplets and shallow convolutional neural networks. In: BMVC, vol. 1, p. 3 (2016)
3. Bay, H., Ess, A., Tuytelaars, T., Van Gool, L.: Speeded-up robust features (SURF). Comput. Vis. Image Underst. **110**(3), 346–359 (2008)

4. Calonder, M., Lepetit, V., Strecha, C., Fua, P.: Brief: binary robust independent elementary features. In: European Conference on Computer Vision, pp. 778–792. Springer (2010)
5. Chen, L., Rottensteiner, F., Heipke, C.: Feature descriptor by convolution and pooling autoencoders. International Archives of the Photogrammetry, Remote Sensing and Spatial Information Sciences-ISPRS Archives 40 (2015), Nr. 3W2 **40**(3W2), 31–38 (2015)
6. Deng, J., Dong, W., Socher, R., Li, L.J., Li, K., Fei-Fei, L.: ImageNet: a large-scale hierarchical image database (2009)
7. He, K., Lu, Y., Sclaroff, S.: Local descriptors optimized for average precision. In: IEEE Conference on Computer Vision and Pattern Recognition (CVPR), June 2018
8. Hinton, G.E., Salakhutdinov, R.R.: Reducing the dimensionality of data with neural networks. Science **313**(5786), 504–507 (2006)
9. Hinton, G.E., Zemel, R.S.: Autoencoders, minimum description length and Helmholtz free energy. In: Advances in Neural Information Processing Systems, pp. 3–10 (1994)
10. Lowe, D.G.: Object recognition from local scale-invariant features. In: ICCV, p. 1150. IEEE (1999)
11. Rublee, E., Rabaud, V., Konolige, K., Bradski, G.: Orb: an efficient alternative to SIFT or SURF (2011)
12. Simo-Serra, E., Trulls, E., Ferraz, L., Kokkinos, I., Fua, P., Moreno-Noguer, F.: Discriminative learning of deep convolutional feature point descriptors. In: Proceedings of the IEEE International Conference on Computer Vision, pp. 118–126 (2015)
13. Springenberg, J.T., Dosovitskiy, A., Brox, T., Riedmiller, M.: Striving for simplicity: the all convolutional net. arXiv preprint arXiv:1412.6806 (2014)
14. Žižakić, N., Ito, I., Pižurica, A.: Efficient local image descriptors learned with autoencoders (in submission)
15. Zagoruyko, S., Komodakis, N.: Learning to compare image patches via convolutional neural networks. In: Proceedings of the IEEE Conference on Computer Vision and Pattern Recognition, pp. 4353–4361 (2015)
16. Zeiler, M.D.: ADADELTA: an adaptive learning rate method. arXiv preprint arXiv:1212.5701 (2012)

The Performance of Three-Hop Wireless Relay Channel in the Presence of Rayleigh Fading

Dragana Krstic[1(✉)], Petar Nikolic[2], and Mihajlo Stefanovic[1]

[1] Faculty of Electronic Engineering, University of Niš, Niš, Serbia
dragana.krstic@elfak.ni.ac.rs
[2] Tigar Tyres, Pirot, Serbia

Abstract. In this paper, the performance of the first and the second order of wireless three-hop relay communication system operating over Rayleigh multipath fading environment are determined. The observed first order performance are: probability density function (PDF), cumulative distribution function, outage probability (OP) and moments. The analyzed second order characteristics are: level crossing rate (LCR) and average fade duration (AFD). A few graphics are plotted and the parameters' influence is studied.

Keywords: Rayleigh fading · The first order performance ·
The second order performance · Three-hop relay system

1 Introduction

The performance of the ratios and products of more random variables (RVs) is much being considered in the last decade. The products of RVs have applications in: channel modeling, multi-hop wireless relay systems, cascaded fading channels, MIMO keyhole systems, quantum physics, signal processing, tensor sensing problem, also in biological and physical sciences, econometrics, classification, ranking and selection [1–3]. In this paper, we will deal with three-hop wireless relay communication system working in the presence of Rayleigh multipath fading. The output signal from the multi-hop relay system is actually product of output signals from all hops (sections) [4]. For such scenario, we will derive the first order system performance (probability density function (PDF), cumulative distribution function (CDF), moments and outage probability (OP)), and the second order system performance (level crossing rate (LCR) and average fade duration (AFD)) [5].

Product of two RVs is considered in many papers. So, using LCR of product of two Nakagami-m RVs evaluated in [6], AFD of wireless relay system working over Nakagami-m fading channel is derived. In [7], PDF of the product of Rayleigh, exponentially, Nakagami-m and Gamma RVs is derived by the Mellin transform.

We started analyzing the product of three RVs. Such product of three RVs and its application in relay telecommunication systems in the presence of

© Springer Nature Switzerland AG 2020
M. Choraś and R. S. Choraś (Eds.): IP&C 2019, AISC 1062, pp. 222–230, 2020.
https://doi.org/10.1007/978-3-030-31254-1_27

multipath fading is presented in [8]. LCR of product of Nakagami-m RV, Rician RV and Rayleigh RV is obtained in closed form in [9]. The statistics of the product of three Nakagami-m RVs is derived in [10].

Further, the papers [11–13] consider multiple RVs with different distributions. PDF for n-Rayleigh distribution, is studied in [11]. The distribution functions are derived by using an inverse Mellin transform, and given in terms of a special function of mathematical physics, the Meijer G-function. An investigation of the second order characteristics of the amplify-and-forward (AF) multi-hop Rayleigh fading channel is done in [12]. This channel is a cascaded one and from source to destination presented as the product of N Rayleigh fading signal envelopes, known as N*Rayleigh channel. The expressions for the LCR and AFD are derived. The PDF, CDF, and moments functions of the N*Nakagami distribution are developed in closed forms using the Meijer's G-function in [13]. The formulas are derived for: OP, amount of fading (AoF), and average bit error probability (BEP), for several modulation schemes. This result is useful for analyzing the cascaded Nakagami-m fading channels.

Unlike previous works, we also obtained results for the second-order characteristics in a closed form. Our paper is organized in four sections. In introduction, the topic is introduced and previously published papers from this area are listed. In Sect. 2, the first order system performance is derived. The derivation of the LCR and AFD of product of three RVs is presented for 3-Rayleigh RV in the third section, with validation of theoretical results by numerical examples and analysis of parameters influence. The last section is conclusion.

2   The First Order Performance of Product of Three Rayleigh Random Variables

2.1   PDF of Product of Three Rayleigh RVs

Let x_i be independent and not necessarily identically distributed (i.n.i.d.) random processes. When the line of sight (LOS) component is absent, the envelope of x_i is Rayleigh distributed and the probability density function (PDF) of x_i is [5]:

$$p_{x_i}(x_i) = \frac{2x_i}{\Omega_i} e^{-\frac{x_i^2}{\Omega_i}}, x_i \geq 0, \tag{1}$$

where $\Omega_i = E\left\{x_i^2(t)\right\}$, $1 \geq i \geq 3$, is the mean power of the i-th RV [12, eq. (1)].

In some wireless systems, the communication channel can be well described as a product of independent RVs. Such a system is, for example, a multi-hop wireless relaying system using AF protocol with fixed-gain amplification factor and statistically independent hops. We consider here the statistics of the AF three-hop Rayleigh fading channel which is modeled as the product of three fading amplitudes. Let the product of three Rayleigh random variables be defined as:

$$x(t) = \prod_{i=1}^{3} x_i(t). \tag{2}$$

Then, it is: $x_1 = x/x_2x_3$. Conditional probability density function of x is:

$$p_{x/x_2x_3}(x/x_2x_3) = \left| \frac{dx_1}{dx} \right| p_{x_1}\left(\frac{x}{x_2x_3} \right),$$

(3)

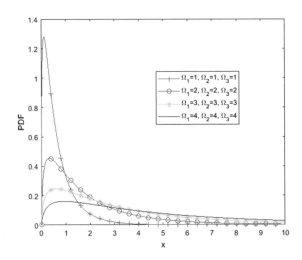

Fig. 1. PDF of product of three Rayleigh random variables for different values of signal powers.

where we apply:

$$\frac{dx_1}{dx} = \frac{1}{x_2x_3}.$$

(4)

After integration, replacements and solving, the PDF of RV x is:

$$p_x(x) = \int_0^\infty dx_2 \int_0^\infty dx_3 \frac{1}{x_2x_3} \frac{x}{x_2x_3} \frac{x_2}{\Omega_2} \frac{x_3}{\Omega_3} \frac{1}{\Omega_1} \cdot e^{-\frac{1}{\Omega_1}\left(\frac{x}{x_2x_3}\right)^2 - \frac{x_2^2}{\Omega_2} - \frac{x_3^2}{\Omega_3}}$$
$$= \frac{8}{\Omega_1\Omega_2\Omega_3} x \cdot \int_0^\infty dx_2 x_2^{-1} \cdot K_0\left(2\sqrt{\frac{1}{\Omega_1\Omega_3}\frac{x^2}{x_2^2}} \right)$$

(5)

where $K_n(x)$ denotes the modified Bessel function of the second kind [14, eq. (1.1)].

Some curves for PDF of x are shown in Fig. 1 depending on the product of three Rayleigh RVs for different values of the signal power Ω_i. It can be noticed from this figure that PDF increases for small signal envelopes x, reaches maximum and start to decrease for higher values of the signal envelope, for all values of the signal powers Ω_i. The influence of the signal envelopes is more significant for their lower values.

2.2   CDF of Product of Three Rayleigh RVs

Cumulative distribution function of x (3*Rayleigh RV) is derived in closed-form:

$$F_x(x) = \int_0^\infty dt p_x(t) = \int_0^x dt \int_0^\infty dx_2 \int_0^\infty dx_3 \frac{1}{x_2 x_3} p_{x_1}\left(\frac{t}{x_2 x_3}\right) p_{x_2}(x_2) p_{x_3}(x_3)$$

$$= \frac{8}{\Omega_1 \Omega_2 \Omega_3} \int_0^x dt t \int_0^\infty dx_2 x_2^{-1} e^{-\frac{x_2^2}{\Omega_2}} \int_0^\infty dx_3 x_3 ^{-1} e^{-\frac{1}{\Omega_1} \frac{t^2}{x_2^2 x_3^2} - \frac{x_3^2}{\Omega_3}}$$

$$= \frac{8}{\Omega_1 \Omega_2 \Omega_3} x^2 \cdot \sum_{j_1=0}^1 \frac{1}{2(j_1)\Omega_1^{j_1}} \cdot \int_0^\infty dx_2 x_2^{-1} e^{-\frac{m_2}{\Omega_2} x_2^2} \cdot \left(\frac{\Omega_3}{\Omega_1} x^2\right)^{-j_1} \cdot K_{-2 j_1}\left(2\sqrt{\frac{x^2}{\Omega_1 \Omega_2 x_2^2}}\right).$$

$$(6)$$

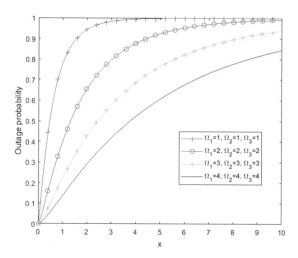

Fig. 2. The outage probability of product of three Rayleigh random variables for different values of signal powers Ω_i.

2.3   Outage Probability of Product of Three Rayleigh RVs

The outage probability is defined as probability that receiver output signal envelope falls below the defined threshold [15, eq. (2.23)]:

$$OP = P(X \le x) = \int_0^\infty dt p_x(t) = F_x(x), \qquad (7)$$

and mathematically is equal to the CDF of x, given by (6).

For some wireless communication cases, the knowledge of the CDF is not sufficient to characterize a system. The frequency and the mean duration of outage events may have equivalent importance, especially in the case of adaptive communication. In this case, the outage probability is characterized not only by the CDF, but also the CDF of the outage duration.

The curves for OP are shown in Fig. 2 for several values of signal envelopes average powers Ω_i at sections. It can be seen from this figure that OP grows with increasing of the signal envelope x and achieves saturation. It is also visible that OP decreases when Ω_i increasing.

2.4   Moment of n-th Order of Product of Three Rayleigh RVs

The moment of n-th order of product of three Raleigh random variables is:

$$m_n = \overline{x^n} = \int_0^\infty dx x^n p_x(x).$$ (8)

For our product of three Rayleigh RVs it is:

$$m_n = \frac{8}{\Omega_1 \Omega_2 \Omega_3} \int_0^\infty dx x^{n+1} \cdot \int_0^\infty dx_2 x_2^{-1} \cdot K_0 \left(2\sqrt{\frac{1}{\Omega_1 \Omega_3} \frac{x^2}{x_2^2}} \right).$$ (9)

3   The Second Order Performance of Product of Three Rayleigh Random Variables

3.1   LCR of Product of Three Rayleigh RVs

The LCR of RV x, equal to triple Rayleigh variable, at given threshold is defined as the rate at which the random variable crosses this given level in the negative (or positive) direction [16]. For obtaining the LCR, it is necessary to determine the joint probability density function (JPDF) of x, $\dot{x}$, and $p_{x\dot{x}}(x\dot{x})$, and apply the Rice's formula [4, Eq. (2.106)]. Then, the LCR of product x of three Rayleigh RVs is [17]:

$$N_X(x) = \int_0^\infty d\dot{x} \dot{x} p_{x\dot{x}}(x\dot{x}).$$ (10)

The first time derivative of x is: $\dot{x} = \dot{x}_1 x_2 x_3 + x_1 \dot{x}_2 x_3 + x_1 x_2 \dot{x}_3$.

It was found that time derivative of i-th envelope is independent from the envelope itself, and has the Gaussian PDF [18]. On that way, RVs $\dot{x}_1$, $\dot{x}_2$ and $\dot{x}_3$ have Gaussian distribution. The average signal level of $\dot{x}$ is:

$$\bar{\dot{x}} = \overline{\dot{x}_1} x_2 x_3 + x_1 \overline{\dot{x}_2} x_3 + x_1 x_2 \overline{\dot{x}_3} = 0,$$ (11)

since: $\dot{x}_1 = \dot{x}_2 = \dot{x}_3 = 0$.

The variance of $\dot{x}$ is: $\sigma_{\dot{x}}^2 = x_2^2 x_3^2 \sigma_{\dot{x}_1}^2 + x_1^2 x_3^2 \sigma_{\dot{x}_2}^2 + x_1^2 x_2^2 \sigma_{\dot{x}_3}^2$, where $\sigma_{\dot{x}_i}^2 = \pi^2 f_m^2 \Omega_i$, and f_m being maximal Doppler frequency [12, eqs. (2)-(4)].

After the appropriate change, the variance becomes:

$$\sigma_{\dot{x}}^2 = \pi^2 f_m^2 (x_2^2 x_3^2 \Omega_1 + x_1^2 x_3^2 \Omega_2 + x_1^2 x_2^2 \Omega_3)$$
$$= \pi^2 f_m^2 x_2^2 x_3^2 \Omega_1 \left(1 + \frac{x^2}{x_2^4 x_3^2} \frac{\Omega_2}{\Omega_1} + \frac{x^2}{x_2^2 x_3^4} \frac{\Omega_3}{\Omega_1} \right).$$ (12)

The joint probability density function of x, $\dot{x}$, x_2 and x_3 is:

$$p_{x\dot{x}x_2x_3}(x\dot{x}x_2x_3) = p_{\dot{x}}(\dot{x}/xx_2x_3) p_x(x/x_2x_3) p_{x_2}(x_2) p_{x_3}(x_3),$$ (13)

with $p_x(x/x_2x_3)$ from (3) and $\frac{dx_1}{dx}$ from (4).

The joint probability density function of x and the first derivative of $\dot{x}$ is:

$$p_{x\dot{x}} = \int_0^\infty dx_2 \int_0^\infty dx_3 p_{\dot{x}}(\dot{x}/xx_2x_3) \frac{1}{x_2 x_3} p_{x_1} \left(\frac{x}{x_2 x_3} \right) p_{x_2}(x_2) p_{x_3}(x_3).$$ (14)

After some replacements, the final LCR of triple Rayleigh RV is:

$$N_X = \frac{1}{\sqrt{2\pi}} \pi f_m \Omega_1^{1/2} x \frac{8}{\Omega_1 \Omega_2 \Omega_3}$$
$$\int_0^\infty dx_2 \int_0^\infty dx_3 e^{-\frac{1}{\Omega_1}\frac{x^2}{x_2^2 x_3^2} - \frac{1}{\Omega_2}x_2^2 - \frac{1}{\Omega_3}x_3^2} \sqrt{1 + \frac{x^2}{x_2^4 x_3^2}\frac{\Omega_2}{\Omega_1} + \frac{x^2}{x_2^2 x_3^4}\frac{\Omega_3}{\Omega_1}} \quad (15)$$

By using Laplace approximation theorem for solution the two-fold integrals, previous double integral can be solved in the process below [19]:

$$\int_0^\infty dx_2 \int_0^\infty dx_3 g(x_2, x_3) e^{\lambda f(x_2, x_3)} = \frac{\pi}{\lambda} \frac{g(x_{20}, x_{30})}{(A(x_{20}, x_{30}))^{1/2}} \quad (16)$$

where A is given by dint of [20]:

$$A(x_{20}, x_{30}) = \begin{vmatrix} \frac{\partial^2 f(x_{20}, x_{30})}{\partial x_{20}^2} & \frac{\partial^2 f(x_{20}, x_{30})}{\partial x_{20}\partial x_{30}} \\ \frac{\partial^2 f(x_{20}, x_{30})}{\partial x_{20}\partial x_{30}} & \frac{\partial^2 f(x_{20}, x_{30})}{\partial x_{20}^2} \end{vmatrix} \quad (17)$$

and x_{20} and x_{30} are solutions of: $\frac{\partial f(x_2, x_3)}{\partial x_2} = \frac{2}{\Omega_1}\frac{x^2}{x_2^3 x_3^2} - \frac{2}{\Omega_2}x_2 = 0$ and $\frac{\partial f(x_2, x_3)}{\partial x_3} = \frac{2}{\Omega_1}\frac{x^2}{x_2^2 x_3^3} - \frac{2}{\Omega_3}x_3 = 0$.

These solutions are introduced in (16) for solving two-fold integral from (15). So, LCR of the product of triple Rayleigh RV is obtained in the closed form.

In Fig. 3, the LCR normalized by f_m is presented depending on signal envelope for some values of signal power Ω_i. LCR increases for low values of the signal envelope, achieves maximum and start to decrease for higher values of the signal envelope. The influence of the signal envelope on the LCR is bigger for small values of these envelopes. For higher values of x, and higher Ω_i, the LCR is higher. The system has better performance when LCR has smaller values.

3.2 AFD of Product of Three Rayleigh RVs

The AFD of RV at threshold x is defined as average time that triple Rayleigh RV stays below the level x after crossing that level in the downward direction. It can be calculated as the ratio of the OP and the LCR by formula [5]: $T_x(x) = F_x(x)/N_x(x)$, where $F_x(x)$ is the CDF of x presented in (6), and $N_x(x)$ is obtained in (15).

In Fig. 4, the normalized AFD $(T_x f_m)$, versus signal envelope x, is shown. It is possible to see from this figure that AFD increases for all values of the signal envelopes. It also can be seen that AFD has lower values for bigger Ω_i.

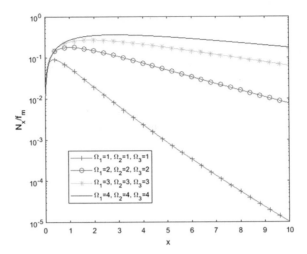

Fig. 3. LCR normalized by f_m versus signal envelope x for different values of signal power Ω_i.

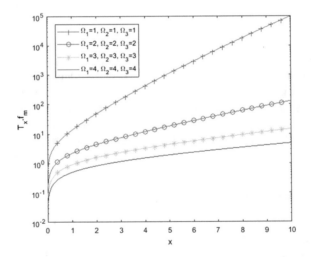

Fig. 4. AFD normalized by f_m versus signal envelope for different values of signal powers Ω_i.

4   Conclusion

The multi-hop communication in relay networks is a very promising approach to improve the transmission coverage of cellular and ad hoc networks. Because of transmit power limitations, multi-hop transmission also leads to remarkable coverage extensions by dividing a total end-to-end transmission into a group of shorter paths. The advantage of multi-hop relaying exists especially for rural areas with low traffic density and rare population [21].

Here, we formed RV as a product of three Rayleigh RVs, so called 3*Rayleigh RV. For this RV, the first and second order performance are considered. We obtained formulas for probability density function, cumulative distribution function, moments, outage probability, level crossing rate and average fade duration in closed forms and analyzed fading parameters influence to them based on some graphs.

Our future papers will be related to the analysis of the product of multiple RVs with other distributions because of its application in multi-hop wireless relay communication systems in the presence of fading. This topic has growing interest in wireless propagation research.

References

1. Chen, Y., Karagiannidis, G.K., Lu, H., Cao, N.: Novel approximations to the statistics of products of independent random variables and their applications in wireless communications. IEEE Trans. Veh. Technol. **61**(2), 443–454 (2012)
2. Nadarajaha, S., Dey, D.K.: On the product and ratio of T random variables. Appl. Math. Lett. **19**, 45–55 (2006)
3. The problem of the random walk. Nature **72**, 318 (1905)
4. Stuber, G.L.: Mobile Communication, 2nd edn. Kluwer Academic Publisher, Dordrecht (2003)
5. Shankar, P.M.: Fading and Shadowing in Wireless Systems. Springer, New York (2012)
6. Zlatanov, N., Hadzi-Velkov, Z., Karagiannidis, G.K.: Level crossing rate and average fade duration of the double Nakagami-m random process and application in MIMO keyhole fading channels. IEEE Commun. Lett. **12**(11), 822–824 (2008)
7. Ahmed, S., Yang, L.L., Hanzo, L.L.: Probability distributions of products of Rayleigh and Nakagami-m variables using Mellin transform. In: IEEE International Conference on Communication, ICC 2011, Kyoto, Japan (2011)
8. Krstic, D., Nikolic, P., Aleksic, D., Minic, S., Vuckovic, D., Stefanovic, M.: Product of three random variables and its application in relay telecommunication systems in the presence of multipath fading. J. Telecommun. Inform. Technol. (JTIT) (1), 83–92 (2019)
9. Krstic, D., Stefanovic, M., Nikoliæ, P.: Level crossing rate of product of Nakagami-m random variable, Rician random variable and Rayleigh random variable. In: IEICE Information and Communication Technology Forum, ICTF 2018, Graz, Austria (2018)
10. Krstic, D., Stefanovic, M., Nikolic, P., Minic, S.: Statistics of the product of three Nakagami-m random variables with applications. In: 26th International Conference on Software Telecommunications and Computer Networks - SoftCOM 2018, Split - Supetar, Croatia, pp. 36–40 (2018)
11. Salo, J., El-Sallabi, H.M., Vainikainen, P.: The distribution of the product of independent Rayleigh random variables. IEEE Trans. Antenn. Propag. **54**(2), 639–643 (2006)
12. Hadzi-Velkov, Z., Zlatanov, N., Karagiannidis, G.K.: Level crossing rate and average fade duration of the multihop Rayleigh fading channel. In: IEEE International Conference on Communications, Beijing, China (2008)
13. Karagiannidis, G., Nikos, S., Takis, M.: N*Nakagami: a novel stochastic model for cascaded fading channels. IEEE Trans. Commun. **55**(8), 1453–1458 (2007)

14. Yang, Z.H., Chu, Y.M.: On approximating the modified Bessel function of the second kind. J. Inequalities Appl. **1**, 41 (2017)
15. Cheikh, D.B.: Outage probability formulas for cellular networks: contributions for MIMO. CoMP and time reversal features. Telecom Paris Tech, English (2012)
16. Jakes, W.C.: Microwave Mobile Communications. IEEE Press, Piscataway (1994)
17. Rice, S.O.: Mathematical analysis of random noise. Bell Syst. Tech. J. **23**, 282–332 (1944)
18. Akki, A.: Statistic properties of mobile-to-mobile land communication channels. IEEE Trans. Veh. Tech. **43**(4), 826–831 (1994)
19. Lopez, J.L., Pagola, P.J.: A simplification of the Laplace method for double integrals. Application second Appell function. Electr. Trans. Numer. Anal. **30**, 224–236 (2008)
20. Abramowitz, M., Stegun, I.A.: Handbook of Mathematical Functions. National Bureau of Standards. Dover Publications, Mineola (1965)
21. Eltahir, I.K., Bilal, K.H., Taha, S.: Evaluate comparative of cooperative relaying protocols in wireless communication. Int. J. Sci. Eng. Res. **4**(8), 112–116 (2013)

Simulation Study of Routing Protocols for Wireless Mesh Networks

Maciej Piechowiak[1(✉)], Piotr Owczarek[2], and Piotr Zwierzykowski[2]

[1] Kazimierz Wielki University, Bydgoszcz, Poland
mpiech@ukw.edu.pl
[2] Poznan University of Technology, Poznan, Poland

Abstract. The article presents the key routing protocols for wireless mesh networks and routing metrics that affect transmission efficiency. Routing protocols are crucial element for performance of mesh networks. However, the specificity of the radio channel and transmission between closely neighboring nodes requires the use of routing techniques not yet implemented in wired networks. As part of the research includes a comparative analysis of representative routing protocols carried out in the OMNeT++ environment. Results of simulation has been presented and compared.

Keywords: Ad-hoc networks · Topology generator · Routing protocols · Multi-hop networks

1 Introduction

The importance of wireless mesh networks (WMNs) has rapidly growing for many years. Their development is particularly evident with the development of the Internet of Things [1]. In general, mesh networks connect stationary or mobile clients and may offer access to the Internet. Unlike the ad-hoc networks, mesh nodes are not mobile – offering access to the network for clients, overcome the limitations of ad-hoc networks such as limited energy resources and significant movement of nodes. However, the energy problem is not always present in WMN networks. The problem is not significant in the last mile networks used in Smart Street Lighting and Smart Metering [2,3].

In contrast to the wireless networks – mesh networks have specific features that distinguish them from other technologies. They share the same frequency spectrum, generate interferences between nodes, operate on higher packet loss levels, create longer packet transmission paths. Abovementioned properties make it impossible to apply the routing protocols used to determine the packet transmission path in wired networks. Mesh networks are composed of multiple nodes exchanging information with each other mainly by radio (where mesh refers to rich interconnection among devices or nodes). The network nodes are acting as client nodes and routing nodes that direct packets to destinations and are responsible for maintaining and managing communication with neighbors.

© Springer Nature Switzerland AG 2020
M. Choraś and R. S. Choraś (Eds.): IP&C 2019, AISC 1062, pp. 231–238, 2020.
https://doi.org/10.1007/978-3-030-31254-1_28

With the use of the WMN network it is possible to reduce the power of transmitters by using directional antennas and precise beam controlling. It allows energy-saving equipment and reduces the inter-channel interference ratio. It is also possible to increase the transmission bandwidth and frequency reuse ratio. Due to redundancy of mesh wireless transmission paths they are characterized by high reliability. This results from the possibility of quick reconfiguration in the event of failure of one or more nodes in the network.

Mesh networks are dynamically self-organizing, so that reduced the need for a central point of network management (although the LoRaWAN technology, which uses long radio links and enables low-speed transmission, is extremely effective in this approach [4]). Customers equipped with access radio card have the ability to connect directly to the mesh network, while others can take advantage of the bridges for the integration of both individual clients as well as entire networks operating in other technology than the mesh network, e.g. Ethernet, WiMAX and UMTS [5].

The article focuses on the efficiency of routing protocols in wireless mesh networks according to the following set of parameters as the measure of the efficiency: packet loss ratio, packet delivery delay, average energy consumption of all network nodes and number of collisions in network and type of routing metrics. The proposed solutions do not apply only to WMN but everywhere where a shared medium is used, e.g. power lines [6,7].

The article consists of five sections. In Sect. 2 representative routing protocols for wireless mesh networks are presented. Section 3 describes the methodology used in the efficiency analysis presented in Sect. 4. Finally, Sect. 5 concludes the article.

2   Routing Protocols

One of the most popular protocol used for both reactive MANET networks and wireless mesh networks is AODV [8]. The AODV protocol is based on the DSDV protocol [9] and similarly uses the sequence numbers to determine the optimal path of the package. During the determination of the path to the destination node it checks the current routing table. If the address of the target device is not in routing table, RREQ message is sent to all neighbors. This procedure continues until the destination node is reached. Each router stores information about the sender of the message and ignore it in case of the copy of the same information. The reply to the source router is passed in a RREP message along the return path.

The AODV protocol manages the entries in the routing tables using HELLO messages, and when it detects a link failure it generates Unsolicited Route Reply message. AODV protocol does not require a broadcast transmission, which is its unquestionable advantage compared to DSDV. It also minimizes the number of nodes that do not participate in the transmission, contributing to the reduction of energy consumption.

The DYMO protocol [10] is based on two basic operations: route discovery and route maintenance. The first of these procedures are initiated when a source node wants to send a packet to the target device, which does not appear in the routing table. For this purpose, the route request message is broadcast on the network. When the destination node receives an addressed message – it transmits feedback information containing the cumulative path. Each intermediate node stores in its routing table the following information: the address of the destination node, the sequence number, the number of hops, next hop address, the name of the interface on the next hop, gateway and storage time of path in the table. The route maintenance is triggered when there is a change in the network topology [11].

The OLSR [12] protocol uses the concept of multipoint relays (MPs), which limit the network traffic and the number of link state updates. Each node operates in a minimum group of MPRs spaced by one hop. The set of MPR is formed by a rule providing availability of each of the nodes spaced apart by two hops. The information necessary for the operation of the network are exchanged only by the nodes that are in the group of MPR – other nodes do not broadcast protocol messages. HELLO messages sent periodically between routers are used to obtain information about nodes spaced about two hops. All devices in the network are periodically informed of the MPR group in Communications Topology Control messages. One of the features of the OLSR protocol is the redundancy of routes.

A key feature of the B.A.T.M.A.N. protocol [13] is a decentralized management system of routing path. A single router stores only the information about the interface, the packet will be sent to the recipient. In this way, the routing process is run dynamically until it is received at the destination. This protocol also detects new network nodes and informs the neighboring routers. The quality of links is determined on the basis of Transmit Link Quality and Receive Quality factors for particular channels (used further in the deciding process of the optimal path routing).

The HWMP hybrid protocol is described in the 802.11s standard for wireless mesh networks. It is a combination of a proactive routing and pro-active tree method. It uses a modified AODV protocol, which defines the metric based on the link parameters of the physical layer. The pro-active tree method builds a spanning tree from the root, which is usually a gateway while AODV protocol finds shortest path between nodes (HWMP supports two kinds of path selection protocols) [14].

3 Simulaton Study

Comparative analysis of routing protocols dedicated to wireless mesh networks were performed with simulation environment OMNeT++, which offers a wide range of configurations and implements the components of each layer of the reference model [15]. The application's GUI allows to create transparent network configuration. The functionality of wireless mesh networks and radio propagation approximating actual transmission conditions have been implemented in the OMNeT++ using INET and INETMANET libraries [16] which offer:

- mobility models,
- propagation models,
- support of multiple radio interfaces,
- data link layer standards,
- routing protocols specific to wireless networks,
- modeling operation on battery power.

The network topology has been arranged on a rectangular plane with dimensions 1000×1000 m. It consists of 20 nodes used to route packets between wireless hosts located on opposite sides of the topology. Each of the nodes has a limited transmit power, so it can only communicate with neighboring devices with one of the three available radio channels in a band of 5.2 GHz according to the IEEE 802.11a standard. The 802.11a standard has not been as widely used as 802.11b so far mainly due to problems with range and higher power consumption, but it has been fully implemented in the INETMANET library.

To measure the parameters of the wireless mesh network, a connectionless UDP transmission was used between hosts located outside the backbone of the network. The process of sending packets started within 20 s after the simulation process was initiated, which in case of using proactive routing protocols gives sufficient time to gather knowledge about available routes. Measurements were made by repeating each scenario ten times, and the results from individual runs were averaged, then presented in the form of diagrams [17].

4 Simulation Results

Figure 1(a) presents the average ratio of packet loss for all devices on the network depending on the routing protocol for wireless mesh network. Among the considered protocols the smallest average value (1.6%) was obtained for BATMAN protocol. At the opposite extreme are reactive protocols AODV, DYMO and HWMP with the loss rate of 2.7–3.7% of all sent packets. It also confirms that the reactive routing protocols should be used in the case when the reliability of delivery in the network is required.

Figure 1(b) shows the distribution of the average packet delivery delay for different routing protocols. The highest value of delay is typical for reactive AODV protocol (about 1.5 ms). The DYMO and HWMP protocols are characterized by a three times smaller delay than in AODV protocol. The value obtained for the OLSR protocol is 0.7 ms and is higher than for the DYMO and HWMP protocols. The expected delay value should be less than those obtained for the two reactive routing protocols, because this protocol has a proactive knowledge of the routing path and does not require additional time for setting routes.

The average energy consumption for all devices on the network is presented on Fig. 1(c). The values in the plot represent the total value of simulation units in which the radio interfaces in the network have been in use. The resulting value for the AODV protocol is highest. The DYMO protocol results are reduced by one third. For other protocols the energy consumption, with respect to the

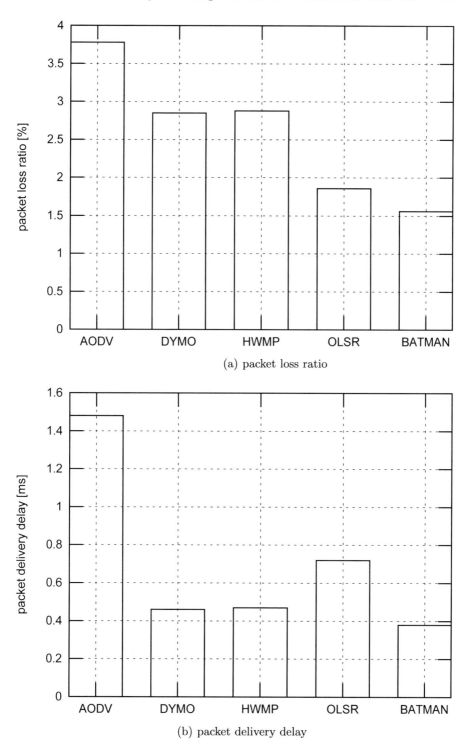

(a) packet loss ratio

(b) packet delivery delay

Fig. 1. Simulation results for WMN routing protocols

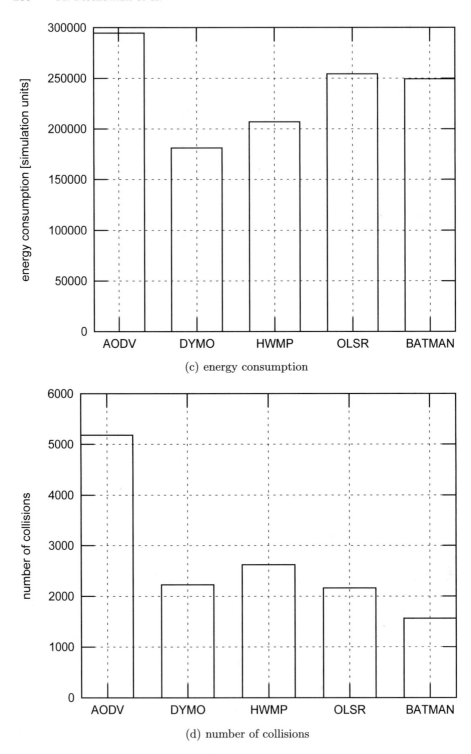

(c) energy consumption

(d) number of collisions

Fig. 1. (*continued*)

AODV, is reduced by 30% for HWMP and 20% for OLSR. The usage of the proactive routing protocols require continuous exchange of information between network nodes, which requires a constant energy consumption, even when the transmission of user data is not carried out. The obtained results show that although the AODV protocol belongs to a group of reactive protocols, the energy consumption on the network is greater than the energy consumption of OLSR proactive protocol.

Figure 1(d) shows the total number of collisions on the network for each routing protocol used in the wireless mesh network. The number of collisions illustrates the effectiveness of transmission path differentiation. The best results were obtained for DYMO and OLSR protocols, for which the total number of collisions in the network does not exceed the value of 2200. The AODV protocol causes twice as many collisions than other considered protocols. The reason for generating more collisions than other protocols is the reactive mode, which calculates routes on demand. The AODV protocol uses a metric based on the number of hops, which means that it always selects the same shortest path. This translates directly to the number of collisions, due to the limited resources of time and spectral.

5   Conclusions

In this article we studied the efficiency of network routing protocols for wireless mesh networks with OMNeT++. For this purpose, a unified model of the network configuration for both, reactive and proactive protocols was create. This will help to develop a consistent methodology of these protocols. The only variable in the simulation process was the routing protocol. The results indicate a lower latency in packet delivery and lower values of packet loss factor using proactive protocols. This is done by higher overhead of data sent over the network and higher energy consumption. However, due to the relatively wide bandwidth and lack of restrictions on battery power as is the case with network ad-hoc networks, mentioned type of routing can be implemented in wireless mesh networks.

References

1. Musznicki, B.: Empirical approach in topology control of sensor networks for urban environment. J. Telecommun. Inf. Technol. **1**, 47–57 (2019)
2. Dubalski, B., Kiedrowski, P.: WSN networks with hot potato protocol for automatic meter reading systems: methods of analysis based on graph theory. Rynek Energii **5**(90), 48–53 (2010)
3. Andrysiak, T., Saganowski, L., Kiedrowski, P.: Anomaly detection in smart metering infrastructure with the use of time series analysis. J. Sens. **2017**, Article ID 878213, pp. 1–15 (2017)
4. Casals, L., Mir, B., Vidal, R., Gomez, C.: Modeling the energy performance of LoRaWAN. Sensors (Basel) **17**(10) (2017)
5. Akyildiz, I.F., Wang, X.: Wireless Mesh Networks. Wiley, Hoboken (2008)

6. Andrysiak, T., Saganowski, L., Kiedrowski, P.: Predictive abuse detection for a PLC smart lighting network based on automatically created models of exponential smoothing. Secur. Commun. Netw. **2017**(1), 1–19 (2017)
7. Kiedrowski, P.: Errors nature of the narrowband PLC transmission in smart lighting LV network. Int. J. Distrib. Sens. Netw. **2016**, Article ID 9592679, pp. 1–9 (2016)
8. Perkins, C., Belding-Royer, E., Das, S.: Ad hoc On-Demand Distance Vector (AODV) Routing. IETF. RFC 3561, July 2003
9. Perkins, C., Bhagwat, P.: Highly dynamic Destination-Sequenced Distance-Vector routing (DSDV) for mobile computers. In: Proceedings of the Conference on Communications Architectures, Protocols and Applications, SIGCOMM 1994, pp. 234–244 (1994)
10. Chakeres, I.D., Perkins, C.E.: Dynamic MANET on demand (DYMO) routing protocol. Internet-Draft, Version 06, IETF, October 2006
11. Bisoyi, S.K., Sahu, S.: Performance analysis of Dynamic MANET On-demand (DYMO) Routing protocol. Spec. Issue IJCCT **1**(2), 3 (2010)
12. Clausen, T., Jacquet, P.: Optimized Link State Routing Protocol (OLSR). RFC 3626, October 2003
13. Neumann, A., Aichele, C., Lindner, M., Wunderlich, S.: Better Approach To Mobile Ad-hoc Networking (B.A.T.M.A.N.). Internet-Draft, April 2008
14. Bari, S.M.S., Anwar, F., Masud, M.H.: Performance study of hybrid wireless mesh protocol (HWMP) for IEEE 802.11s WLAN Mesh Networks. In: 2012 International Conference on Computer and Communication Engineering (ICCCE). IEEE (2012)
15. https://omnetpp.org
16. https://github.com/inetmanet/inetmanet/wiki
17. Wasłowicz, M.: Routing in wireless mesh networks. M.Sc. thesis. Poznan University of Technology (2014)

Call-Level Analysis of a Two-Link Multirate Loss Model with Restricted Accessibility

I. P. Keramidi[1], I. D. Moscholios[1(✉)], P. G. Sarigiannidis[2],
and M. D. Logothetis[3]

[1] Department of Informatics and Telecommunications,
University of Peloponnese, 221 00 Tripolis, Greece
{mst18002,idm}@uop.gr
[2] Department Informatics and Telecommunications Engineering,
University of Western Macedonia, 501 00 Kozani, Greece
psarigiannidis@uowm.gr
[3] WCL, Department of Electrical and Computer Engineering,
University of Patras, 265 04 Patras, Greece
mlogo@upatras.gr

Abstract. We consider a two-link system with restricted accessibility that accommodates Poisson arriving calls from different service-classes and propose a multirate teletraffic loss model for its analysis. In a restricted accessibility system, a new call may be blocked even if available bandwidth units do exist at the time of its arrival. In the two-link system, each particular link has two thresholds which refer to the number of in-service calls in the link. The lowest threshold, named support threshold, defines up to which point the link can support calls offloaded from the other link. The highest threshold, named offloading threshold, defines the point where the link starts offloading calls to the other link. The two-link system with restricted accessibility is modelled as a loss model whose steady state probabilities do not have a product form solution. However, we propose approximate formulas for the calculation of call blocking probabilities. The accuracy of the formulas is verified through simulation and found to be quite satisfactory.

1 Introduction

Quality of Service (QoS) mechanisms are necessary in contemporary communication networks in order to provide the required bandwidth needed by calls. In the case of call-level traffic in a single link, modelled as a loss system, such a QoS mechanism is a bandwidth sharing policy [1]. The simplest bandwidth sharing policy is the Complete Sharing (CS) policy. In the CS policy, a new call has full access to all non-occupied bandwidth units (b.u.) of the link and is accepted in the link if the required bandwidth is available. Otherwise, call blocking occurs. Often the terms 'full accessibility' or 'full availability' are used in the literature

© Springer Nature Switzerland AG 2020
M. Choraś and R. S. Choraś (Eds.): IP&C 2019, AISC 1062, pp. 239–251, 2020.
https://doi.org/10.1007/978-3-030-31254-1_29

instead of the term 'CS policy' [1,2]. However, the term 'availability' may also refer to the proportion of time the link is available [3]. Herein, we prefer the term 'CS policy'.

The simplest teletraffic loss model that adopts the CS policy is the classical Erlang model [1]. In this model, the call arrival process is Poisson while each call requires one b.u. to be accepted in the system. If this b.u. is available then a call remains in the link for a generally distributed service time. Otherwise call blocking occurs. The fact that Call Blocking Probabilities (CBP) are calculated via the classical Erlang B formula has led to numerous extensions of Erlang's model for the call-level analysis of wired (e.g., [4–18]), wireless (e.g., [19–30]), satellite (e.g., [31–33]) and optical networks (e.g., [34–39]).

In the recent paper of [27], a two-link loss system is studied that accommodates Poisson arriving calls from a single service-class. Each link may support and provide service to calls offloaded from the other link. An offloaded call is a new call that arrives in a link but will be served (subject to bandwidth availability) by the other link. This offloading mechanism is achieved with the aid of two thresholds per link, a low and a high threshold. Both thresholds refer to the number of in-service calls in each link. The first (low) threshold is the support threshold while the second (high) one is the offloading threshold. The support threshold refers to the point up to which the link can support offloaded calls from the other link. On the other hand, the offloading threshold refers to the point where call offloading can start from one link to the other.

The model of [27] does not have a Product Form Solution (PFS) for the steady state probabilities. This is due to the fact that the offloading mechanism destroys Local Balance (LB) between adjacent states of the system. In order to calculate CBP, either a linear system of Global Balance (GB) equations should be solved or an approximate method that relies on the independence between the links and the classical Erlang B formula can be adopted.

A potential application of the offloading scheme of [27] is in the area of mobile/WiFi networks. To manage the increasing traffic in mobile networks, traffic can be offloaded to WiFi networks [40,41]. To further increase the available bandwidth of WiFi access links, recent research focuses on the aggregation of backhaul access link capacities and on the bandwidth sharing policies that should be adopted. The impact of such an aggregation to CBP in the case of a single service-class can be well studied by the offloading scheme of [27].

In this paper, we extend the model of [27] to include the case of multirate Poisson traffic, i.e., we consider that the system accommodates calls of different service-classes with different bandwidth requirements. In addition, we incorporate the notion of restricted accessibility. In a restricted accessibility system, a new call may be blocked even if available b.u. do exist at the time of its arrival. Under the general term 'restricted accessibility' one may include:

- bandwidth sharing policies such as the threshold policy [15,17], the probabilistic threshold policy [28,42] or even the bandwidth reservation policy [4,9,13]. In the threshold policy, a new call (of a particular service-class) is not allowed to enter a system if the number of in-service calls (of that service-class) together with the new call exceeds a predefined threshold. In the probabilistic threshold policy, a new call may enter the system with (a state dependent) probability when the number of in-service calls plus the new call exceeds the threshold. In the bandwidth reservation policy, a new call will be blocked when the only available b.u. at the time of its arrival in one link are already reserved for other calls (e.g., of the same or the other link).

- the case where each particular state of the system (except from the initial state where the system is empty) is associated with a blocking probability. This means that a new call may be blocked and lost with a certain blocking probability when the system is in a particular state at the time of its arrival. Such an approach is useful when modeling interference between neighbouring cells (e.g., in CDMA systems) [3,43]. In this paper, we focus on this type of restricted accessibility and study the case where the blocking probability is different for each state. To this end, we provide an approximate method for the CBP determination which is verified via simulation. The CBP calculation in the proposed two-link model is based on the classical Erlang Multirate Loss Model (EMLM) [44,45], which refers to a single link.

The remainder of this paper is organized as follows: In Sect. 2, we review the system of [27]. In Sect. 3, we propose the extension of [27] which includes the case of multirate traffic and the notion of restricted accessibility. In Sect. 4, we provide analytical and simulation CBP results for the proposed model. We conclude in Sect. 5.

2 Review of the Two-Link System for Single-Rate Traffic

We consider a loss system of two links with capacities C_1 and C_2 b.u., respectively. Each link accommodates Poisson arriving calls of a single service-class which require one b.u. in order to be connected in a link. Let λ_l be the arrival rate in link l ($l = 1, 2$). Also let j_l be the occupied b.u. in link l. Then, $0 \leq j_1 \leq C_1$ and $0 \leq j_2 \leq C_2$. Since a call requires one b.u. the value of j_l also expresses the number of in-service calls in link l.

Each link l ($l = 1, 2$) has two different thresholds: the support (low) threshold th_{1l} and the offloading (high) threshold th_{2l}, with $th_{1l} < th_{2l}$ and $0 \leq th_{1l}, th_{2l} \leq 1$. Assuming that $\lfloor x \rfloor$ is the largest integer not exceeding x, then the role of these thresholds, in link l, is described as follows (see Fig. 1):

1. If $0 \leq j_l < \lfloor th_{1l} C_l \rfloor$ then link l is in a *support mode* of operation, i.e., it accepts and serves not only new calls that initially arrive in link l but also new calls offloaded from link m ($m = 1, 2$, $m \neq l$).
2. If $\lfloor th_{1l} C_l \rfloor \leq j_l < \lfloor th_{2l} C_l \rfloor$ then link l is in a *normal mode* of operation, i.e., it does not accept calls offloaded from link m. It only accepts calls that initially arrive in link l.
3. If $\lfloor th_{2l} C_l \rfloor \leq j_l$ then link l operates in *offloading mode* of operation, i.e., a new call that initially arrives in link l will be offloaded to link m. If link m operates in *support mode* (i.e., $0 \leq j_m < \lfloor th_{1m} C_m \rfloor$) then the call will be

accepted in link m. If link m does not operate in *support mode* and at the same time $j_l \leq C_l - 1$, the call will be accepted in link l. Otherwise the call will be blocked and lost without further affecting the system.

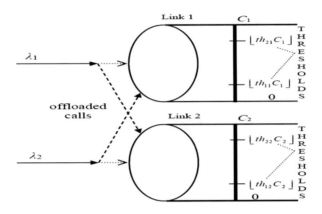

Fig. 1. The single-rate system of the two links.

Based on the above, the call admission of a new call that initially arrives in link l ($l = 1$, 2) is summarized in the following steps:

(1) If $(0 \leq j_l < \lfloor th_{2l}C_l \rfloor)$ then the call is accepted by the link l and remains for a generally distributed service-time with mean μ^{-1}.

(2) If $\lfloor th_{2l}C_l \rfloor \leq j_l$ then: (2a) if $0 \leq j_m < \lfloor th_{1m}C_m \rfloor$ the call is offloaded to link m and remains for a generally distributed service-time with mean μ^{-1}, (2b) if $\lfloor th_{1m}C_m \rfloor \leq j_m$, then link m is in a *normal mode* of operation and does not support offloaded calls from link l. In that case, the call will be handled by link l. In addition, if $j_l \leq C_l - 1$, then the call is accepted in link l and remains for a generally distributed service-time with mean μ^{-1}. Otherwise, call blocking occurs.

Due to the *support* and *offloading modes* of operation of the two links, the corresponding 2-D Markov chain of the system is not reversible and therefore LB between adjacent states is destroyed. Thus, the steady state distribution, $P(j) = P(j_1, j_2)$, of this system cannot be described by a PFS. To determine the values of $P(j_1, j_2)$ (and consequently CBP) there exist two different methods.

The first method provides accurate results compared to simulation. The drawback of the method is that it requires the knowledge of the system's state space and the solution of the set of linear GB equations for each state $j = (j_1, j_2)$ expressed as *rate into state j = rate out of state j*:

$$
\begin{aligned}
&\lambda_1 (j_1 - 1, j_2) P (j_1 - 1, j_2) + \lambda_2 (j_1, j_2 - 1) P (j_1, j_2 - 1) + \\
&(j_1 + 1) \mu P (j_1 + 1, j_2) + (j_2 + 1) \mu P (j_1, j_2 + 1) = \\
&\lambda_1 (j_1, j_2) P (j_1, j_2) + \lambda_2 (j_1, j_2) P (j_1, j_2) + (j_1 \mu + j_2 \mu) P (j_1, j_2)
\end{aligned}
\tag{1}
$$

where:

$$\lambda_l \left(j_1, j_2 \right) \overset{\substack{l=1,2 \\ m \neq l}}{=} \begin{cases} \lambda_l + \lambda_m, & if \ \left(j_l < \lfloor th_{1l}C_l \rfloor \right) \cap \left(j_m \geq \lfloor th_{2m}C_m \rfloor \right) \\ 0, & if \ \left(j_l \geq \lfloor th_{2l}C_l \rfloor \right) \cap \left(j_m < \lfloor th_{1m}C_m \rfloor \right) \\ 0, & if \ \left(j_1, j_2 \right) \ is \ a \ boundary \ state \\ \lambda_l, & otherwise \end{cases} \tag{2}$$

Before we proceed with the CBP determination, via the first method, we emphasize that this method can be quite complex even for a system with links of small capacity.

Having obtained $P(j_1, j_2)$, we can calculate the CBP in each link, P'_{b_1} and P'_{b_2} via (3) and (4), respectively [27]:

$$P'_{b_1} = \sum_{j_2=\lfloor th_{12}C_2 \rfloor}^{C_2} P(C_1, j_2) \tag{3}$$

$$P'_{b_2} = \sum_{j_1=\lfloor th_{11}C_1 \rfloor}^{C_1} P(j_1, C_2) \tag{4}$$

The rationale behind (3) is that a call can be blocked in the first link if there is no available bandwidth in that link (i.e., if $j_1 = C_1$) and the second link is not in support mode (i.e, $\lfloor th_{12}C_2 \rfloor \leq j_2$). A similar description applies to (4).

In addition, we can calculate the total CBP in the two-link system via the following weighted summation [27]:

$$P'_b = \frac{\lambda_1}{\lambda_1 + \lambda_2} P'_{b_1} + \frac{\lambda_2}{\lambda_1 + \lambda_2} P'_{b_2} \tag{5}$$

The second method leads to approximate CBP results (compared to the corresponding simulation results) by assuming that both links operate independently from one another. This means that each link l operates as an independent Erlang loss system of capacity C_l ($l = 1, 2$). Such an assumption simplifies (at the cost of accuracy) the necessary calculations for the CBP determination.

The CBP in the first and the second link can be approximated by (6) and (7), respectively:

$$P_{b_1} = P_1 (C_1) P_2 (j_2 \geq \lfloor th_{12}C_2 \rfloor) \tag{6}$$

$$P_{b_2} = P_2 (C_2) P_1 (j_1 \geq \lfloor th_{11}C_1 \rfloor) \tag{7}$$

where $P_l (C_l)$ refers to the CBP in link l ($l = 1, 2$) which can be determined either via the closed form of the classical Erlang B formula (8a), or via its recurrent form, which is appealing for large values of offered traffic-load a_l and capacity C_l (8b):

$$P_l (C_l) = \frac{\frac{a_l^{C_l}}{C_l!}}{\sum_{i=0}^{C_l} \frac{a_l^i}{i!}}, \quad a_l = \lambda_l / \mu \tag{8a}$$

$$P_l(C_l) = \frac{aP_l(C_l - 1)}{C_l + aP_l(C_l - 1)}, \quad C_l \geq 1, \ P_l(0) = 1 \tag{8b}$$

As far as the values of $P_l(j_l \geq \lfloor th_{1l}C_l \rfloor)$ are concerned, they are given by:

$$P_l(j_l \geq \lfloor th_{1l}C_l \rfloor) = \sum_{j_l = \lfloor th_{1l}C_l \rfloor}^{C_l} P_l(j_l) \tag{9}$$

where $P_l(j_l)$ is determined via the truncated Poisson distribution:

$$P_l(j_l) = \frac{\frac{a_l^{j_l}}{j_l!}}{\sum_{i=0}^{C_l} \frac{a_l^i}{i!}}, \quad a_l = \lambda_l/\mu \tag{10}$$

Finally, the total blocking probability can be determined via (5), where P_{b_1}' and P_{b_2}' are replaced by P_{b_1} and P_{b_2} (calculated via (6) and (7)), respectively.

An alternative recursive way for the determination of $P_l(j_l)$, $j_l = 1, \ldots, C_l$, is based on the link independence assumption. In the Erlang loss model, used to describe each link l, there exist LB between the adjacent states $j_l - 1$ and j_l which has the form [46]:

$$j_l P_l'(j_l) = a_l P_l'(j_l - 1) \tag{11}$$

Based on (11), we can determine the unnormalized values of $P_l'(j_l)$'s considering an initial value of $P_l'(0) = 1$. The normalized values of $P_l'(j_l)$'s are given by:

$$P_l(j_l) = P_l'(j_l) \Big/ \sum_{x=0}^{C_l} P_l'(x) \tag{12}$$

Based on (12), we can calculate P_{b_1}, P_{b_2} and the total CBP, via (6), (7) and (5), respectively.

3   The Proposed Model

In the proposed model, we consider again the loss system of the two links. Each link accommodates Poisson arriving calls of K different service-classes. Calls of service-class k ($k = 1, \ldots, K$) require b_k b.u. in order to be connected in a link. Let λ_{1k} and λ_{2k} be the arrival rates in the first and second link of service-class k calls, respectively. We also denote by j_1 and j_2 the occupied b.u. in the first and second link, respectively. Similar to Sect. 2, each link l ($l = 1, 2$) has a support threshold th_{1l} and an offloading threshold th_{2l}, with $th_{1l} < th_{2l}$ and $0 \leq th_{1l}, th_{2l} \leq 1$. To incorporate the notion of restricted accessibility in the proposed model, we assume that each state j_l of link l, except from the initial state where link l is empty (i.e., when $j_l = 0$) is associated with a blocking probability, $pb_{l,k}(j_l)$ which can be different for each service-class k. Clearly, when there is no available b.u. for calls of service-class k in link l (i.e., when $j_l \geq C_l - b_k + 1$), then $pb_{l,k}(j_l) = 1$. On the same hand, when the system is empty, then $pb_{l,k}(0) = 0$.

The call admission of a new service-class k call that initially arrives in link l ($l = 1, 2$) is summarized in the following steps:

(1) If $(0 \leq j_l < \lfloor th_{2l}C_l \rfloor)$ then the call will be handled by link l. In addition, if $j_l + b_k \leq C_l$, then the call is accepted in link l with probability $1 - pb_{l,k}(j_l)$ and remains for a generally distributed service-time with mean μ_k^{-1}.

(2) If $\lfloor th_{2l}C_l \rfloor \leq j_l$ then:

(2a) if $(0 \leq j_m < \lfloor th_{1m}C_m \rfloor)$ the call is offloaded to link m and assuming that $j_m + b_k \leq C_m$, the call is accepted link m with probability $1 - pb_{m,k}(j_m)$ and remains in link m for a generally distributed service-time with mean μ_k^{-1},

(2b) if $\lfloor th_{1m}C_m \rfloor \leq j_m$, then link m is in a *normal mode* of operation and does not support offloaded calls from link l. In that case, the call will be handled by link l. If $j_l + b_k \leq C_l$, then the call is accepted in link l with probability $1 - pb_{l,k}(j_l)$ and remains for a generally distributed service-time with mean μ_k^{-1}. Otherwise, the call is blocked and lost.

To determine in an approximate but efficient way the CBP of service-class k calls we assume that the two links operate independently from one another. In that case, each independent link behaves as an EMLM system under restricted accessibility [43], and therefore the CBP of service-class k calls in the first and the second link can be approximated by (13) and (14), respectively:

$$P_{res,b_{1k}} = P_{res,1k}(C_1) P_{res,2}(j_2 \geq \lfloor th_{12}C_2 \rfloor) \tag{13}$$

$$P_{res,b_{2k}} = P_{res,2k}(C_2) P_{res,1}(j_1 \geq \lfloor th_{11}C_1 \rfloor) \tag{14}$$

where $P_{res,lk}(C_{lk})$ refers to the CBP of service-class k calls in link l ($l = 1, 2$) and $P_{res,l}(j_l \geq \lfloor th_{1l}C_l \rfloor)$ refers to the probability that link l is not in support mode.

The values of $P_{res,lk}(C_l)$ in (13) and (14) are determined by the formula:

$$P_{res,lk}(C_l) = \sum_{j_l=1}^{C_l} G_l^{-1} q(j_l) pb_{l,k}(j_l) \tag{15}$$

where $q(j_l)$ refers to the unnormalized values of the link occupancy distribution of link l ($l = 1,2$) while $G_l = \sum_{j_l=0}^{C_l} q(j_l)$ is the normalization constant.

In (15), the values of $q(j_l)$ can be recursively determined via the following formula:

$$q(j_l) = \begin{cases} 1, & \text{for } j_l = 0 \\ \frac{1}{j_l} \sum_{k=1}^{K} a_{lk} b_k q(j_l - b_k)(1 - pb_{l,k}(j_l - b_k)), & \\ \quad \text{for } j_l = 1, ..., C_l \end{cases} \tag{16}$$

where: $a_{lk} = \lambda_{lk}/\mu_k$ is the total offered traffic-load of service-class k calls in link l.

As far as the values of $P_{res,l}\,(j_l \geq th_{1l}C_l)$, in (13) and (14), are concerned they can be determined by:

$$P_{res,l}\,(j_l \geq \lfloor th_{1l}C_l \rfloor) = \sum_{j_l=\lfloor th_{1l}C_l \rfloor}^{C_l} G_l^{-1} q(j_l) \qquad (17)$$

where $q(j_l)$ is determined via (16).

Finally, we propose the following formula for the total blocking probability of service-class k calls in the system of the two links:

$$P_{res,b_k} = \frac{\lambda_{1k}}{\lambda_{1k} + \lambda_{2k}} P_{res,b_{1k}} + \frac{\lambda_{2k}}{\lambda_{1k} + \lambda_{2k}} P_{res,b_{2k}} \qquad (18)$$

4   Numerical Examples – Evaluation

In this section, we present an application example and provide analytical and simulation CBP results of the proposed model. Simulation results are mean values of 7 runs and are based on the Simscript III simulation language [47]. In each simulation run a total of ten million calls is generated. To account for a warm-up period, the first 5% of these calls are not taken into account in the CBP results. Furthermore, to increase the readability of figures we do not present reliability ranges. The latter are less than two order of magnitude.

As an application example, consider a two-link system of capacities $C_1 = 24$ b.u. and $C_2 = 20$ b.u. that accommodates $K = 2$ service-classes whose calls require $b_1 = 1$ and $b_2 = 2$ b.u., respectively. For the first link, let: $\lambda_{11} = 9$ calls/min and $\lambda_{12} = 1$ calls/min. Similarly, for the second link let $\lambda_{21} = 7$ calls/min and $\lambda_{22} = 1$ calls/min. Also let $\mu_1^{-1} = \mu_2^{-1} = 1.0$ min. As far as the values of the thresholds are concerned, we assume that the offloading thresholds are equal to $th_{21} = th_{22} = 0.7$ and consider two different sets of support thresholds: (1) $th_{11} = th_{12} = 0.05$ and (2) $th_{11} = th_{12} = 0.25$. Regarding the restricted accessibility blocking probability factors for each link, we consider two sets: (1) $pb_{l,1}\,(j_l) = pb_{l,2}\,(j_l) = (j_l/C_l)^5$ and (2) $pb_{l,1}\,(j_l) = pb_{l,2}\,(j_l) = (j_l/C_l)^7$ where $l = 1,\,2$.

In the x-axis of Figs. 2, 3, 4 and 5, λ_{11} and λ_{21} increase in steps of 1.0 and 0.5, respectively. So, point 1 is: ($\lambda_{11} = 9.0$, $\lambda_{12} = 1.0$, $\lambda_{21} = 7.0$, $\lambda_{22} = 1.0$) while point 7 is: ($\lambda_{11} = 15.0$, $\lambda_{12} = 1.0$, $\lambda_{21} = 10.0$, $\lambda_{22} = 1.0$).

In Figs. 2 and 3, we present the CBP for the first service-class calls in each link, respectively. In Figs. 4 and 5, we present the corresponding CBP results for calls of the second service-class. Figures 2, 3, 4 and 5 show that the analytical CBP results: (a) are close to the simulation results especially when the values of the support thresholds th_{11} and th_{12} are kept within a reasonable level (e.g., 0.05 to 0.25). Depending on the two-link system, higher values of th_{11} and th_{12} may increase the error between analytical and simulation CBP results. This behavior can also be observed in the model of [27] and is anticipated due to the fact that (6), (7) and consequently (13), (14) imply that both links work independently

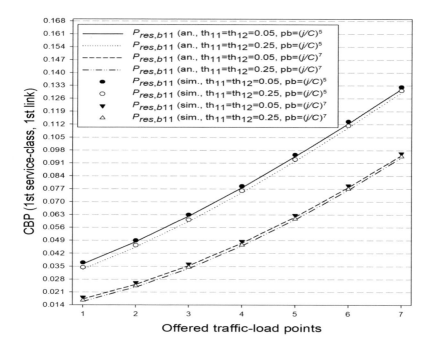

Fig. 2. 1st service-class - CBP in the first link.

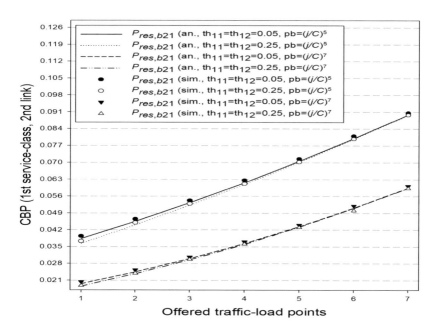

Fig. 3. 1st service-class - CBP in the second link.

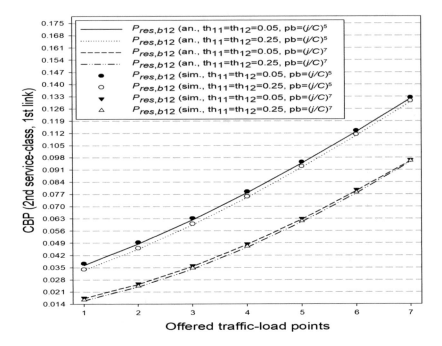

Fig. 4. 2nd service-class - CBP in the first link.

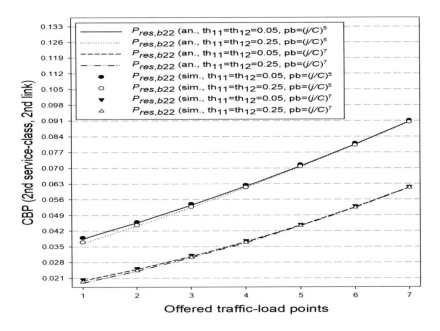

Fig. 5. 2nd service-class - CBP in the second link.

from one another. (b) The choice of $pb_{l,k}(j_l)$ greatly affects CBP. The higher values of set 1 $\left(pb_{l,k}(j_l) = (j_l/C_l)^5\right)$ result in much higher CBP compared to the lower values of set 2 $\left(pb_{l,k}(j_l) = (j_l/C_l)^7\right)$.

5   Conclusion

We propose a new multirate loss model for the call-level analysis of a two-link system with restricted accessibility that accommodates Poisson arriving calls of different service-class with different bandwidth-per-call requirements. In the two-link system, each link may support and provide service to calls offloaded from the other link. The proposed model does not have a PFS for the steady state probabilities due to the offloading mechanism and the existence of restricted accessibility. However, we show that an approximate method does exist for the determination of CBP that achieves satisfactory accuracy compared to simulation. As a future work, we intend to use the proposed model as the springboard for the analysis of interference between the two links.

References

1. Stasiak, M., Glabowski, M., Wisniewski, A., Zwierzykowski, P.: Modeling and Dimensioning of Mobile Networks. Wiley, West Sussex, UK (2011)
2. Iversen, V.: Teletraffic Engineering and Network Planning. DTU Photonic, Denmark (2015)
3. Iversen, V.: Modelling restricted accessibility for wireless multi-service systems. LNCS, vol. 3883, pp. 93–102 (2006)
4. Stasiak, M., Glabowski, M.: A simple approximation of the link model with reservation by a one-dimensional Markov chain. Perf. Eval. **41**(2–3), 195–208 (2000)
5. Moscholios, I., Logothetis, M., Kokkinakis, G.: Connection dependent threshold model: a generalization of the Erlang multiple rate loss model. Perform. Eval. **48**(1–4), 177–200 (2002)
6. Glabowski, M., Stasiak, M.: Point-to-point blocking probability in switching networks with reservation. Ann. Telecommun. **57**(7–8), 798–831 (2002)
7. Rácz, S., Gerő, B., Fodor, G.: Flow level performance analysis of a multi-service system supporting elastic and adaptive services. Perform. Eval. **49**(1–4), 451–469 (2002)
8. Glabowski, M., Stasiak, M.: Multi-rate model of the limited availability group with finite source population. In: Proceedings of the 10th Asia-Pacific Conference on Communication, Beijing, China (2004)
9. Moscholios, I., Logothetis, M.: Engset multirate state-dependent loss models with QoS guarantee. Int. J. Commun. Syst. **19**(1), 67–93 (2006)
10. Vassilakis, V., Moscholios, I., Logothetis, M.: Call-level performance modelling of elastic and adaptive service-classes with finite population. IEICE Trans. Commun. **E91-B**(1), 151–163 (2008)
11. Huang, Q., Ko, K.-T., Iversen, V.: Approximation of loss calculation for hierarchical networks with multiservice overflows. IEEE Trans. Commun. **56**(3), 466–473 (2008)

12. Stasiak, M., Sobieraj, M., Weissenberg, J., Zwierzykowski, P.: Analytical model of the single threshold mechanism with hysteresis for multi-service networks. IEICE Trans. Commun. **E95-B**(1), 120–132 (2012)
13. Moscholios, I., Vardakas, J., Logothetis, M., Koukias, M.: A quasi-random multirate loss model supporting elastic and adaptive traffic under the bandwidth reservation policy. Int. J. Adv. Netw. Serv. **6**(3&4), 163–174 (2013)
14. Yan, S., Razo, M., Tacca, M., Fumagalli, A.: A blocking probability estimator for the multi-application and multi-resource constraint problem. In: Proceedings of the ICNC, Honolulu, Hawaii (2014)
15. Moscholios, I., Logothetis, M., Vardakas, J., Boucouvalas, A.: Performance metrics of a multirate resource sharing teletraffic model with finite sources under both the threshold and bandwidth reservation policies. IET Netw. **4**(3), 195–208 (2015)
16. Huang, Y., Rosberg, Z., Ko, K., Zukerman, M.: Blocking probability approximations and bounds for best-effort calls in an integrated service system. IEEE Trans. Commun. **63**(12), 5014–5026 (2015)
17. Moscholios, I., Logothetis, M., Vardakas, J., Boucouvalas, A.: Congestion probabilities of elastic and adaptive calls in Erlang-Engset multirate loss models under the threshold and bandwidth reservation policies. Comput. Netw. **92**(1), 1–23 (2015)
18. Glabowski, M., Sobieraj, M.: Analytical modelling of multiservice switching networks with multiservice sources and resource management mechanisms. Telecommun. Syst. **66**(3), 559–578 (2017)
19. Widjaja, I., Roche, H.: Sizing X2 bandwidth for inter-connected eNBs. In: Proceedings of the IEEE VTC Fall, Anchorage, Alaska, USA (2009)
20. Stasiak, M., Zwierzykowski, P., Parniewicz, D.: Modelling of the WCDMA interface in the UMTS network with soft handoff mechanism. In: Proceedings of the IEEE GLOBECOM, Hawaii (2009)
21. Renard, B., Elayoubi, S., Simonian, A.: A dimensioning method for the LTE X2 interface. In: Proceedings of the IEEE WCNC, Shanghai, China (2012)
22. Stasiak, M., Parniewicz, D., Zwierzykowski, P.: Traffic engineering for multicast connections in multiservice cellular network. IEEE Trans. Ind. Inform. **9**(1), 262–270 (2013)
23. Moscholios, I., Kallos, G., Katsiva, M., Vassilakis, V., Logothetis, M.: Call blocking probabilities in a W-CDMA cell with interference cancellation and bandwidth reservation. In: Proceedings of the IEICE ICTF, Poznan, Poland (2014)
24. Moscholios, I., Kallos, G., Vassilakis, V., Logothetis, M.: Congestion probabilities in CDMA-based networks supporting batched Poisson input traffic. Wirel. Pers. Commun. **79**(2), 1163–1186 (2014)
25. Khedr, M., Makki Hassan, R.: Opportunistic call admission control for wireless broadband cognitive networks. Wirel. Netw. **20**(1), 105–114 (2014)
26. Machado de Medeiros, A., Yacoub, M.: BlockOut: blocking and outage in a single performance measure. IEEE Trans. Veh. Tech. **63**(7), 3451–3456 (2014)
27. Burger, V., Seufert, M., Hossfeld, T., Tran-Gia, P.: Performance evaluation of backhaul bandwidth aggregation using a partial sharing scheme. Phys. Commun. **19**, 135–144 (2016)
28. Moscholios, I., Vassilakis, V., Logothetis, M., Boucouvalas, A.: A probabilistic threshold-based bandwidth sharing policy for wireless multirate loss networks. IEEE Wirel. Commun. Lett. **5**(3), 304–307 (2016)
29. Vassilakis, V., Moscholios, I., Logothetis, M.: Uplink blocking probabilities in priority-based cellular CDMA networks with finite source population. IEICE Trans. Commun. **E99-B**(6), 1302–1309 (2016)

30. Vassilakis, V., Moscholios, I., Logothetis, M.: Quality of service differentiation of elastic and adaptive services in CDMA networks: a mathematical modelling approach. Wirel. Netw. **24**(4), 1279–1295 (2018)
31. Wang, Z., Mathiopoulos, P., Schober, R.: Performance analysis and improvement methods for channel resource management strategies of LEO-MSS with Multiparty Traffic. IEEE Trans. Veh. Tech. **57**(6), 3832–3842 (2008)
32. Yiltas, D., Zaim, A.: Evaluation of call blocking probabilities in LEO satellite networks. Int. J. Satell. Commun. **27**(2), 103–115 (2009)
33. Wang, Z., Mathiopoulos, P., Schober, R.: Channel partitioning policies for multi-class traffic in LEO-MSS. IEEE Trans. Aerosp. Electron. Syst. **45**(4), 1320–1334 (2009)
34. Vardakas, J., Moscholios, I., Logothetis, M., Stylianakis, V.: An analytical approach for dynamic wavelength allocation in WDM-TDMA PONs servicing ON-OFF traffic. IEEE/OSA J. Opt. Commun. Netw. **3**(4), 347–358 (2011)
35. Deng, Y., Prucnal, P.: Performance analysis of heterogeneous optical CDMA networks with bursty traffic and variable power control. IEEE/OSA J. Opt. Commun. Netw. **3**(6), 487–492 (2011)
36. Vardakas, J., Moscholios, I., Logothetis, M., Stylianakis, V.: On code reservation in multi-rate OCDMA passive optical networks. In: Proceedings of the CSNDSP, Poznan, Poland (2012)
37. Vardakas, J., Moscholios, I., Logothetis, M., Stylianakis, V.: Performance analysis of OCDMA PONs supporting multi-rate bursty traffic. IEEE Trans. Commun. **61**(8), 3374–3384 (2013)
38. Casares-Giner, V.: Some teletraffic issues in optical burst switching with burst segmentation. Electron. Lett. **52**(11), 941–943 (2016)
39. Guan, Y., Jiang, H., Gao, M., Bose, S., Shen, G.: Migrating elastic optical networks from standard single-mode fibers to ultra-low loss fibers: strategies and benefits. In: Proceedings of the Optical Fiber Communication Conference, Los Angeles, USA (2017)
40. Mamatas, L., Psaras, I., Pavlou, G.: Incentives and algorithms for broadband access sharing. In: Proceedings of the ACM SIGCOMM Workshop on Home Networks, New Delhi, India (2010)
41. Psaras, I., Mamatas, L.: On demand connectivity sharing: queuing management and load balancing for user-provided networks'. Comput. Netw. **55**(2), 399–414 (2011)
42. Moscholios, I., Vassilakis, V., Logothetis, M., Boucouvalas, A.: State-dependent bandwidth sharing policies for wireless multirate loss networks. IEEE Trans. Wirel. Commun. **16**(8), 5481–5497 (2017)
43. Iversen, V., Stepanov, S., Kostrov, A.: Dimensioning of multiservice links taking account of soft blocking. LNCS, vol. 4003, pp. 3–10. Springer (2006)
44. Kaufman, J.: Blocking in a shared resource environment. IEEE Trans. Commun. **29**(10), 1474–1481 (1981)
45. Roberts, J.: A service system with heterogeneous user requirements. In: Pujolle, G. (ed.) Performance of Data Communication Systems and Their Applications, North Holland, Amsterdam, pp. 423–431 (1981)
46. Pantelis, S., Moscholios, I., Papadopoulos, S.: Call-level evaluation of a two-link single rate loss model for Poisson traffic. In: Proceedings of the IEICE ICTF, Poznan, Poland, pp. 4–6 (2017)
47. Simscript III. http://www.simscript.com. Accessed July 2019

Performance Metrics in OFDM Wireless Networks Under the Bandwidth Reservation Policy

P. I. Panagoulias[1], I. D. Moscholios[1(✉)], and M. D. Logothetis[2]

[1] Department of Informatics and Telecommunications,
University of Peloponnese, 221 00 Tripolis, Greece
{panagoulias,idm}@uop.gr
[2] WCL, Department of Electrical and Computer Engineering,
University of Patras, 265 04 Patras, Greece
mlogo@upatras.gr

Abstract. In this paper we study the downlink of an Orthogonal Frequency Division Multiplexing (OFDM) based cell that accommodates calls from different service-classes with different resource requirements. We assume that calls arrive in the system according to a Poisson process while the call admission is based on the Bandwidth Reservation (BR) policy. The BR policy is used for the reservation of subcarriers in favor of calls of high subcarrier requirements. To determine the most important performance metrics, i.e., Call Blocking Probabilities (CBP) and resource utilization in this system, we model it as a multirate loss model and propose recursive formulas which reduce the complexity of the calculations. The accuracy of the formulas is verified via simulation and found to be quite satisfactory.

1 Introduction

The teletraffic modelling is important in the call-level performance evaluation of contemporary Orthogonal Frequency Division Multiplexing (OFDM) wireless networks that service calls from different service-classes with different Quality of Service (QoS) requirements. In such a multidimensional call-level traffic environment, it is required to have resource sharing policies capable of treating differently some calls (of a particular service-class) from other calls (of another service-class).

The most common resource sharing policy is the Complete Sharing (CS) policy. It is characterized as complete, since the only restriction is the "complete" system capacity [1]. In the CS policy, a new call is accepted in the system whenever the available system resources are greater than (or at least equal to) the call's required resources. The CS policy is an easy to apply policy and can be taken as the default policy, but it can lead to an unfair resource allocation among service-classes. On the other side of the CS policy is the complete partitioning policy, where a part of the system resources is allocated to each service-class.

© Springer Nature Switzerland AG 2020
M. Choraś and R. S. Choraś (Eds.): IP&C 2019, AISC 1062, pp. 252–263, 2020.
https://doi.org/10.1007/978-3-030-31254-1_30

The only benefit of this policy is to guarantee a certain QoS to a service-class. However, due to the fixed partition of resources, this policy leads to a decreased resource utilization.

In [2–4], the CS policy is considered for the determination of Call Blocking Probabilities (CBP) in OFDM wireless networks. More specifically, in [2], Paik and Suh (P-S) consider the downlink of an OFDM-based cell that accommodates Poisson arriving calls generated by multiservice classes. The system is described via a multirate loss model, i.e., new calls are not allowed to wait in a queue if their required resources are not available. Instead, in case of resource unavailability, calls are blocked and lost. Contrary to [3] and [4], where the acceptance of a new call in the cell depends only on the availability of subcarriers (i.e., the subcarriers are the only system resource), in the P-S model both the subcarriers and power are modelled as system resources and thus participate in call admission.

The P-S model is significant since power is a limited resource in OFDM wireless networks and should be taken into consideration in call admission. In addition, the steady-state probabilities in the P-S model can be described via a Product Form Solution (PFS). The latter is quite important in teletraffic modelling since it usually results in efficient formulas for the determination of the various performance measures. However, the calculation of CBP and resource utilization in the P-S model is based on highly complexed formulas which are not attractive for network planning and dimensioning procedures.

To solve this problem, a recursive formula which reduces substantially the complexity of the calculations of the P-S model is proposed in [5]. In addition, according to [5], the proposed formula can be used for the analysis of the Multiple Fractional Channel Reservation (MFCR) policy [6–8]. We name this model, P-S/MFCR model. The MFCR policy allows the reservation of subcarriers in order to favor calls of high subcarrier requirements. In that sense, and contrary to the CS policy, the MFCR policy can provide a certain QoS to calls of certain service-classes.

In this paper, we propose a recursive formula for the CBP determination in OFDM wireless networks that accommodate Poisson arriving calls generated by multiservice classes under the Bandwidth Reservation (BR) policy. We name the proposed model, P-S/BR model. The BR policy is a variant of the MFCR policy since it allows the reservation of an integer number of resource units [9–12]. On the other hand, the MFCR policy permits the reservation of a real (not integer) number of resource units.

This paper is organized as follows: In Sect. 2, we provide a review of the P-S model. In Sect. 3, we present the recursive formula for the determination of CBP and resource utilization in the P-S model. In Sect. 4, we review the P-S/MFCR model. In Sect. 5, we propose the P-S/BR model. In Sect. 6, we compare the analytical with simulation results for the P-S and the P-S/BR models. The comparison verifies the accuracy of the proposed formulas. We conclude in Sect. 7.

2   Review of the P-S Model

To describe the P-S model of [2], consider the downlink of an OFDM-based cell that has M subcarriers and let R, P and B be the average data rate per subcarrier, the available power in the cell and the system's bandwidth, respectively. We assume that the entire range of channel gains or signal to noise ratios per unit power is partitioned into K consecutive and non-overlapping intervals and denote as $\gamma_k, k = 1, \ldots, K$ the average channel gain of the kth interval. By further assuming L subcarrier requirements and K average channel gains, the cell accommodates a total of LK service-classes. A new call of service-class (k, l) call $(k = 1, \ldots, K$ and $l = 1, \ldots, L)$ requires b_l subcarriers in order to be accepted in the cell. This means that the call has a data rate requirement $b_l R$. In addition, it has an average channel gain γ_k. If these subcarriers are not available at the time of the call's arrival, then call blocking occurs. Otherwise, the call remains in the cell for a generally distributed service time with mean μ^{-1}. To calculate the power p_k required to achieve the data rate R of a subcarrier assigned to a call whose average channel gain is γ_k, we use the Shannon theorem: $R = (B/M) \log_2(1 + \gamma_k p_k)$.

Assuming that service-class (k, l) calls follow a Poisson process with rate λ_{kl} and that n_{kl} is the number of in-service calls, then the system can be described as a multirate loss model whose steady-state probabilities $\pi(\boldsymbol{n})$ have the following PFS [2]:

$$\pi(\boldsymbol{n}) = G^{-1} \left(\prod_{k=1}^{K} \prod_{l=1}^{L} p_{kl}^{n_{kl}} / n_{kl}! \right) \tag{1}$$

where: $\boldsymbol{n} = (n_{11}, \ldots, n_{k1}, \ldots, n_{K1}, \ldots, n_{1L}, \ldots, n_{kL}, \ldots, n_{KL})$, G is the normalization constant, $G = \sum_{\boldsymbol{n} \in \boldsymbol{\Omega}} \left(\prod_{k=1}^{K} \prod_{l=1}^{L} p_{kl}^{n_{kl}} / n_{kl}! \right)$, $\boldsymbol{\Omega}$ is the state space of the system denoted as $\boldsymbol{\Omega} = \{\boldsymbol{n} : 0 \leq \sum_{k=1}^{K} \sum_{l=1}^{L} n_{kl} b_l \leq M, \; 0 \leq \sum_{k=1}^{K} \sum_{l=1}^{L} p_k n_{kl} b_l \leq P\}$ and $p_{kl} = \lambda_{kl}/\mu$ is the offered traffic-load (in erlang) of service-class (k, l) calls.

The derivation of the PFS requires that the available power in the cell, P, and the power p_k (required to achieve the data rate R of a subcarrier) are integers. This can be achieved by multiplying both P and p_k by a constant in order to have an equivalent representation of the constraint $\sum_{k=1}^{K} \sum_{l=1}^{L} p_k' n_{kl} b_l \leq P'$ (that appears in the definition of $\boldsymbol{\Omega}$), where P' and p_k' are integers [2]. Thus, without loss of generality, it is assumed that P and p_k are integers.

According to [2], all performance metrics (such as CBP) are based on the calculation of $\pi(\boldsymbol{n})$'s via (1). As an example, the CBP $B(k, l)$ of service-class (k, l) calls is given by:

$$B(k, l) = 1 - G(P - p_k b_l, M - b_l)/G(\boldsymbol{\Omega}) \tag{2}$$

However, since the cardinality of Ω grows as $(MP)^{KL}$, the applicability of (1) is limited to systems of moderate size and therefore is not recommended for network planning and dimensioning procedures.

In [2], Paik and Suh propose the algorithms of [13] and [14] for the determination of $G(P - p_k b_l, M - b_l)$ (and consequently for the calculation of the CBP $B(k, l)$) without providing explicit details. The algorithms of [13] and [14] are proposed in the literature for the CBP determination in circuit-switched networks (see e.g., [15] and [16]). The algorithms of [13] are based on z-transforms and mean-value analysis. On the other hand, the algorithm of [14] is based on numerical inversion of generating functions which is a quite complex approach (e.g., [17] and [18]). Both algorithms: i) are applied to loss models whose steady-state probabilities have a PFS and (ii) are less general than the classical Kaufman-Roberts (K-R) recursive formula [19, 20]. The latter provides an efficient way for the CBP determination in a multirate loss system that accommodates Poisson arriving calls. Due to the effectiveness of the K-R formula, there exist an extensive list of applications not only in PFS but also in non-PFS models (e.g., [21–38]).

3   A Recursive Formula for the Determination of Performance Metrics

To circumvent the complexity problem of (1), a recursive yet efficient formula that resembles the K-R formula is proposed in [5]. To present this formula, the following notation is necessary: let $j_1 = \sum_{k=1}^{K} \sum_{l=1}^{L} n_{kl} b_l$ be the occupied subcarriers, i.e., $j_1 = 0, \ldots, M$ and $j_2 = \sum_{k=1}^{K} \sum_{l=1}^{L} p_k n_{kl} b_l$ the occupied power in the cell, i.e., $j_2 = 0, \ldots, P$. Also, let $q(\boldsymbol{j}) = q(j_1, j_2)$ be the occupancy distribution, given by:

$$q(\boldsymbol{j}) = q(j_1, j_2) = \sum_{n \in \Omega_j} \pi(\boldsymbol{n}) \tag{3}$$

where: Ω_j is the set of states in which the occupied subcarriers and the occupied power in the cell is j_1 and j_2, respectively.

The recursive determination of the unnormalized values of $q(j_1, j_2)$'s is based on the following formula [5]:

$$q(j_1, j_2) = \begin{cases} 1, & \text{for } j_1 = j_2 = 0 \\ \frac{1}{j_1} \sum_{k=1}^{K} \sum_{l=1}^{L} p_{kl} b_l q(j_1 - b_l, j_2 - p_k b_l) \\ \text{for } j_1 = 1, ..., M \text{ and } j_2 = 1, ..., P \end{cases} \tag{4}$$

The recursive form of (4) and its lower computational complexity, in the order of $O(MPLK)$, makes (4) attractive for network planning and dimensioning procedures. Contrary to the formulas proposed in [2] (or the algorithms of [13, 14]), (4) can be used as the springboard for the analysis of complicated subcarrier sharing policies, e.g., the MFCR policy [5], the BR policy (presented herein) or the threshold policy (see e.g., [30,35]). In the threshold policy, a call of a certain service-class can be blocked if the number of in-service calls (of the same service-class) plus the new call exceeds a predefined threshold. This means that a new call can be blocked, even if available resources do exist, when the threshold is exceeded.

Having obtained $q(j_1, j_2)$ we can calculate $B(k, l)$ via the formula:

$$B(k,l) = \sum_{\{(j_1+b_l>M)\,\cup\,(j_2+p_k b_l>P)\}} G^{-1} q(j_1, j_2) \tag{5}$$

and the mean number of in-service calls of service-class (k, l), $E(k, l)$, via the formula:

$$E(k,l) = p_{kl}(1 - B(k,l)) \tag{6}$$

where: G is the normalization constant, determined via the formula $G = \sum_{j_1=0}^{M} \sum_{j_2=0}^{P} q(j_1, j_2)$.

Having determined the values of $E(k, l)$, we can also calculate the entire system Blocking Probability (BP), the Subcarrier Utilization (SU) and the Power Utilization (PU), via the formulas:

$$BP = \sum_{k=1}^{K}\sum_{l=1}^{L} B(k,l)\lambda_{k,l} \bigg/ \Lambda, \quad \Lambda = \sum_{k=1}^{K}\sum_{l=1}^{L}\lambda_{k,l} \tag{7}$$

$$SU = \sum_{k=1}^{K}\sum_{l=1}^{L} E(k,l)b_l \bigg/ M \tag{8}$$

$$PU = \sum_{k=1}^{K}\sum_{l=1}^{L} p_k E(k,l)b_l \bigg/ P \tag{9}$$

4   Review of the P-S Model Under the MFCR Policy

In the MFCR policy, a service-class (k, l) call requests b_l subcarriers and has a reservation parameter t_l that expresses the number of subcarriers reserved to benefit calls of all other service-classes except for (k, l). The reservation of t_l subcarriers is achieved since $\lfloor t_l \rfloor +1$ subcarriers are reserved with probability $t_l - \lfloor t_l \rfloor$ while $\lfloor t_l \rfloor$ subcarriers are reserved with probability $1 - (t_l - \lfloor t_l \rfloor)$. To describe the call admission mechanism in the P-S/MFCR model we proceed as follows [5]: Let j_1 be the occupied subcarriers and j_2 be the occupied power in the cell when a new service-class (k, l) call arrives in the cell. Then, if: (a) $(M - j_1 - \lfloor t_l \rfloor > b_l) \cap (j_2 + p_k b_l \le P)$ the call is accepted in the cell, (b) $(M - j_1 - \lfloor t_l \rfloor = b_l) \cap (j_2 + p_k b_l \le P)$ the call is accepted with probability $1 - (t_l - \lfloor t_l \rfloor)$, (c) $(M - j_1 - \lfloor t_l \rfloor < b_l) \cup (j_2 + p_k b_l > P)$ the call is blocked and lost.

The P-S/MFCR model does not have a PFS for the steady state probabilities. Based on [5], it can be proved that the unnormalized values of $q(j_1, j_2)$'s can be determined by an approximate but recursive formula:

$$
q(j_1, j_2) = \begin{cases} 1, & \text{for } j_1 = j_2 = 0 \\ \frac{1}{j_1} \sum\limits_{k=1}^{K} \sum\limits_{l=1}^{L} p_{kl}(j_1 - b_l) b_l q(j_1 - b_l, j_2 - p_k b_l) \\ & \text{for } j_1 = 1, \ldots, M \text{ and } j_2 = 1, \ldots, P \end{cases} \tag{10}
$$

where:

$$
p_{kl}(j_1 - b_l) = \begin{cases} p_{kl}, & \text{for } j_1 < M - \lfloor t_l \rfloor \\ (1 - (t_l - \lfloor t_l \rfloor)) \, p_{kl}, & \text{for } j_1 = M - \lfloor t_l \rfloor \\ 0, & \text{for } j_1 > M - \lfloor t_l \rfloor \end{cases} .
$$

Having obtained $q(j_1, j_2)$ we calculate $B(k, l)$ via:

$$
\begin{aligned}
B(k, l) = & \sum\nolimits_{\{(j_1 + b_l + \lfloor t_l \rfloor > M) \cup (j_2 + p_k b_l > P)\}} G^{-1} q(j_1, j_2) \\
& + (t_l - \lfloor t_l \rfloor) \sum\nolimits_{\{(j_1 + b_l + \lfloor t_l \rfloor = M) \cap (j_2 + p_k b_l \le P)\}} G^{-1} q(j_1, j_2)
\end{aligned} \tag{11}
$$

while the values of $E(k, l)$, BP, SU and PU can be determined via (6), (7), (8) and (9), respectively.

5   The P-S Model Under the BR Policy

In the BR policy, a new service-class (k, l) call requests b_l subcarriers and has a reservation parameter t_l similar to the MFCR policy. The call admission mechanism in the proposed P-S/BR model is as follows: (a) if $(M - j_1 - t_l \ge b_l) \cap (j_2 + p_k b_l \le P)$ then the service-class (k, l) call is accepted in the cell, (b) if $(M - j_1 - t_l < b_l) \cup (j_2 + p_k b_l > P)$ then the service-class (k, l) call is blocked and lost.

The steady-state probabilities in the P-S/BR model do not have a PFS, since the BR policy destroys local balance between the adjacent states $\boldsymbol{n}_{kl}^-$ and $\boldsymbol{n}$, where: $\boldsymbol{n}_{kl}^- = (n_{11}, \ldots, n_{k1}, \ldots, n_{K1}, \ldots, n_{1l}, \ldots, n_{kl} - 1, \ldots, n_{Kl}, n_{1L}, \ldots, n_{kL}, \ldots, n_{KL})$.

However, based on Sect. 3 and [5], it can be proved that the unnormalized values of $q(j_1, j_2)$'s are determined via (10) where: $p_{kl}(j_1 - b_l) = \begin{cases} p_{kl}, & \text{for } j_1 \le M - t_l \\ 0, & \text{for } j_1 > M - t_l \end{cases} .$

Having obtained $q(j_1, j_2)$ we calculate $B(k, l)$ via:

$$B(k, l) = \sum_{\{(j_1 + b_l + t_l > M)\} \cup (j_2 + p_k b_l > P)} G^{-1} q(j_1, j_2) \qquad (12)$$

while the values of $E(k, l)$, BP, SU and PU can be determined via (6), (7), (8) and (9), respectively.

6   Evaluation

We consider the downlink of an OFDM-based cell and provide analytical and simulation CBP results for the P-S and the P-S/BR models. The input parameters for the abovementioned models are those of [2]: $B = 20$ MHz, $P = 25$ Watt, $M = 256$, $R = 329.6$ kbps, $L = 64$, $b_l = l$, $l = 1, \ldots, 64$, and the values of b_l are uniformly distributed. In addition, let $K = 3$ which results in $LK = 192$ service-classes. Let the integer representations of p_k $(k = 1, 2, 3)$ and P be [2]: $p_1 = 6$, $p_2 = 10$, $p_3 = 16$, $P' = 2500$. The values of p_k' require: $p_1 \approx 0.06$, $p_2 \approx 0.01$, $p_3 \approx 0.16$ achieved by $\gamma_1 = 24.679$ dB, $\gamma_2 = 22.460$ dB, $\gamma_3 = 20.419$ dB. We further assume that the probability an arriving call has an average channel gain γ_k is given by two different sets: (1) set 1: $a_k = 1/3$ $(k = 1, 2, 3)$ and (2) set 2: $a_1 = 1/4$, $a_2 = 1/4$, $a_3 = 1/2$. Also, let $\lambda_{kl} = \Lambda a_k / L$ be the arrival rate of Poisson arriving service-class (k, l) calls, where Λ is the total arrival rate in the cell, $\Lambda = pM\mu/\hat{g}$, p is the traffic intensity of the cell, $\mu = 0.00625$ and $\hat{g} = 32.5$ is the average subcarrier requirement of a new call. Note that the value of $\hat{g} = 32.5$ is based on the fact that $b_1 = 1$, $b_{64} = 64$ subcarriers and the values of b_l are uniformly distributed. As far as the values of the BR parameters are concerned, let $t_l = 64$-l, $l = 1, \ldots, 64$, so that $b_1 + t_1 = \ldots = b_{64} + t_{64}$.

In the x-axis of Figs. 1, 2 and 3, the value of p increases from 0.2 to 1.0. Simulation CBP results, based on Simscript III [39], are mean values of 7 runs, while each run is based on the generation of 10 million calls. To account for a warm-up period, the blocking events of the first 3% of these generated calls are not considered in the results.

In Figs. 1 and 2, we consider the case of $\alpha_k = 1/3$, $(k = 1, 2, 3)$. Figure 1 shows the analytical and simulation CBP of service-classes (3, 64), (2, 64) and (1, 64) which require the highest number of subcarriers ($l = 64$). We see that the BR policy reduces the CBP of these service-classes compared to the values of the P-S model. Figure 2 shows the analytical and simulation CBP of service-classes (3, 48), (2, 48) and (1, 48). We see that, in most of the cases, the BR policy increases the CBP of these service-classes compared to the values of the P-S model. The same behavior (CBP increase) appears in most of the service-classes whose calls require less than 64 subcarriers.

On the other hand, the BP of the entire system increases for both sets of α_k (Fig. 3) since the t_l parameters are chosen to benefit service-classes with high subcarrier requirements. In all figures, analytical results are quite close compared to the corresponding simulation results.

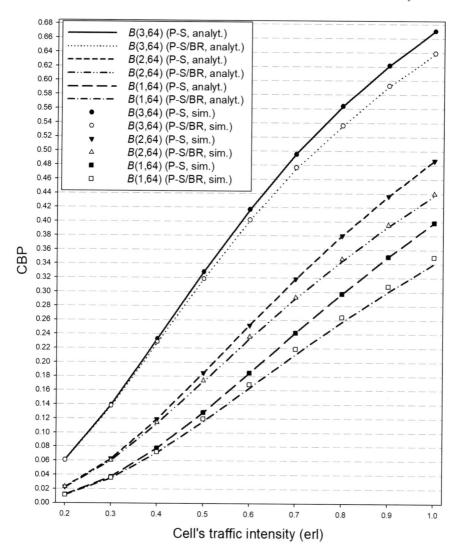

Fig. 1. CBP of service-classes (1, 64), (2, 64) and (3, 64).

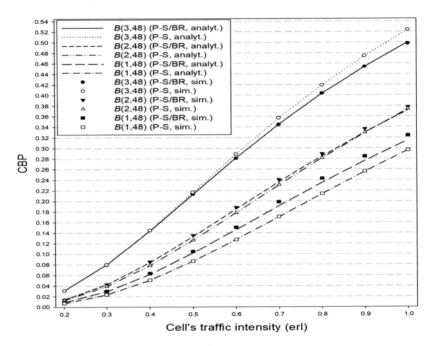

Fig. 2. CBP of service-classes (1, 48), (2, 48) and (3, 48).

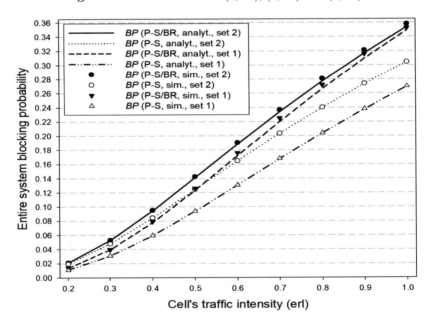

Fig. 3. BP of the entire system.

7 Conclusion

We propose recursive formulas for the determination of performance measures in the downlink of an OFDM cell that accommodates multirate traffic of Poisson arrivals under the CS, the MFCR and the BR policies. The proposed formulas are quite accurate compared to simulation and can be used in network dimensioning procedures for the CBP and resource utilization calculation. As a future work we intend to study call arrival processes that are more peaked and 'bursty' than the Poisson process, such as the batched Poisson process, where calls arrive in batches and the batch arrival process is Poisson [27,40].

References

1. Moscholios, I., Logothetis, M.: Efficient Multirate Teletraffic Loss Models Beyond Erlang. Wiley & IEEE Press, West Sussex, UK (2019)
2. Paik, C., Suh, Y.: Generalized queueing model for call blocking probability and resource utilization in OFDM wireless networks. IEEE Commun. Lett. $15(7)$, 767–769 (2011)
3. Pla, V., Martinez-Bauset, J., Casares-Giner, V.: Comments on call blocking probability and bandwidth utilization of OFDM subcarrier allocation in next-generation wireless networks. IEEE Commun. Lett. $12(5)$, 349 (2008)
4. Chen, J.C., Chen, W.S.E.: Call blocking probability and bandwidth utilization of OFDM subcarrier allocation in next-generation wireless networks. IEEE Commun. Lett. $10(2)$, 82–84 (2006)
5. Moscholios, I., Vassilakis, V., Panagoulias, P., Logothetis, M.: On call blocking probabilities and resource utilization in OFDM wireless networks. In: Proceedings of the CSNDSP, Budapest, Hungary (2018)
6. Cruz-Pérez, F., Vázquez-Ávila, J., Ortigoza-Guerrero, L.: Call blocking probability and bandwidth utilization of OFDM subcarrier allocation in next-generation wireless networks. IEEE Commun. Lett. $8(10)$, 629–631 (2004)
7. Moscholios, I.: Congestion probabilities in Erlang-Engset multirate loss models under the multiple fractional channel reservation policy. Image Proces. Commun. $21(1)$, 35–46 (2016)
8. Moscholios, I., Vassilakis, V., Logothetis, M.: Call blocking probabilities for Poisson traffic under the multiple fractional channel reservation policy. In: Proceedings of the CSNDSP, Prague, Czech Republic (2016)
9. Stasiak, M., Glabowski, M.: A simple approximation of the link model with reservation by a one-dimensional Markov chain. Perform. Eval. $41(2–3)$, 195–208 (2000)
10. Glabowski, M., Stasiak, M.: Point-to-point blocking probability in switching networks with reservation. Ann. Telecommun. $57(7–8)$, 798–831 (2002)
11. Moscholios, I., Vardakas, J., Logothetis, M., Koukias, M.: A quasi-random multirate loss model supporting elastic and adaptive traffic under the bandwidth reservation policy. Int. J. Adv. Netw. Serv. $6(3\&4)$, 163–174 (2013)
12. Moscholios, I., Logothetis, M., Vardakas, J., Boucouvalas, A.: Congestion probabilities of elastic and adaptive calls in Erlang-Engset multirate loss models under the threshold and bandwidth reservation policies. Comput. Netw. $92(P1)$, 1–23 (2015)
13. Pinsky, E., Conway, A.: Computational algorithms for blocking probabilities in circuit-switched networks. Ann. Oper. Res. $35(1)$, 31–41 (1992)

14. Choudhury, G., Leung, K., Whitt, W.: An algorithm to compute blocking probabilities in multi-rate multi-class multi-resource loss models. Adv. Appl. Probab. **27**(1), 1104–1143 (1995)
15. Caro, F., Simchi-Levi, D.: Optimal static pricing for a tree network. Ann. Oper. Res. **196**(1), 137–152 (2012)
16. Wang, M., Li, S., Wong, E., Zukerman, M.: Performance analysis of circuit-switched multi-service multi-rate networks with alternative routing. J. Lightwave Technol. **32**(2), 179–200 (2014)
17. Iversen, V.: Review of the convolution algorithm for evaluating service integrated systems. COST-257. Leidschendam, The Netherlands (1997)
18. Beard, C., Frost, V.: Prioritized resource allocation for stressed networks. IEEE/ACM Trans. Netw. **9**(1), 618–633 (2001)
19. Kaufman, J.: Blocking in a shared resource environment. IEEE Trans. Commun. **29**(10), 1474–1481 (1981)
20. Roberts, J.: A service system with heterogeneous user requirements. In: Performance of Data Communications Systems and Their Applications, North Holland, Amsterdam, pp. 423–431 (1981)
21. Wang, Z., Mathiopoulos, P., Schober, R.: Channel partitioning policies for multiclass traffic in LEO-MSS. IEEE Trans. Aerosp. Electron. Syst. **45**(4), 1320–1334 (2009)
22. Stasiak, M., Glabowski, M., Wisniewski, A., Zwierzykowski, P.: Modeling and Dimensioning of Mobile Networks. Wiley, West Sussex (2011)
23. Deng, Y., Prucnal, P.: Performance analysis of heterogeneous optical CDMA networks with bursty traffic and variable power control. IEEE/OSA J. Opt. Commun. Netw. **3**(6), 487–492 (2011)
24. Vardakas, J., Moscholios, I., Logothetis, M., Stylianakis, V.: On code reservation in multi-rate OCDMA passive optical networks. In: Proceedings of the CSNDSP, Poznan, Poland (2012)
25. Stasiak, M., Parniewicz, D., Zwierzykowski, P.: Traffic engineering for multicast connections in multiservice cellular network. IEEE Trans. Ind. Inform. **9**(1), 262–270 (2013)
26. Khedr, M., Makki Hassan, R.: Opportunistic call admission control for wireless broadband cognitive networks. Wirel. Netw. **20**(1), 105–114 (2014)
27. Moscholios, I., Kallos, G., Vassilakis, V., Logothetis, M.: Congestion probabilities in CDMA-based networks supporting batched Poisson input traffic. Wirel. Pers. Commun. **79**(2), 1163–1186 (2014)
28. Huang, Y., Rosberg, Z., Ko, K., Zukerman, M.: Blocking probability approximations and bounds for best-effort calls in an integrated service system. IEEE Trans. Commun. **63**(12), 5014–5026 (2015)
29. Casares-Giner, V.: Some teletraffic issues in optical burst switching with burst segmentation. Electron. Lett. **52**(11), 941–943 (2016)
30. Moscholios, I., Vassilakis, V., Logothetis, M., Boucouvalas, A.: A probabilistic threshold-based bandwidth sharing policy for wireless multirate loss networks. IEEE Wirel. Commun. Lett. **5**(3), 304–307 (2016)
31. Vassilakis, V., Moscholios, I., Logothetis, M.: Uplink blocking probabilities in priority-based cellular CDMA Networks with finite source population. IEICE Trans. Commun. **E99-B**(6), 302–1309 (2016)
32. Moscholios, I.: A multirate loss model for quasi-random traffic under the multiple fractional channel reservation policy. In: Proceedings of the IEICE ICTF, Patras, Greece (2016)

33. Glabowski, M., Kaliszan, A., Stasiak, M.: Modelling overflow systems with distributed secondary resources. Comput. Netw. **108**(10), 171–183 (2016)
34. Guan, Y., Jiang, H., Gao, M., Bose, S., Shen, G.: Migrating elastic optical networks from standard single-mode fibers to ultra-low loss fibers: strategies and benefits. In: Proceedings of the Optical Fiber Communication Conference, Los Angeles, USA (2017)
35. Moscholios, I., Vassilakis, V., Logothetis, M., Boucouvalas, A.: State-dependent bandwidth sharing policies for wireless multirate loss networks. IEEE Trans. Wirel. Commun. **16**(8), 5481–5497 (2017)
36. Moscholios, I., Logothetis, M., Shioda, S.: Performance evaluation of multirate Loss systems supporting cooperative users with a probabilistic behavior. IEICE Trans. Commun. **E100-B**(10), 1778–1788 (2017)
37. Glabowski, M., Sobieraj, M.: Analytical modelling of multiservice switching networks with multiservice sources and resource management mechanisms. Telecommun. Syst. **66**(3), 559–578 (2017)
38. Hanczewski, S., Stasiak, M., Weissenberg, J.: Non-full-available queueing model of an EON node. Opt. Switching Netw. (2018, accepted)
39. Simscript III. http://www.simscript.com. Accessed July 2019
40. Moscholios, I., Logothetis, M.: The Erlang multirate loss model with batched Poisson arrival processes under the bandwidth reservation policy. Comput. Commun. **33**(Suppl. 1), S167–S179 (2010)

Traffic Modeling for Industrial Internet of Things (IIoT) Networks

Mariusz Głąbowski[1], Sławomir Hanczewski[1], Maciej Stasiak[1],
Michał Weissenberg[1(✉)], Piotr Zwierzykowski[1],
and Vito Bai[2]

[1] Faculty of Electronics and Telecommunications,
Poznan University of Technology (PUT), Poznań, Poland
{michal.weissenberg,piotr.zwierzykowski}@put.poznan.pl
[2] Huawei Technologies Co., Ltd., Shenzhen, Guangdong, China

Abstract. This article presents an engineering proposal of methods to model typical traffic sources to be found in industrial networks. To develop these methods, the ON/OFF model was used. On the basis of this particular model, a number of methods are proposed that can be used to model typical types of traffic that occur in industrial networks, such as Time-Triggered (TT) traffic, Audio-Video Bridging (AVB) traffic and Best Effort (BE) traffic. The article presents four traffic models and a method for their application in modeling a number of exemplary types of traffic used in industrial networks.

1 Introduction

Over the past years we have been experiencing a rapid and dynamic increase in the number of available internet services based on the IP protocol and Ethernet technology. One of the latest areas in which new services are being implemented are the solutions that have been introduced to the so-called *Internet of Things* and its industrial version IIoT - *Industrial Internet of Things* [1,2]. The more and more common use of IIoT is getting even more evident with the increase of the popularity of the concept of Industry 4.0. IIoT networks will be capable of transmitting traffic streams that are known from industrial networks as well as types of traffic that have been associated until now with computer networks. Transmission of such a diverse array of traffic over just one network should be taken into consideration as early as the designing and dimensioning stages. Appropriate modeling of traffic is also important from the point of view of the analysis of traffic management mechanisms that are typically used in such networks. These evolutionary changes that are being introduced to industrial networks also involve changes in communications and control systems. Industrial network standards that were offered by the leading manufacturers and producers of network devices and industrial automation drivers, initially closed, tend now to become open standards. A large number of them is based on Ethernet standards and use the Ethernet or its modifications as a transport network [3].

M. Choraś and R. S. Choraś (Eds.): IP&C 2019, AISC 1062, pp. 264–271, 2020.
https://doi.org/10.1007/978-3-030-31254-1_31

To widen our knowledge on the nature of traffic generated in industrial networks based on the Ethernet, it is necessary to know the parameters of traffic in the network that is generated by all sources. These parameters for each type of traffic include:

- intensity of packet generation;
- distribution of gaps between single packets for a given traffic stream;
- intensity of stream generation for a given type;
- time interval between successive streams in a given service.

The above values make it possible to parametrize appropriately models of traffic sources based on the ON/OFF source concept.

The present article is divided into five chapters. The introduction starts the article. Chapter Two presents a description of the proposed methods for modeling traffic sources that are to be found in IIoT networks. The models that are presented in the chapter can be used to model three basic types of traffic that are used in modern industrial networks, i.e. Time-Triggered (TT) traffic, traffic characteristic for Audio-Video Bridging (AVB) and Best Effort (BE) traffic. With the obtained results of measurements, it was possible to develop approximate models of traffic sources that were based on the ON/OFF concept. These models are presented in Chapter Three. Chapter Four is devoted to the types of traffic in industrial networks and their properties and characteristic features. This chapter also includes a table in which a proposal to use these traffic models to model different types of traffic is presented. A short summary concludes the article.

2   Modeling of Traffic Sources in Industrial Networks

An important element of modeling traffic phenomena in industrial networks is to describe the sources that occur in them. Therefore, this chapter presents a classification of traffic sources to be found in industrial networks and describes the traffic source models for the Time-Triggered (TT) traffic and traffic typical for Audio-Video Bridging (AVB).

2.1   Traffic Sources in Industrial Networks

Modern industrial networks are assumed to include different types of traffic. Beside traffic typical for industrial networks, called Time-Triggered (TT), Audio-Video Bridging (AVB) and Best Effort (BE), there is all traffic typical for computer networks, considered in [3]. Because of the significance of the TT traffic in the process of manufacturing or other production assignments, this type of traffic has the highest priority in the network. The way traffic of this type is generated is also characteristic on account of a deterministic behavior of traffic sources. A TT traffic source can be any device or sensor that participates in the production process. In general, TT traffic sources can be divided into the following groups:

- **passive sources** - characterized by regular packet generation. A good example of such sources is the vast array of different sensors that send results of measurements regularly to controlling devices;
- **active sources** - characterized by generation of both regular and irregular messages. Regular messages can result from, for example, sending reports on the current state of the operation of devices, while irregular messages can be sent, for example, as a reply to a demand sent by a controller;
- **maintenance messages sources** - that generate messages necessary to sustain effective production process. Activity of these sources results from control algorithms that control performance of individual tasks in a production process.

TT traffic sources can be modelled using suitably designed analytical models known from the literature. The present authors believe that the most appropriate for this purpose are ON/OFF traffic source models. Over years, these sources have been widely used in modeling traffic in packet networks [5].

2.2   Time-Triggered Traffic

One of the basic TT traffic sources models in industrial networks is the ON/OFF traffic source model. This model reflects well the real traffic sources that are characterized by alternating repeated active (ON) and idle (OFF) states. During the active state the source generates packets, whereas during the idle state the source is switched off. The simplest example of a source of this type is the telephone subscriber in a telephone conversation. The time in which he or she speaks corresponds to the active state, whereas the time of listening corresponds to the idle state. The basic ON/OFF traffic source model [6] assumes that duration times in states ON/OFF will have exponential distribution with the parameters respectively $1/\alpha$ (ON state) and $1/\beta$ (OFF state). The parameter α is the intensity of transition from state ON to state OFF, whereas the parameter β is the intensity of transition from state OFF to state ON. The operation of the source is represented by the two-state Markov process shown in Fig. 1.

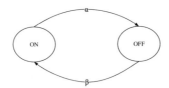

Fig. 1. Two-state Markov chain

The model of traffic source under consideration assumes that in state ON packets are generated with constant speed equal to $1/T$, where T denotes the fixed time interval between successive packets. In state OFF, packets are not generated by the source. The presented ON/OFF source model can also be viewed as a simple example of the modulating process that is a combination of two different processes (Fig. 2).

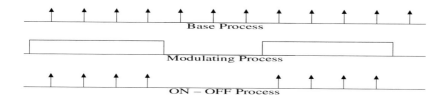

Fig. 2. ON/OFF traffic as an example of a simple modulating process.

The *Base Process* is characterized by constant speed in generating packets $(1/T)$. The process is modulated by *Modulating Process*, whose Markov chain is presented in Fig. 1. As a result of the modulating process, packets are not generated in those periods in which the modulating process is in state OFF. Considering the nature of traffic generated by TT sources, it is possible to model them using appropriately parametrized ON/OFF sources. The operation of passive traffic sources is limited to sending messages in regular time intervals. To determine the parameters of an ON-OFF source, that is an analytical counterpart of a real source, it is necessary to determine the intensity of transitions between states ON and OFF first. Taking into consideration sustained and uninterrupted working time of industrial networks, the duration time of state ON is usually far more higher than the duration time of state OFF, which means that the intensity of transition to state of total switch off (α) is very low (total switch off state is a marginal state). A traffic source can be switched off only when all system is switched off, during a system failure or, alternatively, at a demand of a driver. The speed at which messages are generated in state ON depends on the parameters of a real device (sensor).

2.3 Audio-Video Bridging Traffic

The most widely used models to model traffic sources of the type AVB are MMPP models (*Markov Modulated Poisson Process*) that allow variability in a packet stream in time to be taken into consideration, and in particular the rapid nature of this traffic. The MMPP process is called double stochastic, because both the base process and modulating processes are Markov process. The simplest, two-state 2-MMPP model [7] is characterized by constant alternate active states (ON1) and (ON2). During state (ON1), the source generates packets that create a Poisson traffic stream with the intensity α_1, whereas during state (ON2) a Poisson stream with the intensity α_2. Duration times of states (ON1) and (ON2) have exponential distribution with the parameters ω_1, ω_2, respectively. The parameter ω_1 is the intensity of transition from state (ON1) to state (ON2), while the parameter ω_2 is the intensity of transition from state (ON2) to state (ON1). Figure 3 shows a Markov process divided into the base and modulating processes that define the operation of a 2-MMPP source.

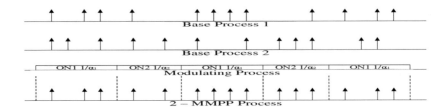

Fig. 3. 2-MMPP traffic.

3   Proposed ON/OFF Models

This chapter provides basic information on ON/OFF traffic surce model. During the study, four variants of the ON/OFF traffic source models were distinguished. The differences included:

- *method for a determination of frame length*: frame length can be fixed or variable (distribution of length can be exponential or normal);
- *method for a determination of the distance between successive frames*: gaps can be interpolated between each of the frames, duration time of frames can be either constant or random with required distribution (normal or exponential);
- *the way duration time for ON state is generated*: it can be constant or random with required distribution (normal or exponential);
- *the way duration time for Off state is generated*: it can be constant or random with required distribution (normal or exponential).

The above parameters influence the duration time of ON state. The OFF state period can be established or its length can be a random variable with required average value and selected distribution. Table 1 shows the four selected configurations of the variants of ON/OFF traffic source models.

Table 1. ON/OFF sources in the testbed

	M1	M2	M3	M4
Duration time of ON state	Constant	Constant	Exponential	Exponential
Duration time of OFF state	Constant	Exponential	Exponential	Exponential
Frame length	Constant	Constant	Constant	Exponential
Gap length between frames	Constant	Constant	Constant	Exponential

Each of the variants of ON/OFF source models can be further modified, which will make it possible to determine randomly the type of each of the successive events in state ON. In state ON, the type of an event is determined randomly (i.e. the frame or the gap between frames, which is a random variable). Then, service time for the event is chosen randomly (frame duration time or gap time between frames). Hence, to the names of ON/OFF models, in which the event type is a random variable, the letter R is added, just as in, for example, M2-R.

3.1   Models

Model M1 - duration time for state ON is pre-defined, whereas the number of frames generated within this time is constant (CON). Duration time for state OFF is also CON, just as the length of frames and gap time between them. In the boundary case, the length of gaps can be equal to zero, i.e. frames are generated without any gaps between them.

Model M2 - duration time for state ON is pre-defined, whereas the number of frames generated within this time is CON. Duration time for state OFF is determined by a random variable with the exponential distribution (ED). The lengths of frames and gap time between them are CON.

Model M3 - duration time for state ON is defined using a random variable with the ED. The size of frames is CON, whereas their number is proportional to the duration time of state ON. Duration time for state OFF is determined by a random variable with an exponential distribution. The lengths of frames and gap time between them is CON.

Model M4 - duration time for state ON is determined using a random variable with the ED. The size of frames is determined using a random variable with the ED and their number is proportional to the duration time of state ON. Duration time for state OFF is determined by a random variable with the ED. The length of frames and gap time between them are also determined by a random variable with the ED.

Modification R - this modification is based on the fact that events in state ON have random characteristics. This means that each event can represent a frame or a gap between frames depending on the random variable. This modification can be used to the models 1–4.

Modification D - this modification is based on the fact that part of frames and gaps between frames can have constant length or random length, i.e. the one that is determined by a random variable with any distribution (exponential or normal). The assumption in the "D" modification is that the time characteristics of a variable of states ON and OFF can be described by an exponential or normal distribution.

Modification T - in the M1–M4 models, duration time for state ON took on a constant value or was determined by a random variable with required distribution. The assumption in this modification is that duration time for state ON is strictly defined by the number of events (frames and gaps between frames) with required time characteristics (constant or random). On account of the possibility of random duration time of all events in state ON, duration time for this state will be determined as the sum of all its component events. Duration time for state ON will be deemed terminated when all events defined as component parts of state ON are serviced. The number of events in state ON will always be constant.

4   Types of Traffic in Industrial Networks

On the basis of measurements carried out in one of the largest IIoT test networks, launched in Austin, eight types of traffic can be distinguished [4]: Isonchronic (I), Cyclic (C), Alarm and Events (AaE), Configuration and diagnostics (CaD), Network Control (NC), Best effort (BE), Video (V) and Audio/Voice (A/V). Each of the traffic types is characterized by a number of different features that can be classified as follows:

- *Periodicity* (P) - this parameter describes two types of transmission: cyclic and acyclic (e.g. controlled by events);
- *Cycle length* (CL) - the parameter refers to only periodical data streams and describes planned data transmission periods in the application layer;
- *Network synchronization* (NS) - the parameter describes whether a traffic stream is synchronized with network time in the application layer;
- *Data delivery guarantee* (DDG) - the parameter denotes limitations in application delivery by the network for intact operations. To select an appropriate QoS mechanism, the range of this particular feature is limited to the data transmission demands for an application. Three levels of this parameters are defined: time limitation, delay (latency) and bit rate. For traffic without any particular requirements, this parameter is replaced by the value: n.a.
- *Interference tolerance* (IT) - the parameter refers only to periodic data streams and describes the tolerance of an application to jitter;
- *Frame loss tolerance* (FLT) - the parameter describes tolerance of an application to the loss of successive packets in a stream;
- *Data size* (DS) - the parameter describes the size of data transmitted in Ethernet frames;
- *Criticality* (C) - the parameter describes the data criticality level for the operation necessary for part of the system and is used to determine appropriate QoS/TSN mechanisms for them.

Table 2 shows which of the proposed models and with what parameters can be used to model particular types of traffic distinguished during measurements in the test network.

Table 2. Juxtaposition of the values of the parameters that describe different types of traffic and their corresponding models

Type	P	CL	NS	DDG	IT	FLT	DS	C	Proposed model
I	Yes	0.12–2 ms	Yes	Time constraint	No	No	30–100	High	M1 with R
C	Yes	2–200 ms	No	Delay	Yes	Yes	50–1000	High	M2–M4
AaE	No	—	No	Delay	—	Yes	50–1500	High	M1–M4 with R, D
CaD	No	—	No	Bitrate	—	Yes	500–1500	Avg	M1–M4
NC	No	50 ms–1 s	No	Bitrate	Yes	Yes	50–500	High	M1–M4 with R
BE	No	—	No	N.a.	—	Yes	30–1500	Low	M1–M4
V	No	—	No	Delay	—	Yes	1000–1500	Low	M1–M4
A/V	No	—	No	Delay	—	Yes	1000–1500	Low	M1–M4

5 Conclusions

The analyses carried out by the authors to analyze traffic management mechanisms in industrial Ethernet networks prove a particular relationship between generating traffic and packet delays in the network. This fact has made the authors launch works on models of traffic sources for this particular type of network. The goal of the present study is to propose a number of models that would render the characteristics and properties of real traffic as precise as possible. These characteristics can be then successfully used in a test environment to evaluate effectiveness of traffic management mechanisms. Research has been conducted under project: *Research on network and protocol architecture for industry4.0/IIoT* on PUT in cooperation with Huawei Technologies Co., Ltd.

References

1. Al-Fuqaha, A., Guizani, M., Mohammadi, M., Aledhari, M., Ayyash, M.: Internet of Things: a survey on enabling technologies, protocols, and applications. IEEE Commun. Surv. Tutorials **17**(4), 2347–2376 (2015)
2. Da, X.L., He, W., Li, S.: Internet of Things in industries: a survey. IEEE Trans. Industr. Inf. **10**(4), 2233–2243 (2014)
3. Time-sensitive networking task group (2019). IEEE website
4. Time sensitive networks for flexible manufacturing testbad – description of converged traffic types. In: Industrial Interent Consortium (2018)
5. Staehle, D., Leibnitz, K., Tran-Gia, P.: Source traffic modeling of wireless applications. Würzburg, Report no. 261(2000)
6. Mandelbrot, B., Ness, J.: Long-run linearity, locally Gaussian processes, H-spectra and infinite variances. Int. Econ. Rev. **10**, 82–111 (1969)
7. Heffes, H., Lucantoni, D.: A Markov modulated characterization of packetized voice and data traffic and related statistical multiplexer performance. IEEE J. Sel. Areas Commun. **4**(6), 856–868 (1986)

Modelling of Switching Networks with Multi-service Sources and Multicast Connections in the Last Stage

Maciej Sobieraj$^{(\boxtimes)}$, Maciej Stasiak$^{(\boxtimes)}$, and Piotr Zwierzykowski$^{(\boxtimes)}$

Chair of Communication and Computer Networks,
Poznan University of Technology, ul. Polanka 3, 60-965 Poznań, Poland
{maciej.sobieraj,maciej.stasiak,piotr.zwierzykowski}@put.poznan.pl

Abstract. This article proposes a new calculation methods for a determination of traffic characteristics in switching networks with multi-service traffic sources carrying multicast traffic. In the switching networks were implemented two scenarios of selecting subsequent links which belongs to a given multicast connection. The analytical method described in the paper can be used to determine the point-to-point blocking probability in switching networks with multicast connections. In this method, each traffic source can generate calls of different service classes. The proposed calculation method is approximate, therefore the results of calculations are compared with the results of simulation experiments. The simulation results confirm acceptable accuracy of the proposed analytical method.

1 Introduction

Modern telecommunication networks support multi-service traffic, which consist of a mixture of traffic streams with different bitrates, including real-time streams generated, for example, by VoIP (Voice over Internet Protocol), DVOIP (Digital Video over Internet Protocol) or IPTV (Internet Protocol Television). Many streams related to the real-time services are implemented on the basis of multicast connections [1,2]. The multicast connection is based on duplication of transmitted streams in those nodes of the network that are closest to the recipients and, as a consequence, have a small impact on the network load. Therefore, in the paper is assumed that the multicast switches are placed in the last stage of the switching network (Fig. 1).

2 Analytical Model of the Switching Network

In the paper we assume that for an analytical determination of the traffic characteristics of multi-service switching networks we use models based on the approximation of a multidimensional Markov process by one-dimensional Markov chain

M. Choraś and R. S. Choraś (Eds.): IP&C 2019, AISC 1062, pp. 272–278, 2020.
https://doi.org/10.1007/978-3-030-31254-1_32

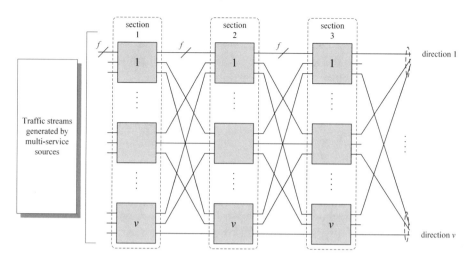

Fig. 1. Structure of switching network with multi-service sources

proposed in [3–6]. The switching networks described in the paper (Fig. 1) operates in accordance with the point-to-point algorithm, thus for determination of blocking probability for unicast traffic we will use PPBMT method extended for multi-service sources [3].

Let us assume that the switching network services a mixture of different multi-service traffic stream. In the model the following assumptions and parameters have been introduced [3,6]:

- The switching network carries three types of streams of Erlang, Engset and Pascal traffic.
- Each stream is generated by sources that belong to an appropriate set of traffic sources $\mathbb{Z}_{Er,i}$, $\mathbb{Z}_{En,j}$ and $\mathbb{Z}_{Pa,k}$.
- Sources that belong to the set $\mathbb{Z}_{Er,i}$ can generate Erlang call streams from the set $\mathbb{C}_{Er,i} = 1, 2, ..., c_{Er,i}$.
- Sources that belong to the set $\mathbb{Z}_{En,j}$ can generate Engset call streams from the set $\mathbb{C}_{En,j} = 1, 2, ..., c_{En,j}$.
- Sources that belong to the set $\mathbb{Z}_{Pa,k}$ can generate Pascal call streams from the set $\mathbb{C}_{Pa,k} = 1, 2, ..., c_{Pa,k}$.
- The total number of the sets of traffic sources in the system is $S = s_I + s_J + s_K$, where s_I are sets of traffic sources generating Erlang traffic streams, s_J are sets of traffic sources generating Engset traffic streams and s_K are sets of traffic sources generating Pascal traffic streams.
- The system services m traffic classes which belong to the set $\mathbb{M} = 1, 2, ..., m$.
- Each class c is defined by the number t_c of demanded AUs necessary to set up a new connection of class c.
- The parameter μ_c define the exponential distribution of the service time for calls of class c.

- The participation of class c (from the set $\mathbb{M}$) in the structure of traffic generated by sources from set $\mathbb{Z}_{\mathrm{Er},i}$, $\mathbb{Z}_{\mathrm{En},j}$ and $\mathbb{Z}_{\mathrm{Pa},k}$ is described by the parameter $\eta_{\mathrm{Er},i,c}$, $\eta_{\mathrm{En},j,c}$ and $\eta_{\mathrm{Pa},k,c}$, which for particular sets of traffic sources fulfils the following dependence:

$$\sum_{c=1}^{c_{\mathrm{Er},i}} \eta_{\mathrm{Er},i,c} = 1, \quad \sum_{c=1}^{c_{\mathrm{En},j}} \eta_{\mathrm{En},j,c} = 1, \quad \sum_{c=1}^{c_{\mathrm{Pa},k}} \eta_{\mathrm{Pa},k,c} = 1. \tag{1}$$

3   Multicast Connections

The article assumes that the multicast connection requires a setting-up of connection to all q desired output directions. If this is not possible, the connection cannot be serviced [7,8]. The total blocking probability for calls of a given class (E_T) consists of the internal blocking probability (E_{in}) and the external blocking probability (E_{ex}).

The internal blockage occurs when a new connection cannot be set-up in the switching network, while the external blockage appears when lack of free available resources in the required outgoing directions for servicing a call of a given class. The value of internal blocking probability in the proposed model can be calculated for both unicast and multicast calls on the basis of the PPBMT method [3]. The method of external blocking probability calculation depends on the adopted scenario of selecting subsequent links which belongs to a given multicast connection. In the first scenario the control algorithm of the switching network selects the appropriate output link in the required direction, after setting-up subsequent unicast connection belonging to a given multicast connection. This means that all selected output links in the required directions must belong to the same last stage switch. Therefore, the probability of external blocking of each subsequent component connection will be defined as the probability of blocking one output link (in the selected switch). Thus, the value of the external blocking probability can be calculated on the basis of model of limited availability group [3].

In the second scenario the control algorithm first selects the switches of the last stage of the switching network in which is possible to set-up all q_c component connections of a multicast connection of class c. In the next step the algorithm try to set-up connections to the subsequent switches of the last stage of switching network. The external blockage occurs when we cannot find the switch in the last stage of the switching network, which have no less then q_c free outgoing links for servicing a call of a given class. In accordance with the adopted blocking definition, the external blocking probability for a multicast call of class c can be expressed in the following form:

$$E_{ex} = 1 - P(c, q_c), \tag{2}$$

where $P(c, q_c)$ can be determined by the distribution of links in the limited-availability group [3].

4   Numerical Evaluation

The analytical method proposed in the paper is an approximate method. There-fore, the results of the calculation of the blocking probability in the switching network with multi-service sources and multicast connections were compared with the results of the simulation experiments. The simulations experiments were conducted for an exemplary 3-stage switching network with the Clos struc-ture which were composed of switches that had $v \times v$ links (Fig. 1) [9].

The results of calculation of blocking probability in the switching network with multi-service sources and unicast and multicast connections have been com-pared with the results of simulation experiments. The simulations experiments were conducted for an exemplary 3-stage switching networks in the Clos struc-ture (Fig. 1)[1].

The research has been conducted for the following two structures of switching networks:

System 1:
- $k = 4$, $f = 30$ AU, $V = 120$ AU (capacity of the system),
- structure of offered traffic (traffic classes): $m = 4$, $t_1 = 1$ AU, $\mu_1^{-1} = 1$, $t_2 = 3$ AU, $\mu_2^{-1} = 1$, $t_3 = 5$ AU, $\mu_3^{-1} = 1$, $t_4 = 7$ AU, $\mu_4^{-1} = 1$.
- structure of offered traffic (sets of traffic sources): $S = 2$, $\mathbb{C}_{\text{Er},1} = 1, 2, 4$, $\eta_{\text{Er},1,1} = 0.5$, $\eta_{\text{Er},1,2} = 0.4$, $\eta_{\text{Er},1,4} = 0.1$, $\mathbb{C}_{\text{En},2} = 1, 3, 4$, $\eta_{\text{En},2,1} = 0.6$, $\eta_{\text{En},2,2} = 0.3$, $\eta_{\text{En},2,3} = 0.1$, $N_{\text{En},2} = 800$;
- unicast: classes 1, 2 and 3;
- multicast: class 4, $q_4 = 2$.

System 2:
- $k = 4$, $f = 32$ AU, $V = 120$ AU (capacity of the system),
- structure of offered traffic (traffic classes): $m = 4$, $t_1 = 1$ AU, $\mu_1^{-1} = 1$, $t_2 = 4$ AU, $\mu_2^{-1} = 1$, $t_3 = 6$ AU, $\mu_3^{-1} = 1$, $t_4 = 8$ AU, $\mu_4^{-1} = 1$.
- structure of offered traffic (sets of traffic sources): $S = 2$, $\mathbb{C}_{\text{Er},1} = 1, 2, 3$, $\eta_{\text{Er},1,1} = 0.5$, $\eta_{\text{Er},1,2} = 0.3$, $\eta_{\text{Er},1,3} = 0.2$, $\mathbb{C}_{\text{Pa},2} = 1, 3, 4$, $\eta_{\text{Pa},2,1} = 0.6$, $\eta_{\text{Pa},2,2} = 0.3$, $\eta_{\text{Pa},2,3} = 0.1$, $N_{\text{Pa},2} = 900$;
- unicast: classes 1, 2 and 3;
- multicast: class 4, $q_4 = 2$.

Figures 2, 3, 4 and 5 show the results of the calculation and the simulation for two structures of the exemplary switching networks and for two scenarios of selecting subsequent links which belongs to a given multicast connection. Simulation results are presented in the form of charts with confidence intervals stated at the 95% confidence level. The confidence intervals have been calculated in accordance with the t-Student distribution for 5 series with 1,000,000 calls of each class.

---

[1] Similar switching networks are used in electronic and optical switching.

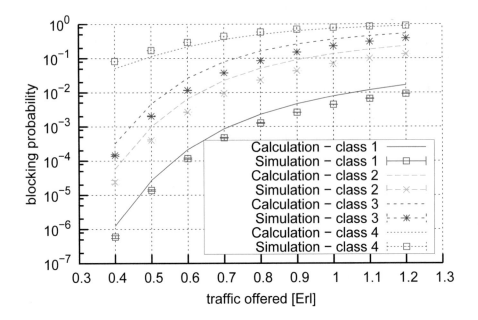

Fig. 2. Blocking probability for calls of particular traffic classes in System 1 and the first scenario

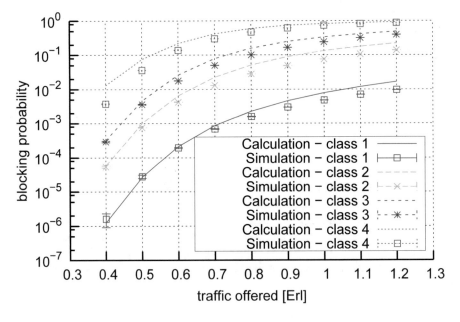

Fig. 3. Blocking probability for calls of particular traffic classes in System 1 and the second scenario

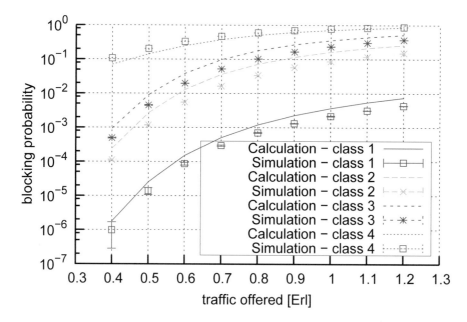

Fig. 4. Blocking probability for calls of particular traffic classes in System 2 and the first scenario

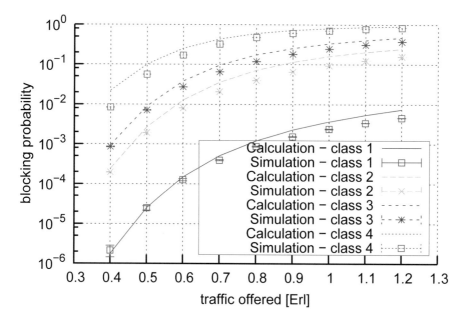

Fig. 5. Blocking probability for calls of particular traffic classes in System 2 and the second scenario

5   Conclusion

In the article the new analytical method for point-to-point blocking probability in switching networks with multi-service traffic sources and unicast and multicast connections has been proposed. The proposed method allows us to determine blocking probability in the switching network with two scenarios of selecting subsequent links which belongs to a given multicast connection. The evaluation of the accuracy of the proposed method was done by a comparison of the analytical results with the results of the simulation experiments. The results indicated fair accuracy of the proposed method.

References

1. Parniewicz, D., Stasiak, M., Zwierzykowski, P.: Analytical model of the multi-service cellular network servicing multicast connections. Telcommun. Syst. **52**(2), 1091–1100 (2013)
2. Stasiak, M., Zwierzykowski, P.: Analytical model of ATM node with multicast switching. In: Proceedings of MELECON 1998 - 9th Mediterranean Electrotechnical Conference, vols. 1 and 2, pp. 683–687 (1998). https://doi.org/10.1109/MELCON.1998.699302
3. Głąbowski, M., Sobieraj, M.: Analytical modelling of multiservice switching networks with multiservice sources and resource management mechanisms. Telecommun. Syst. **66**(559), 559–578 (2017). https://doi.org/10.1007/s11235-017-0305-4
4. Głąbowski, M., Sobieraj, M., Stasiak, M.: Analytical modeling of multi-service systems with multi-service sources. In: Proceedings of 16th Asia-Pacific Conference on Communications, pp. 285-290 (2010)
5. Głąbowski, M., Sobieraj, M., Stasiak, M.: A full-availability group model with multi-service sources and threshold mechanisms. In: Proceedings of the 8th International Symposium on Communication Systems, Networks & Digital Signal Processing, Poznan, Poland, pp. 1–5 (2012)
6. Głąbowski, M., Sobieraj, M., Stasiak, M.: Modeling switching networks with multi-service sources and point-to-group selection. In: Asia-Pacific Conference on Communications (APCC), Jeju Island, pp. 686–691 (2012). https://doi.org/10.1109/APCC.2012.6388282
7. Parniewicz, D., Stasiak, M., Zwierzykowski, P.: Traffic engineering for multicast connections in multiservice cellular networks. IEEE Trans. Ind. Inf. **9**(1), 262–270 (2013)
8. Stasiak, M., Zwierzykowski, P.: Point-to-group blocking in theswitching networks with unicast and multicast switching. Perform. Eval. **48**(1/4), 247–265 (2002)
9. Kleban, J., Sobieraj, M., Weclewski, S.: The modified MSM Closswitching fabric with efficient packet dispatching scheme. In: Workshop on High Performance Switching and Routing, pp. 1–6 (2007)

The Analytical Model of Complex Non-Full-Availability System

Sławomir Hanczewski[✉], Maciej Stasiak, and Michał Weissenberg

Faculty of Electronics and Telecommunications,
Poznan University of Technology, Poznań, Poland
slawomir.hanczewski@put.poznan.pl

Abstract. This paper presents an analytical model of a complex non-full-availability system. In systems of this type, non-full availability resources consist of a certain number of component elements, each of which being a non-full-availability system. During the study, the results obtained on the basis of the proposed model were compared with the results obtained in simulation experiments, which confirmed the correctness of the adopted theoretical assumptions. The model can be used to model complex networks/systems, such as cloud computing systems or data centers.

1 Introduction

The development of network technologies and solutions over recent years has been truly remarkable and enormously facilitated access to networks and, what follows, to a wide range of offered services that can now be accessed from nearly any place. Whether wired or wireless networks are used in the process is not relevant any more. From the point of view of the user, the most rapid advance of network technologies is to be observed in mobile networks. The reports published by institutions that are engaged in analysing the network market as well as those of the producers of network devices are at one when they acknowledge this trend [1,2]. Wide and easy availability of network access makes the user independent of his or her location that from now has ceased to impose any limitations. The user can have access to a server that is located virtually in any part of the world, while the required QoS (Quality of Service) parameters will still be provided. Thanks to this, a construction of distributed web-based systems has become feasible to carry out. For these solutions to function properly, however, appropriate mechanisms that would control the infrastructure in such a way as to provide available resources to be used in as optimum way as possible are needed. To achieve this goal, virtualization is used to make available resources to as large group of users as possible on the one hand, while the application of algorithms for load equalization is used, on the other. The aforementioned mechanisms, though improve the effectiveness of the use of network resources (available transmission speed, server resources) make systems become, from the user's perspective, non-full-availability systems, as the user can have access to

© Springer Nature Switzerland AG 2020
M. Choraś and R. S. Choraś (Eds.): IP&C 2019, AISC 1062, pp. 279–286, 2020.
https://doi.org/10.1007/978-3-030-31254-1_33

only parts of available resources and not to all offered resources. A good example of the above is a demand to create a new virtual machine. The admission of a call of this type for service involves reservation of appropriate resources: necessary disk space, the number of processors, RAM memory and an access link. The physical infrastructure of a cloud computing system can have far more resources than those needed for a call to be serviced. There is, however, the essential condition for access to resources, i.e. the resources to be utilized have to be available in one physical server. This problem is visually presented in Fig. 1. A new call that demands a creation of a new virtual machine requires a certain defined amount of resources. The available resources of servers are represented in the figure by empty rectangles, whereas the resources occupied by the virtual machines that have been created earlier are denoted by VM rectangles. In total, the servers have resources capable of creating two such machines, but in fact it is only one server (the middle one) that has enough resources for a virtual machine to be created.

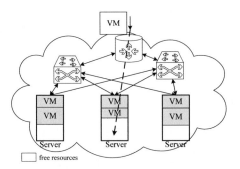

Fig. 1. Available resources in cloud computing infrastructure

Non-full-availability network systems have been addressed by researchers in traffic engineering for a large number of years. The literature of the subject offers a number of analytical models of these systems. They include Limited Availability Group – LAG model [3] and Erlang's Ideal Grading – EIG [4,5]. The limited-availability group is an analytical model of a group of separated network links and was successfully used to model, for example, a group of cells or multi-service switching networks [6]. The EIG group, in turn, was a starting point in the development of such systems as switching networks [7], overflow systems [8] and self-optimizing mobile networks [9].

From the point of view of traffic engineering, non-full-availability systems belong to the group of state-dependent systems. Systems of this type are subject to intensive scrutiny in modern research. The results of relevant studies make it possible to construct analytical models that provide more and more complicated and complex network solutions. These solutions are already available to users of present-day networks or will become available in the near future.

This paper is structured as follows. Section 2 presents topical issues related to traffic offered in modern telecommunications and computer network. Section 3 includes basic information on modeling state-dependent systems. The next section (Sect. 4) presents the architecture and the analytical model of a complex non-full-availability systems. Section 5 provides a number of examples of the results that were obtained using the analytical model proposed in the paper. A brief summary of the paper is presented in Sect. 6.

2   Traffic in Telecommunications and Computer Networks

The initial assumption adopted in our investigations, the results of which are presented in this paper, was that a non-full-availability system was offered m call classes. Calls of individual call classes create a Poisson stream with the parameters $\lambda_1, ..., \lambda_i, ...\lambda_m$. Service time is described by the exponential distribution with the parameters $\mu_1, ..., \mu_i, ...\mu_m$. Traffic offered by calls of class i is then equal to:

$$A_i = \lambda_i/\mu_i. \tag{1}$$

Moreover, each class demanded an appropriate number of *allocation units* (AU): $t_1, ..., t_i, ..., t_m$ specific for every class. The allocation unit is a unit of the capacity of a system and the demands specific for classes that are offered to a system under consideration and is determined on the basis of a discretization process [10]. In addition, each class has a related parameter, i.e. availability $d_1, ..., d_i, ..., d_m$. This is a parameter that is characteristic for Erlang's Ideal Grading and its significance is explained in Sect. 4.2.

3   State-Dependent Systems

In traffic engineering, a state-dependent system is defined as such a system in which admission of new calls for service is made dependent on the number of occupied allocation units. The dependence can be a result of a limited number of traffic sources – in this case, we refer to them as systems with state-dependent call stream. This dependence can also result from the structure of a system or from the algorithm used for the operation of the call admission function. Such systems are called systems with state-dependent service stream. Multi-service state-dependent systems, similarly as in the case of full-availability systems, can be approximated by a one-dimensional Markov process. In the case of full-availability systems, the transition between two states depends only on the intensity of the call stream of a given class. In state-dependent systems this transition additionally depends on the so-called conditional transition coefficient $\sigma_i(n)$ (where i is the number of a call class, whereas n is the number of occupied AUs in the system). It is proved in [11] that in the case of systems in which state-dependence is related to a number of different sources and these sources are independent from one another, the total conditional transition coefficient can be determined using the following equation:

$$\sigma_i(n) = [\sigma_i(n)]^R[\sigma_i(n)]^A[\sigma_i(n)]^P, \tag{2}$$

where:

- $[\sigma_i(n)]^R$ – conditional transition coefficient of class i in occupancy state n resulting from the properties of a call stream of this class,
- $[\sigma_i(n)]^A$ – conditional transition coefficient of class i in occupancy state n resulting from the structure of a system,
- $[\sigma_i(n)]^P$ – conditional transition coefficient of class i in occupancy state n resulting from the function of call admission control for in the system.

Knowing the values of all parameters $\sigma_i(n)$, the occupancy distribution in the system can be determined using the following formula [11]:

$$n[P_n]_V = \sum_{i=1}^{m} a_i t_i [\sigma_i(n)]_V [P_{(n-t_i)}]_V, \tag{3}$$

where:

- V is the capacity of the system, expressed in AU,
- $[P_n]_V$ is the occupancy probability n AU in the system.

4   Complex Non-Full-Availability System

The non-full-availability system under consideration is composed of k independent resources (eg. links). Each of the resources, with the capacity of f AUs, shows the properties of an EIG group. The total capacity of the system is:

$$V = kf. \tag{4}$$

A simplified diagram of the structure of the system is presented in Fig. 2a. It is easily observable that the system is a combination of two non-full-availability systems known from the literature, i.e. a Limited-Availability Group (LAG) [3] and an Erlang's Ideal Grading (EIG) [4]. The explanation of the operation of the proposed model should be, however, preceded with a short description of the properties of both groups.

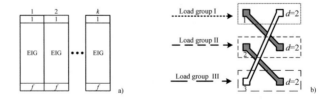

Fig. 2. Model of complex non-full-availability system (a) structure, (b) structure of Erlang's Ideal Grading ($V = 3$ AUs, $d = 2$ AUs)

4.1 Limited-Availability Group

The LAG is a model of separated full-availability network resources with a given capacity f AUs to which call streams with common source population are offered. Calls can be admitted for service only when they can be serviced by a single resource. This means that even when the number of free resources in the group is equal or higher than the number of AUs demanded by a call, this does not guarantee that this call can be admitted for service. It is so because unoccupied AUs can be distributed in individual separated resources in such an unfavorable way that in each of the resources there is not a sufficient number of AUs that are demanded by this new call. Therefore, this fact has to be taken into consideration in a formula that determines the conditional transition coefficient (Eq. (2)). In keeping with [3], the parameter $[\sigma_i(n)]_V^{\mathrm{LAG}}$ for this type of group can be determined on the basis of the following formula:

$$[\sigma_i(n)]_V^{\mathrm{LAG}} = \frac{F(V - n, k, f) - F(V - n, k, t_i - 1)}{F(V - n, k, f)}, \tag{5}$$

where the function $F(x, k, f)$ is given by the formula below:

$$F(x, k, f) = \sum_{i=0}^{\lfloor \frac{x}{f+1} \rfloor} (-1)^i \binom{k}{i} \binom{x + k - 1 - i(f+1)}{k-1}. \tag{6}$$

4.2 Erlang's Ideal Grading

In the case of the EIG, calls have no access to all f AUs in a group. Calls have access to a certain number of AU that is called availability d_i. Calls that have access to the same AUs in the group are called the load group. The structure and concept of availability in the EIG group is shown in Fig. 2b. Similarly as in the case of the LAG group, the number of free AUs that is equal to or higher than the number of AUs demanded by a given call does not guarantee this call to be admitted for service. For calls to be admitted for service, unoccupied AUs have to be available to a new arriving call. By taking this fact into consideration, the conditional transition coefficient can be determined according to the following formula [5]:

$$[\sigma_i(n)]_f^{\mathrm{EIG}} = 1 - \sum_{d_i - t_i + 1}^{z} \binom{d_i}{x} \binom{V - d_i}{n - x} \Big/ \binom{V}{n}, \tag{7}$$

where:
$z = n - t_i$, if $(d_i - t_i + 1) \le (n - t_i) < d_i$,
$z = d_i$, if $(n - t_i) \ge d_i$.

4.3 Analytical Model of a Complex Non-Full-Availability System

In keeping with the adopted structure of the proposed non-full-availability system, a new call will be admitted for service only when the conditions for admittance resulting from the properties of LAG and EIG groups will be satisfied concurrently. This means then, that a decision as to whether a new call can be admitted or not will be taken in two steps. The first step involves checking if the number of unoccupied AUs in at least one of the subsystems is equal or higher than the number of demanded AUs required for the call under investigation to be serviced. If such a subsystem does exist, then the following step involves determining whether free AUs belong to the load group relevant to the new oncoming call. From the point of view of the service process that occurs in the system, the transition between the states resulting from the admission of a new call for service will be influenced by both the properties of the LAG group and the EIG group. By taking into account the conclusions proposed in [11], we can state that the total conditional transition coefficient in the considered system can be determined in the following way:

$$[\sigma_i(n)]_V = [\sigma_i(n)]_V^{LAG}[\sigma_i(\lfloor n/k \rfloor)]_f^{EIG}. \tag{8}$$

The difference in the state number for which calculations of the parameters $[\sigma(n)_i]_V$ in LAG and EIG are made results from the fact that the subsystems that behave like EIG groups have the capacity equal to f. Therefore, it is necessary to rescale the occupancy state in the whole of the system n to the occupancy state in a single sub-system with the capacity f. The assumption of uniform loads of resources in the system means that the occupancy n AUs in the system corresponds to and represents the occupancy n/k AUs in a single component resource.

When we know the values of the parameters σ for all call classes and states in the system, it is possible to determine the occupancy distribution using Eq. (3). Once we know the occupancy distribution, the blocking probability in the system can be determined using the following formula:

$$E_i = \sum_{n=0}^{V-t_i}[P_n]_V[1-[\sigma_i(n)]_V] + \sum_{n=V-t_i+1}^{V}[P_n]_V. \tag{9}$$

5   Numerical Examples

To verify the operation of the analytical model proposed in the paper the results obtained with the model were compared with the results obtained in a digital simulation. For this purpose, a simulation model of the considered non-full-availability system was developed and implemented in the C++ language. The event scheduling approach was used in the simulator. To determine a single measurement, 5 series of simulation were conducted, each with 1,000,000 calls of the class that demanded the highest number of AU for service. The obtained results are plotted in the function of traffic offered to a single AU in the system:

$$a = \sum_{i=1}^{m} a_i t_i / V. \tag{10}$$

Another assumption was that the total offered traffic was divided between individual call classes in the following proportions: $a_1 t_1 : ... : a_m t_m = 1 : ... : 1$.

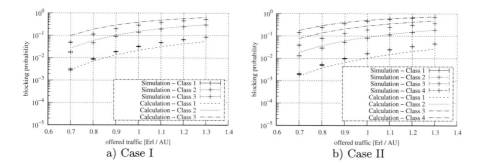

Fig. 3. Blocking probability in complex non-full-availability system

Figure 3 shows the results for the system: $k = 3$, $f = 20$ AU. In the first case, the system was offered $m = 3$ call classes with the parameters: $t_1 = 1$ AU, $d_1 = 15$ AUs, $t_2 = 2$ AUs, $d_2 = 15$ AUs, $t_3 = 3$ AUs, and $d_3 = 15$ AUs. In the second case shows the results for the case $m = 4$, $t_1 = 1$ AU, $d_1 = 15$ AUs, $t_2 = 2$ AUs, $d_2 = 15$ AUs, $t_3 = 3$ AUs, $d_3 = 15$ AUs, $t_4 = 4$ AUs, and $d_4 = 15$ AUs. The proposed analytical model is an approximate model. However, the presented results of the investigation clearly indicate that the model approximates the non-full-availability system under consideration considerably well. In the calculations, the most significant error was made for the class that demanded the highest number of AUs. This error lies in a certain underestimation of the blocking probability. The perceived repeatability of this error, however, creates feasible grounds for researchers to develop of an appropriate correction. The confirmed accuracy of the model makes it possible to use it to model real non-full-availability systems, such as cloud computing. This will be possible only when appropriate equations that would allow availability in such systems to be determined properly will be proposed. Further research on this will be undertaken by the present authors.

6   Conclusions

This paper proposes an architecture of a complex non-full-availability system and its corresponding analytical model. A comparison of the results obtained using the analytical model and those obtained during a digital simulation show high accuracy of the proposed model. The errors in approximation introduced by the model are of fixed nature, which provides good grounds for future development that would include a correction to the model and would make adjustments to obtained results. The present authors are of the opinion that the presented model can be used to model real systems such as those operating in a computational cloud [12]. The model itself is easy in its implementation, while the time needed for calculations is indeed very low.

Acknowledgments. The presented work has been funded by the Polish Ministry of Science and Higher Education within the status activity task "Structure, analysis and design of modern switching system and communication networks" (08/82/SBAD/8229) in 2019.

References

1. CISCO: Cisco virtual networking index: forecast and methodology, 2015–2020 white paper. Technical report (2016)
2. Internet Society: Internet society global internet report 2015 mobile evolution and development of the internet. Technical report (2015)
3. Stasiak, M.: Blocking probability in a limited-availability group carrying mixture of different multichannel traffic streams. Ann. Telecommun. **48**(1–2), 71–76 (1993)
4. Brockmeyer, E., Halstrom, H., Jensen, A.: The life and works of A.K. Erlang. Acta Polytechnika Scand. **6**(287), 138–155 (1960)
5. Głąbowski, M., Hanczewski, S., Stasiak, M., Weissenberg, J.: Modeling Erlang's ideal grading with multi-rate BPP traffic. Math. Probl. Eng. **2012**, 35 (2012). Art ID 456910
6. Głąbowski, M.: Recurrent calculation of blocking probability in multiservice switching networks. In: Asia-Pacific Conference on Communications, Busan, pp. 1–5 (2006)
7. Hanczewski, S., Sobieraj, M., Stasiak, M.: The direct method of effective availability for switching networks with multi-service traffic. IEICE Trans. Commun. **E99–B**(6), 1291–1301 (2016)
8. Głąbowski, M., Hanczewski, S., Stasiak, M.: Modelling of cellular networks with traffic overflow. Math. Probl. Eng. **2015**, 15 (2015). Art ID 286490
9. Głąbowski, M., Hanczewski, S., Stasiak, M.: Modelling load balancing mechanisms in self-optimising 4G mobile networks with elastic and adaptive traffic. IEICE Trans. Commun. **E99–B**(8), 1718–1726 (2016)
10. Roberts, J., Mocci, U., Virtamo, J.: Broadband network teletraffic. In: Final Report of Action COST 242. Commission of the European Communities. Springer, Berlin (1996)
11. Głąbowski, M., Kaliszan, A., Stasiak, M.: Modeling product-form state-dependent systems with BPP traffic. Perform. Eval. **67**, 174–197 (2010)
12. Hanczewski, S., Weissenberg, M.: Concept of an analytical model for cloud computing infrastructure. In: 11th International Symposium on Communication Systems, Networks & Digital Signal Processing, Budapest, pp. 1–4 (2018)

Model of a Multiservice Server
with Stream and Elastic Traffic

Sławomir Hanczewski[✉], Maciej Stasiak, and Joanna Weissenberg

Faculty of Electronics and Telecommunications,
Poznan University of Technology, Poznań, Poland
slawomir.hanczewski@put.poznan.pl

Abstract. This paper discusses an approximate analytical model of a
telecommunications system to which a mixture of stream and elastic
traffic is offered. These two types of traffic are characteristic for modern
networks. The first one is typical for services that require fast and reliable
data transfer. The other is characteristic for typical public services known
from the Internet.

1 Introduction

A present-day multiservice networks are packet networks in which packet streams
that are matched with appropriate services available in the network can undergo
the influence of different network mechanisms, such as threshold and threshold-
less compression, resource reservation, traffic overflow and call redirection, and
priorities for selected call classes and services [1]. In networks that are based on
the TCP/IP protocols, the traffic compression mechanisms is widely used. This
mechanism is activated when all resources of a given system become occupied.
In these circumstances, an arrival of a new packet stream triggers activation of
a given mechanism and is followed by a decrease in bitrate of currently serviced
streams, which in consequence leads to a release of free resources capable of
servicing this new stream. Generally, to transmit all data, a decrease in bitrate
is followed by an increase in service time for a given stream. Traffic shaped by
the operation of the compression mechanism in the way as described above is
defined as elastic traffic (e.g. FTP services, email or the like services that use
the TCP protocol). If a decrease in bitrate is not accompanied by an increase
in service time, then this traffic is defined as adaptive traffic (e.g. real time services that use the RTP protocol). In the majority of cases real network systems
have to handle both concurrently, i.e. simultaneously elastic traffic, that under-
goes compression, and stream traffic, which is not influenced by the compression
mechanism. Stream traffic is typical for those services that require constant
bitrate in data transmission, and, what follows, low latency, e.g. remote surgery
and examination.

Multiservice systems are mainly analyzed at a certain level of generality,
called the call level (with other notion used as well, such as stream or flow, e.g.
[7,15]). A call is defined at this level as a packet stream with constant bitrate

© Springer Nature Switzerland AG 2020
M. Choraś and R. S. Choraś (Eds.): IP&C 2019, AISC 1062, pp. 287–294, 2020.
https://doi.org/10.1007/978-3-030-31254-1_34

(CBR) that can be determined on the basis of the maximum bitrate of real packet streams with variable bitrate (VBR), or on the basis of equivalent bitrates EB (Equivalent Bandwidth) [2,3]. A choice of these constant bitrates results from an adopted method of network dimensioning and has no influence on the structure of a model. It is pointed out in the literature that it is possible to approximate traffic at the call level in TCP/IP networks by Poisson streams, e.g. [3].

The problem of the analysis of multiservice servers (called in traffic engineering groups or full-availability resources) with bitrate compression has been addressed widely in the literature over the recent years. All those publications have broached the subject of stream traffic and elastic traffic separately. The first model for multiservice systems with stream traffic was proposed in [4,5]. [6] proposes the first model of a multiservice server with finite compression for elastic traffic. The notion of finite compression means that the bit rate of a given call can be decreased exclusively within certain borders (limits). The distribution of serviced resources of a server is determined in the model on the basis of the algorithm of balanced fairness resource allocation [16]. The "balanced fairness" algorithm results from the reversible Markov process, and in consequence leads to a simple description of a full-availability server with elastic traffic at the macrostate [6,8] and microstate level [10]. [12] analyses a server with infinite compression for elastic traffic in which, in the case of a lack of free resources in the server, calls always decrease their bitrate. In this system the phenomenon of blocking in servers never occurs, since, with large loads in the server, the bit rate of traffic streams tends towards zero. In [7], the model [6] is generalized to include the case of elastic and adaptive traffic service. The occupancy distribution proposed in [7] also provides a basis for a large number of models with elastic and adaptive traffic that have been developed for different traffic streams and a variety of limitations (constraints) imposed on calls that are handled in the network [13,14]. [9] and [10] propose a number of multiservice queuing models for stream traffic, while [11] proposes a multiservice model of a queuing system that services elastic and adaptive traffic.

This paper presents a new approximate model of a multiservice server that services a mixture of elastic and stream traffic. The model is based on an approximation of the service process of elastic and stream traffic by a reversible two-dimensional Markov process.

The paper is structured as follows. Section 2 proposes a model of a system that services elastic and stream traffic, while Sect. 3 presents a comparison of the results of the analytical modelling with the results of a digital simulation. In Sect. 4, which is a summary of the article, the most important conclusions that result from the study are formulated.

2   Model of a Server with Elastic and Stream Traffic

In considerations devoted to multiservice networks, e.g. Internet networks, traffic is most frequently expressed in AUs (Allocation Units) [3]. Such an approach leads, however, to a certain disregard for packet streams with variable bit rate

VBR, and in consequence the latter are represented by calls, i.e. packet streams with constant bitrate CBR, determined mainly on the basis of the maximum bitrates of VBR calls or on the basis of other heuristic algorithms employed [3]. With CBRs that have been determined earlier, it is possible to determine the value of AU. This value is defined as such value of bitrate that a given set of CBRs for calls of individual classes (called demands in traffic engineering) is its multiple number. The maximum value of the allocation unit can be calculated as the GCD (Greatest Common Divisor) of all CBRs of calls offered to the system.

Let us assume that a server is offered M_1 call classes of stream traffic and M_2 call classes of elastic traffic. In the case of a lack of compression, offered traffic can be described by the following parameters:

- M the number of all traffic classes of calls offered in the system under consideration ($M = M_1 + M_2$),
- λ_i call stream intensity of class i ($1 \leq i \leq M$),
- μ_i service intensity of calls of class i ($1 \leq i \leq M$),
- c_i demanded bitrate (CBR) for a call of class i
- c_{AU} maximum bitrate of AU:

$$c_{AU} = \text{GCD}(c_1, c_2, ..., c_M). \tag{1}$$

- t_i demanded number of AUs necessary for a call of class i, in occupancy state of the system n AUs, to be executed:

$$t_i = \lceil c_i / c_{AU} \rceil \tag{2}$$

- A_i average traffic intensity for traffic of class i (in relation to calls):

$$A_i = \lambda_i / \mu_i, \tag{3}$$

- C the total bitrate of the server (capacity of the system in bps),
- V the total capacity of the server in AUs:

$$A_i = \lfloor C / c_{AU} \rfloor. \tag{4}$$

Let us consider now a multiservice server [8] with the real capacity C_r and virtual capacity C_v. Service process in the server can be described as follows: if there is not enough resources in the server to service a new call, then all currently serviced elastic calls are compressed so the new call can be serviced.

To simplify the analysis of the system, the notion of total, virtual capacity is introduced V_v ($V_v > V_r$). The assumption is that calls of the elastic type can be compressed for as long as the number of occupied AUs in the server (understood to be the sum of non-compressed demands of calls of all classes) exceeds the virtual capacity. Otherwise, a new call will be rejected. These occupancy states n, in which $V_r < n \leq V_v$, determine the area of potential compression of elastic traffic (Fig. 1). The adopted visualisation of the system means that in the real area ($0 \leq n \leq V_r$) calls of elastic and stream classes can be serviced, whereas in the virtual area ($V_r < n \leq V_v$) only calls of elastic classes can be serviced. The

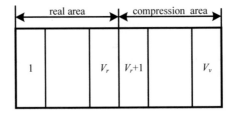

Fig. 1. Visualisation of the multiservice server with elastic and stream traffic

initial assumption is that the server services stream traffic only. In this case, the occupancy distribution in the system with the capacity V_r can be determined as follows [4,5]:

$$n_1[P(n_1)]_{V_r} = \sum_{i=1}^{M_1} A_i t_i [P(n_1 - t_i)]_{V_r}, \tag{5}$$

where $[P(n_1)]_{V_r}$ is the occupancy probability n_1 AUs. Our further assumption is that the server services elastic traffic only. The occupancy distribution in the system with the capacity V_v can be determined as follows [6]:

$$(n_2)^*[P(n_2)]_{V_v} = \sum_{i=1}^{M_2} A_i t_i [P(n_2 - t_i)]_{V_v}, \tag{6}$$

where $[P(n_2)]_{V_v}$ is the occupancy probability n_2 AUs. The parameter expresses the following condition:

$$(n_2)^* = \min(n_2, V_r). \tag{7}$$

Our next assumption now is that the server services a mixture of stream and elastic traffic, whereas the service process is a reversible process. In these circumstances, we can write, in line with the assumptions adopted in [6] the following occupancy distribution:

$$(n_1 + n_2)^*[P(n_1, n_2)]_{V_v} =$$
$$\sum_{i=1}^{M_1} A_i t_i [P(n_1 - t_i, n_2)]_{V_r} + \sum_{j=1}^{M_2} A_j t_j [P(n_1, n_2 - t_j)]_{V_v}, \tag{8}$$

where $[P(n_1, n_2)]_{V_v}$ is the occupancy distribution in which n_1 AUs are occupied by calls of the stream classes and n_2 AUs are occupied by calls of the elastic classes. The parameter expresses the following condition:

$$(n_1 + n_2)^* = \min(n_1 + n_2, V_r). \tag{9}$$

The probabilities will be written in the strength of the adopted reversibility on the right side of Eq. (8) in the product form:

$$(n_1 + n_2)^* [P(n_1, n_2)]_{V_v} =$$

$$[P(n_2)]_{V_v} \sum_{i=1}^{M_1} A_i t_i [P(n_1 - t_i)]_{V_r} + [P(n_1)]_{V_r} \sum_{j=1}^{M_2} A_j t_j [P(n_2 - t_j)]_{V_v}, \quad (10)$$

Taking into consideration (5) and (6), the distribution (10) can eventually be written in the following form:

$$[P(n_1, n_2)]_{V_v} = \frac{n_1 + (n_2)^*}{(n_1 + n_2)^*} [P(n_1)]_{V_r} [P(n_2)]_{V_v}. \quad (11)$$

Since the distributions $[P(n_1)]_{V_r}$, $[P(n_2)]_{V_v}$ have different length, then the distribution has to be normalised:

$$[P(n_1, n_2)]_{V_v} = \frac{[P(n_1, n_2)]_{V_v}}{1 - \sum_{n_1+n_2>V_v}^{V_r} [P(n_1, n_2)]_{V_v}}. \quad (12)$$

Following the determination of the occupancy distribution $[P(n_1, n_2)]_{V_v}$ the blocking probability for each call class of the stream type can be determined:

$$E(i) = \sum_{(n_1,n_2)=\Omega(i)} [P(n_1, n_2)]_{V_v}, \quad (13)$$

and, likewise, the blocking probability for each class of the elastic type:

$$E(j) = \sum_{(n_1,n_2)=\Omega(j)} [P(n_1, n_2)]_{V_v}, \quad (14)$$

where $\Omega(i)$, $\Omega(j)$ are sets of blocking states for calls of class i of the stream type and call of class j of the elastic type:

$$\Omega(i) = \{(n_1, n_2) \in \Omega : n_1 + n_2 \geq V_v - t_i + 1\} \cup$$
$$\cup \{(n_1, n_2) \in \Omega : \frac{n_1}{t_1} \in I \wedge n_2 = 0\} \cup$$
$$\cup \{(n_1, n_2) \in \Omega : \frac{n_1}{t_1} \in I \wedge n_1 \geq V_r - t_i \wedge n_2 \neq 0\}, \quad (15)$$

$$\Omega(j) = \{(n_1, n_2) \in \Omega : n_1 + n_2 \geq V_v - t_j + 1\} \cup$$
$$\cup \{(n_1, n_2) \in \Omega : \frac{n_1}{t_1} \in I \wedge n_2 = 0\}, \quad (16)$$

where Ω is the set of all possible states in the system.

The presented model is an approximate model, since the probability distributions for particular types of traffic in (11) are determined independently. This means that in the case of the traffic classes that undergo compression the occupancy probabilities $[P(n_2)]_{V_v}$ are determined with the assumption that the real capacity of the system V_r is constant. Such an approach causes the probability $[P(n_1, n_2)]_{V_v}$ of such a state that $(n_1 + n_2 > V_r)$ to be determined on the basis of a normalised probability product in the proposed model, in which the normalisation coefficient and the probability of state occupancy by calls that undergo compression do not take into consideration the number of resources occupied by calls of stream classes, and in consequence any changes in the number of available real resources of the system for the traffic classes that undergo compression. The approximate character of the proposed model stems indeed from this particular fact.

3   Numerical Example

The proposed model of a multiserver that services a mixture of elastic and stream traffic is an approximate model. To determine its usefulness in modelling network systems the results of the analytical calculations were compared with the results of simulations for a selected number of parameters of the server. A system with the following parameters was chosen for the analytic modelling:

- real server capacity: $C_r = 20$ Mbps, $V_r = 20$ AUs, $c_{AU} = 1$ Mbps,
- virtual server capacity: $C_v = 40$ Mbps, $V_v = 40$ AUs, $c_{AU} = 1$ Mbps,
- classes of stream traffic: $M_1 = 1$, $t_1 = 1$ AU,
- classes of elastic traffic: $M_2 = 2$, $t_2 = 1$ AU, $t_3 = 2$ AUs,
- proportions of offered mixture of traffic (no compression): $A_1 t_1 : A_2 t_2 : A_3 t_3 = 1 : 1 : 1$.

Figure 2 shows the results of modelling the server under investigation for different virtual capacities. The results of the simulation are presented in the figures in the form of dots with the 95% confidence intervals, determined on the basis of the t-Student distribution for 10 series, each with 100,000 calls of the oldest class in each series. The results of the analytical modelling are shown as solid lines.

This study confirms high accuracy of the proposed model of a multiservice server. All the results are presented in relation (and in dependence) to the value of offered traffic per one AU of the real server capacity:

$$a = \sum_{i=1}^{M} A_i t_i / V_r. \tag{17}$$

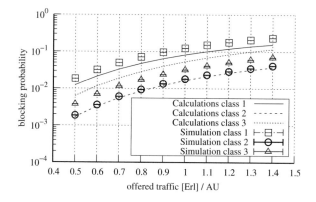

Fig. 2. Blocking probability in system with stream and elastic traffic

4 Conclusions

This article proposes a new, approximate method for modelling multiservice servers to which a mixture of elastic and stream traffic is offered. The presented method is based on an approximation of a multi-dimensional service process by a two-dimensional reversible Markov process. The proposed model corresponds to real-life scenarios for traffic stream services in TCP/IP networks and hence can be employed to solve problems in dimensioning and optimisation of network systems, the more so that the results of the simulation experiments confirm high accuracy of the proposed analytical method. The study carried out by the present authors show that the accuracy of the method depends neither on the number of call classes nor on the degree of compression of elastic traffic, expressed by the value of the virtual capacity.

Acknowledgments. Paper is supported by National Science Center as a part of project 2016/23/B/ ST7/03925 "Modelling and service quality evaluation of Internet-based services".

References

1. Stasiak, M., Głąbowski, M., Zwierzykowski, P., Wiśniewski, A.: Modeling and Dimensioning of Mobile Networks, From GSM to LTE. Wiley, Hoboken (2010)
2. Kelly, F.: Notes on effective bandwidth. University of Cambridge, Technical report (1996)
3. Roberts, J., Mocci, V., Virtamo, I. (eds.): Broadband network teletraffic, final report of action COST 242. Commission of the European communities. Springer, Berlin (1996)
4. Kaufman, J.: Blocking in a shared resource environment. IEEE Trans. Commun. **29**(10), 1474–1481 (1981)
5. Roberts, J.: A service system with heterogeneous user requirements-application to multi-service telecommunications systems. In: Pujolle, G., (ed.) Proceedings

of Performance of Data Communications Systems and their Applications, North Holland, Amsterdam, pp. 423–431 (1981)

6. Stamatelos, G., Koukoulidis, V.: Reservation-based bandwidth allocation in a radio ATM network. IEEE/ACM Trans. Netw. **5**(3), 420–428 (1997)

7. Rácz, S., Gerő, B.P., Fodor, G.: Flow level performance analysis of a multi-service system supporting elastic and adaptive services. Perform. Eval. **49**(1–4), 451–469 (2002)

8. Stasiak, M.: Queuing systems for the internet. IEICE Trans. Commun. **E99–B**(6), 1224–1242 (2016)

9. Hanczewski, S., Stasiak, M., Weissenberg, J.: The queueing model of a multiservice system with dynamic resource sharing for each class of calls. In: Kwiecień, A., Gaj, P., Stera, P. (eds.) Computer Networks. Communications in Computer and Information Science, vol. 370, pp. 436–445. Springer, Berlin (2013)

10. Hanczewski, S., Stasiak, M., Weissenberg, J.: A queueing model of a multi-service system with state-dependent distribution of resources for each class of calls. IEICE Trans. Commun. **E97–B**(8), 1592–1605 (2014)

11. Hanczewski, S., Stasiak, M., Weissenberg, J.: Queueing model of a multi-service system with elastic and adaptive traffic. Comput. Netw. **147**, 146–161 (2018)

12. Bonald, T., Virtamo, J.: A recursive formula for multirate systems with elastic traffic. IEEE Commun. Lett. **9**(8), 753–755 (2005)

13. Moscholios, I.D., Logothetis, M.D., Vardakas, J.S., Boucouvalas, A.C.: Congestion probabilities of elastic and adaptive calls in Erlang-Engset multirate loss models under the threshold and bandwidth reservation policies. Comput. Netw. **92**, 1–23 (2015)

14. Moscholios, I.D., Vardakas, J.S., Logothetis, M.D., Boucouvalas, A.C.: Congestion probabilities in a batched poisson multirate loss model supporting elastic and adaptive traffic. Ann. Telecommun. **68**(5), 327–344 (2013)

15. Bakmaz, B.M., Bakmaz, M.R.: Solving some overflow traffic models with changed serving intensities. AEU - Int. J. Electron. Commun. **66**(1), 80–85 (2012)

16. Bonald, T., Massoulié, L., Proutière, A., Virtamo, J.: A queueing analysis of max-min fairness, proportional fairness and balanced fairness. Queueing Syst. **53**(1), 65–84 (2006)

The Analytical Model of 5G Networks

Sławomir Hanczewski[(✉)], Alla Horiushkina, Maciej Stasiak,
and Joanna Weissenberg

Faculty of Electronics and Telecommunications,
Poznan University of Technology, Poznań, Poland
slawomir.hanczewski@put.poznan.pl

Abstract. 5G networks are to be fully broadband access networks, in which network resources are to be divided in to slices. The concept of slice is based on virtualisation and Software Define Network. Since the effective use of radio interface resources is crucial for the implementation of the services offered at the appropriate QoS level, it is necessary to develop an analytical model of 5G networks. The paper contains a proposal of an analytical model, which in the Authors view can become a tool supporting an engineers responsible for designing and maintaining the 5G networks.

1 Introduction

Every year enormous amount of data is transferred over mobile networks thanks to the more and more expanding throughputs of the radio interface. In LTE networks (4G), it is $300\,\mathrm{Mb/s}$ (downlink) as yet and $50\,\mathrm{Mb/s}$ (uplink). However, the throughput of the radio interface itself is not enough to guarantee appropriate efficacy of the network. Therefore, network operators make use of a number of appropriate traffic shaping mechanisms in their networks, such as reservation [1, 2], priorities [3,4], compression [5–9], traffic overflow [10,11] or buffering [12,13].

Since a typical infrastructure of mobile networks is not very flexible, two previously developed solutions are proposed to be implemented into 5G networks, namely virtualisation and Software Defined Network (SDN). Such an approach will make it possible for the 5G networks infrastructure to construct dedicated networks depending on particular needs. Each of these networks is called a slice. The slices do not involve the radio interface only, but also elements of the backbone network (including servers) that are necessary for the execution of services. As it is easy to notice, the control functions CP (Control Plane) and those responsible for data transmission UP (User Plane) are separated in line with the adapted assumptions for the SDN networks. The slices differ from one another, and hence it is already in the radio interface of the 5G networks that different technologies for their service will be used for the execution of transmission - RAT (Radio Access Technology). Since services offered within the slices are to be easily available by the radio interface, a determination

© Springer Nature Switzerland AG 2020
M. Choraś and R. S. Choraś (Eds.): IP&C 2019, AISC 1062, pp. 295–302, 2020.
https://doi.org/10.1007/978-3-030-31254-1_35

of its traffic characteristics will be of paramount significance for the offering of services. Despite the fact that 5G networks are treated as broadband access networks, the thing is their radio interface still proves to be the bottleneck of the whole of the network. Its appropriate dimensioning will then make it possible to use the available resources in the most optimum way and, as a result, available services provided to the maximum number of users.

5G networks will be broadband access networks, and as a result will be servicing very diversified, in terms of service demands, data streams. Most frequently, the demands that refer to particular streams are described by the two parameters: transmission bitrate and acceptable latency. With reference to latency, streams can be divided into: real-time service streams, that are characterised by very low acceptable latency, and streams for which the same requirements as those for real-time services do not apply. A number of exemplary parameters of data streams are presented in [14].

In the paper we proposed approximated analytical model of 5G network with slices offered real-time and non-real-time services. The model is based on Kaufman-Roberts recursion and SDFIFO model.

The paper is structured as follows. Section 2 proposes an analytical method that allows the values of QoS parameters for a 5G networks to be determined. Exemplary results of the analytical and simulation modelling are presented in Sect. 3, while Sect. 4 summarises and provides conclusions of the study.

2   Modelling of 5G Network

2.1   Distribution of Resources

Activation of each of the slices is related to the allocation of appropriate resources of the physical infrastructure of the networks, including the radio interface. These resources have to be sufficient enough to transfer data with required transmission bitrate. Therefore, the resources of the radio interface can be expressed in bits transmitted in a time unit. Thanks to the above, the distribution of the resources of the radio interface into slices (i.e. a division of available transmission bitrate) is at the same time a distribution of appropriate physical resources of the network. Each slice has C_j bps ($1 \leq j \leq p$) at its disposal, where p is the number of slices activated in the resources of a 5G networks (Fig. 1).

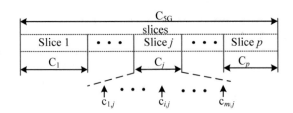

Fig. 1. 5G network resource distribution between slices

If the slices are grouped according to the time of their activation in the network, then it turns out that the best mechanism to guarantee the assumed service level is static reservation [15]. This type of reservation assumes a fixed distribution of available resources of the system. Then, part of the resources of the 5G networks will be distributed permanently between long-time slices, i.e. those where the duration time of operation is long. The allocated resources must guarantee of course the assumed quality of service for a given slice. Second type of slices are short-time slices but these slices are not considered in this paper. Each long-time slice will service m_j different call classes, where j denotes the slice number $1 \leq j \leq p$. The assumption is that calls of individual classes form a call stream with the exponential distribution with the parameter $\lambda_{i,j}$ $(1 \leq i \leq m_j)$. The service time of calls admitted for service also has exponential distribution with the parameter $\mu_{i,j}$ $(1 \leq i \leq m_j)$. Therefore, the intensity of traffic offered by calls of individual classes is equal to:

$$A_{i,j} = \lambda_{i,j}/\mu_{i,j}. \tag{1}$$

Calls of individual call classes that are serviced by a given slice differ in the number of resources demanded for service $c_{i,j}$ (bps). Thanks to discretisation, slices capacity and packet streams bitrates can be represented by allocation units (AU) [16].

Slices service calls with similar characteristics. This means that in a slice where streams of the best effort type are transmitted, streams related to video transmission will not be transmitted. Therefore, each slice can be dimensioned independently and appropriately to the type of currently serviced streams. It should be noted that in the simplest case, a slice can service only one type of traffic stream.

2.2 Services Without Real-Time Constraint

Streams of this type, on top of the delay that results from processing in network devices, can also be subject to additional delays resulting from their buffering. This fact does not mean that the streams will be buffered in base station. This delay may result from waiting for data on the user's device for the resources of a slice to be released. To determine whether the resources of a given slice are sufficient for streams to be serviced at the assumed level, the SDFIFO queueing model can be used [12]. This model is good approximation of considered static slice with non real-time traffic. The model allows, with an assumed capacity of the slice, the probability of the absence of a possibility to service a new stream and the average waiting time for the service to be determined. The system is composed of a server with the capacity V_j AUs (where the index j relates to the number of the slice) and a buffer with the capacity U_j AUs. The system is offered m_j call classes.

Service of calls according to the SDFIFO discipline assumes that virtual queues for each call class will be created in the buffer. In addition, the model assumes that calls of each virtual queue can be serviced even at the expense of service speed (duration of a service) of the calls that are already in the server (being serviced by the slice). This system is discussed in detail in [12].

In [12] it is proved that the occupancy distribution $[P(n)]_{V_j+U_j}$ in a SDFIFO system can be determined recurrently in the following way:

$$[P(n)]_{V_j,U_j} = \begin{cases} \frac{1}{n} \sum_{i=1}^{m_j} A_{i,j} t_{i,j} [P(n-t_{i,j})]_{V_j,U_j} & \text{for } 0 \leq n < V_j, \\ \frac{1}{V_j} \sum_{i=1}^{m_j} A_{i,j} t_{i,j} [P(n-t_{i,j})]_{V_j,U_j} & \text{for } V_j \leq n \leq V_j + U_j, \end{cases} \tag{2}$$

where n is the number of busy AUs in the slice (server).

When we know the occupancy distribution $[P(n)]_{V_j,U_j}$, we are in position to determine the blocking probability, i.e. the situation in which admission of a new call of class i for service is not possible:

$$E_{i,j} = \sum_{n=V_j+U_j-t_{i,j}+1}^{V_j+U_j} [P(n)]_{V_j,U_j} . \tag{3}$$

In the SDFFIFO system, the average waiting time for service is determined by the following formula:

$$T_{i,j} = \frac{1}{\lambda_{i,j} t_{i,j}} Q_{i,j} = \frac{1}{\lambda_{i,j} t_{i,j}} \sum_{n=V_j}^{V_j+U_j} [x_{i,j}(n) - y_{i,j}(n)] [P(n)]_{V_j,U_j}, \tag{4}$$

where:

$y_{i,j}(n)$ is the average number of occupied AUs by calls of class i in the slice j (server j) that is in the occupancy state n AUs,

$x_{i,j}(n)$ is the number of occupied AUs by calls of class i in the system j (the slice and the buffer) that is in the occupancy state n AUs.

The value of the parameter $y_{i,j}(n)$ can be determined with the following dependence:

$$y_{i,j}(n) = \begin{cases} \frac{A_{i,j} t_{i,j} [P(n-t_{i,j})]_{V_j,U_j}}{[P(n)]_{V_j,U_j}} & \text{for } t_{i,j} \leq n \leq V_j + U_j \\ 0 & \text{otherwise,} \end{cases} \tag{5}$$

while the value of the parameter $x_{i,j}(n)$ is approximated by the following dependence:

$$x_{i,j}(n) = \begin{cases} \frac{A_{i,j} [P(n-t_{i,j})]^*_{V_j+U_j,0}}{[P(n)]^*_{V_j+U_j,0}} & \text{for } t_{i,j} \leq n \leq V_j + U_j \\ 0 & \text{otherwise,} \end{cases} \tag{6}$$

where $[P(n)]^*_{V_j+U_j,0}$ is the occupancy distribution in the system with zero buffer and capacity equal to $V_j + U_j$, described by the following formula [17,18]:

$$n[P(n)]^*_{V_j+U_j,0} = \sum_{i=1}^{m_j} A_{i,j} t_{i,j} [P(n-c_i)]^*_{V_j+U_j,0}, \text{ for } 0 \leq n \leq V_j + U_j. \tag{7}$$

2.3   Slices with Real-Time Service Streams

In the case of real-time services, data should not be delayed, due to delay constraints, more than it results from their processing process. Because of the absence of buffers, the properties of such a slice can be determined on the basis of the recurrence dependence (2), with the assumption that the capacity of the buffer is 0. Equation (2) can be reduced then in its essence to the following form:

$$n[P(n)]_{V_j} = \sum_{i=1}^{m_j} A_{i,j} t_{i,j} [P(n - t_{i,j})]_{V_j}, \text{ for } 0 \le n \le V_j. \tag{8}$$

The loss probability can be determined on the basis following formula:

$$E_{i,j} = \sum_{n=V_j - t_{i,j}+1}^{V_j} [P(n)]_{V_j}. \tag{9}$$

3   Exemplary Results

The proposed model was used to model resources of the 5G radio interface with the throughout $C_{5G} = 10$ Gbps. Available bit rate is divided between static slices in the following way [14]:

- Slice 1: Autonomous vehicle control, $C_1 = 300$ Mbps,
- Slice 2: Media on demand, $C_2 = 4$ Gbps,
- Slice 3: Large outdoor event – the rest radio interface capacity is used by thus slice $C_3 = 5{,}7$ Gbps, but results for it are not presented in the paper.

The obtained results are presented in graphs in the function of traffic offered per one AU of the slice j:

$$a = \sum_{i=1}^{m} A_{i,j} / V_j. \tag{10}$$

The results obtained on the basis of presented in the paper model were compared with the results of simulation experiments. In order to do so, an slices in 5G networks simulator was developed. The results comparison confirmed the acceptable accuracy of the proposed model.

3.1   Autonomous Vehicle Control Slice

This slice allows control of autonomous vehicles. Let us assume, that each autonomous vehicle receives and sends data stream (message) with the same time duration ($m_1 = 1$). The demanded bit rate is equal to 2 Mbps. In addition, transmission to and from vehicles should not be buffered. Because the number of serviced autonomous vehicles is high, the traffic generated in this slice can be modeled using the one-service Erlang model. Thus, the blocking probability in this slice can be calculated on the basis of formulae (8) and (9). The slice capacity and streams demands expressed in AUs are equal to: $V_1 = 150$ AUs and $t_{1,1} = 1$ AU. In Fig. 2 are presented results of blocking probability. Adopted slice capacity allows to ensure the value of blocking probability at 0.01 for offered traffic equal to 0.85 of Erlang/AU. To obtain a lower blocking probability for the higher values of the offered traffic, the transmission rate available in the patch should be increased.

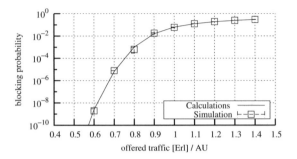

Fig. 2. Blocking probability in slice 1

3.2   Media on Demand Slice

The services offered in this slice, make it possible for users to access various types of content. It was assumed, the slice capacity is equal to $C_2 = 4$ Gbps. Each user, depending on the demands, can send data with one of three bit rates: $c_1, 2 = 10$ Mbps, $c_{2,2} = 20$ Mbps and $c_{3,2} = 80$ Mbps. Due to the user can wait for the data transfer, this slice can be approximated by the model presented in the Sect. 2.2. The capacity of this slice expressed in allocation units is equal to $V_2 = 400$ AUs, while the requested bit rates are equal to $t_{1,2} = 1$ AU, $t_{2,2} = 2$ AUs and $t_{3,2} = 8$ AUs. It was assumed that the capacity of the fictional buffer to determine the average waiting time for service requests is equal to $U_2 = 40$ AUs. In Fig. 3a, the results of the probability of blocking are presented, whereas in Fig. 3b the average expectation for the start of service is presented. Delays associated with transmission are not considered. In the case of this slice, the reduction in the probability of blocking can be achieved at the expense of increasing the waiting time for handling or increasing the available bit rate in the slice.

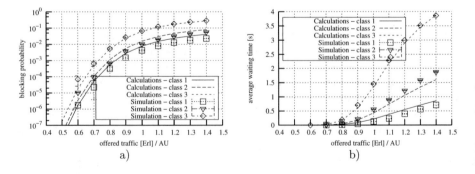

Fig. 3. Blocking probability and average waiting time in slice 2

4 Conclusions

The concept of dividing 5G networks into slices, which are isolated virtual networks, significantly facilitates the process of modelling these networks. The reason is, although the 5G networks itself have a complex infrastructure, the slices themselves are very simple networks that service data streams with similar parameters. Therefore, it is possible to model the slices simply. The models presented in the article can be used not only to calculate, e.g. the blocking probability in the slice, but also on this basis it is possible to determine the parameters of the slice (bit rate), which will provide a better QoS parameters. Due to the presented models, it is possible to prepare tools for engineers dealing with the dimensioning of the slices, which will make this process much simpler. The method can be used to implement a tool that would support engineers responsible for the operation of networks, from the designing stage related to the distribution of resources to optimisation of their use.

Acknowledgements. The presented work has been funded by the Polish Ministry of Science and Higher Education within the status activity task "Structure, analysis and design of modern switching system and communication networks" (08/82/SBAD/8229) in 2019.

References

1. Roberts, J.: Teletraffic models for the Telcom 1 integrated services network. In: Proceedings of 10th International Teletraffic Congress, Montreal, paper 1.1.2 (1983)
2. Głąbowski, K.A., Stasiak, M.: Asymmetric convolution algorithm for blocking probability calculation in full-availability group with bandwidth reservation. IET Circuits Devices Syst. **2**(1), 87–94 (2008)
3. Subramaniam, K., Nilsson, A.A.: Tier-based analytical model for adaptive call admission control scheme in a UMTS-WCDMA system. In: Proceedings of 2005 IEEE 61st Vehicular Technology Conference, Stockholm (2005)
4. Hanczewski, S., Stasiak, M., Zwierzykowski, P.: Modelling of the access part of a multi-service mobile network with service priorities. EURASIP J. Wirel. Commun. Netw. **2015**(1), 1–14 (2005)
5. Stamatelos, G., Koukoulidis, V.: Reservation-based bandwidth allocation in a radio ATM network. IEEE/ACM Trans. Netw. **5**(3), 420–428 (1997)
6. Rácz, S., Gerő, B.P., Fodor, G.: Flow level performance analysis of a multi-service system supporting elastic and adaptive services. Perform. Eval. **49**(1–4), 451–469 (2002)
7. Moscholios, I.D., Logothetis, M.D., Kokkinakis, G.: Connection-dependent threshold model: a generalization of the Erlang multiple rate loss model. J. Perform. Eval. **48**, 177–200 (2002)
8. Sobieraj, M., Stasiak, M., Weissenberg, J., Zwierzykowski, P.: Analytical model of the single threshold mechanism with hysteresis for multi-service networks. IEICE Trans. Commun. **E95–B**(1), 120–132 (2012)
9. Moscholios, I.D., Logothetis, M.D., Boucouvalas, A.C.: Blocking probabilities of elastic and adaptive calls in the Erlang multirate loss model under the threshold policy. Telecommun. Syst. **62**(1), 245–262 (2016)

10. Wilkinson, R.I.: Theories of toll traffic engineering in the USA. Bell Syst. Tech. J. **40**, 421–514 (1956)

11. Głabowski, M., Kaliszan, A., Stasiak, M.: Modelling overflow systems with distributed secondary resources. Comput. Netw. **108**, 171–183 (2016)

12. Hanczewski, S., Stasiak, M., Weissenberg, J.: A queueing model of a multi-service system with state-dependent distribution of resources for each class of calls. IEICE Trans. Commun. **E97–B**(8), 1592–1605 (2014)

13. Hanczewski, S., Kaliszan, A., Stasiak, M.: Convolution model of a queueing system with the cFIFO service discipline. Mobile Inf. Syst. **2016**, 15 (2016). Art. ID 2185714

14. Ercsson: Ericsson White Paper, 5G systems. Ericsson, Technical report (2017)

15. Głąbowski, M., Zwierzykowski, P., Wiśniewski, A.: Modeling and Dimensioning of Mobile Networks, From GSM to LTE. Wiley, Hoboken (2010)

16. Roberts, J., Mocci, V., Virtamo, I. (eds.): Broadband Network Teletraffic, Final Report of Action COST 242. Commission of the European Communities. Springer, Berlin (1996)

17. Kaufman, J.: Blocking in a shared resource environment. IEEE Trans. Commun. **29**(10), 1474–1481 (1981)

18. Roberts, J.: A service system with heterogeneous user requirements-application to multi-service telecommunications systems. In: Pujolle, G. (ed.) Proceedings of Performance of Data Communications Systems and their Applications, North Holland, Amsterdam, pp. 423–431 (1981)

Simulation Studies of a Complex Non-Full-Availability Systems

Sławomir Hanczewski[✉] and Michał Weissenberg

Faculty of Electronics and Telecommunications, Poznan University of Technology,
Poznań, Poland
{slawomir.hanczewski,michal.weissenberg}@put.poznan.pl

Abstract. This article presents the results of a simulation study on multi-service complex non-full-availability systems. To carry out the necessary research and perform research assignments, simulations were done using a simulator developed by the authors, while its implementation constituted the first stage in the study on the systems. In further stages of the study the authors will attempt to develop appropriate analytical models of complex non-full-availability systems and propose a set of new call distribution algorithms in these systems. The developed simulator has the advantage of making verification of obtained results possible.

1 Introduction

The concomitant effect following the increase in the speed and quality of data transmission in telecommunications networks is the perceived increase in the number of offered services and the number of online users. This situation imposes the necessity of application of more and more effective solutions to facilitate optimum usage of available resources. The literature of the subject abounds in presentations of a huge number of traffic manage mechanisms that aim at achieving and maintaining as high QoS level (Quality of Service) in networks as possible, such as: queueing [4], priorities [5] or traffic overflow [3]. Yet another challenge to face, however, is optimization of systems and processes that are constantly being developed and whose complexity becomes more and more extended with every new year. A good example of such systems are cloud systems solutions [12]. As it is apparent from the research studies carried out in recent years, a significant number of present-day systems have the characteristics of non-full-availability systems. In these systems calls that occur at inputs have access to a determined set of resources that is lower than the system's total capacity. These limitations frequently result from a particular structure of these systems (e.g. switching networks) or from a particular applied traffic control mechanism (e.g. systems with reservation). Cloud systems that comprise interconnected myriad of devices designed to process data, make use of elaborate and sophisticated load equalizing mechanisms [7,12] and are capable of employing virtualization within devices [9,11]. Access to physical devices that can be dynamically divided into dedicated

© Springer Nature Switzerland AG 2020
M. Choraś and R. S. Choraś (Eds.): IP&C 2019, AISC 1062, pp. 303–310, 2020.
https://doi.org/10.1007/978-3-030-31254-1_36

machines for specific services and users clearly indicates non-full-availability of cloud systems.

A condition necessary for any mechanisms for the improvement in resource usage to be successfully applied is their proper parametrization. The parametrization can be performed following observations of real systems, on the basis of analytical models or simulation experiments. Due to relatively brief time needed for results to be obtained, as well as a possibility of investigating new systems, quite frequently an application of a digital simulation is the only available solution with this respect. Therefore, the authors made a decision to develop a simulator of a system of interconnected EIG groups of their own design.

The remaining part of the article is structured as follows: Sect. 2 discusses the concept of the complex non-full-availability system. Sect. 3 provides essential information on the purpose-made simulator developed for the study. Sect. 4 is devoted to a presentation of a number of exemplary, illustrative results of the study. Sect. 5 summarizes and concludes the article.

2   Complex Non-Full-Availability System

2.1   Description of the System

The complex non-full-availability system presented in Fig. 1 case a comprises k subsystems with the total capacity V AUs[1]. The concept of the division is illustrated in Fig. 1 case b. Each of the subsystems can additionally be divided to be capable of admitting further successive calls. This system is offered a mixture of m multi-service traffic classes. Individual call classes demand respectively $t_1, ..., t_i, ..., t_m$ AUs for service. Calls of different classes that appear at the input to the system form Poisson call streams with the intensities $\lambda_1, ..., \lambda_i, ..., \lambda_m$. Service stream for each traffic class is described by the exponential distribution with the parameters $\mu_1, ..., \mu_i, ..., \mu_m$. Therefore, traffic offered to this system is equal to:

$$A_i = \lambda_i/\mu_i. \tag{1}$$

2.2   Non-Full-Availability Subsystem

The assumption in the course of the simulation study was that calls that arrived in a subsystem were serviced according to the algorithm that corresponds to call service in an Erlang's Ideal Grading (EIG), whose model was proposed by Erlang in 1917 [1,2]. In this solution, arriving calls of class i have access to a limited number of resources of all the group only. The number of these resources

---

[1] The Allocation Unit (AU) is used in *multi-service* systems and is defined as the highest common divisor of the amount of resources demanded by all traffic streams offered to a system. In systems in which the offered packet streams are characterized by variable bitrate, the so-called bandwidth discretization is used to determine the AU. The bandwidth discretization consists in a replacement of the variable bitrate of a bit stream with a constant bitrate, termed the *equivalent bandwidth*.

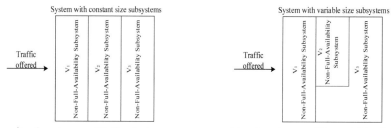

Fig. 1. Complex non-full-availability system

d_i is called availability. Traffic sources that have access to a uniform subset of all resources form the so-called load group and their number is equal to the combination d_i, each with V AUs:

$$g_i = \binom{V}{d_i}. \tag{2}$$

Each of the subsystems can be characterized by a different capacity, whereas each of the call classes that are offered to the system can have different availability within each of the subsystems.

2.3 Applied Algorithms for the Choice of Subsystems

The initial assumption in the study was that a number of allocation disciplines for calls to be allocated to the subsystems were to be implemented in the simulator. These were the following disciplines:

- random discipline - in this algorithm from the subsystems, one subsystem (with equal probability) is chosen randomly. If the number of free (unoccupied) resources in the load group in which a call arrives in the subsystem is not sufficient, the call is relocated to another random subsystem;
- minimum discipline - in this algorithm, each time the subsystems are checked out starting from the subsystem with the lowest capacity, the check is carried out until a subsystem that is capable of servicing a given call is found or the list of available subsystems is exhausted;
- maximum discipline - in this algorithm, each time the subsystems are checked out starting from the subsystem with the highest capacity, the check is carried out until a subsystem that is capable of servicing a given call or the list of available subsystems is exhausted.

3 Simulator of a Complex Non-Full-Availability System

A block diagram of the developed simulator of a complex non-full-availability system is shown in Fig. 2.

The simulator is divided into 5 modules.

Fig. 2. The block diagram of the simulator of a complex non-full-availability system

- *Input data* - the module allows all necessary data to be introduced to the simulator, i.e. the capacity of a system, the number and size of subsystems, the selected call service discipline, etc.;
- *Call stream* - generates calls according to the assumed traffic parameters for each of the classes offered to the system;
- *System structure* - stores data on the occupancy of the resources of the system, the number of calls, the occupancy of individual subsystems during a simulation and other vital parameters;
- *Call service* - is responsible for servicing calls according to the adopted mechanism for calls to be transmitted to individual subsystems;
- *Output data* - calculates the average usage of resources in each of the subsystems, losses in the whole of the system and in each of the subsystems for all classes of calls offered to the system, based on the obtained results, and then saves the results to a file.

The simulator was written in the C++ language based on an event scheduling approach [13] and is an extension to the work discussed in [6]. In order to determine results, each time 10 simulation series were performed. Each series was divided into two basic stages: transitional stage (lasting until 10,000 calls of the class that demand the highest number of AUs appear in the system) and the principal stage (lasting until 100,000 calls of the class that demands the highest number of AUs appear in the system). The obtained results were then processed statistically to determine confidence intervals based on the t-Student distribution.

4   Exemplary Results

Figures 3, 4, 5, 6 and 7 show a number of exemplary illustrative results of the simulation experiments for complex non-full-availability systems. All the obtained results are presented in the function of offered traffic per a single AUs of all of the system.

Figure 3 shows the values of the loss probability and the average occupancy of each of the subsystems for the system with the total capacity $V = 60$ AUs, divided into 3 subsystems with the capacities $V_1 = V_2 = V_3 = 20$ AUs, respectively (System I), obtained in the simulation experiment. The resources were allocated according to the random algorithm. The system was offered $m = 3$ call classes that demanded 1, 2 and 4 AUs, respectively, for service. The availability parameters were equal to: $d_1 = d_2 = d_3 = 6$ AUs. The proportions of traffic offered by individual classes satisfied the condition $A_1t_1 : A_2t_2 : A_3t_3 = 1 : 1 : 1$.

With uniform sizes of the subsystems and availabilities for each call classes, the random algorithm, allocating calls with equal probability to each of the subsystems, leads to an equalization of the usage of each of them.

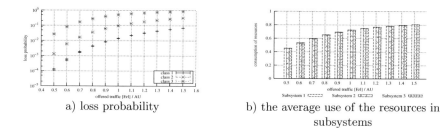

a) loss probability

b) the average use of the resources in subsystems

Fig. 3. Loss probability and the average use of the resources in subsystems (System I) with the application of the random algorithm

Figure 4 shows analogous results for System I, in which the sequential algorithm was used to place calls in the subsystems.

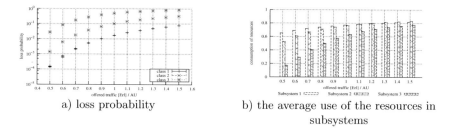

a) loss probability

b) the average use of the resources in subsystems

Fig. 4. Loss probability and the average use of the resources in subsystems (System I) with the application of the sequential algorithm

Figure 5 shows the values of loss probabilities obtained in the study in a system with the capacity $V = 80$ AUs, divided into 3 subsystems with the capacities $V_1 = 15, V_2 = 25, V_3 = 40$ AUs, respectively (System II). To allocate resources to each of the subsystems, the following algorithms were used sequentially: the random algorithm (Fig. 5 case a), MIN algorithm (Fig. 5 case b) and MAX algorithm (Fig. 5 case c). This system was offered $m = 3$ call classes that demanded 1, 2 and 7 AUs for service, respectively. The availability parameters were equal for the subsystems: subsystem 1: $d_1 = 3, d_2 = 5, d_3 = 10$ AUs, subsystem 2: $d_1 = 4, d_2 = 7, d_3 = 15$ AUs, subsystem 3: $d_1 = 5, d_2 = 9; d_3 = 20$ AUs. The proportions of traffic offered by individual classes satisfied the condition $A_1 t_1 : A_2 t_2 : A_3 t_3 = 1 : 1 : 1$.

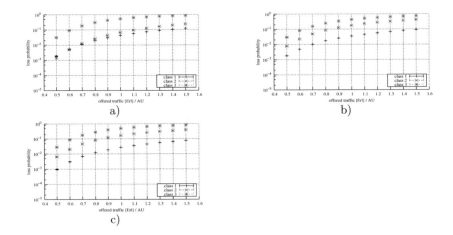

Fig. 5. Loss probability in System II

The occupancies of individual subsystems of System II, with the application of all three algorithms, are presented in Fig. 6.

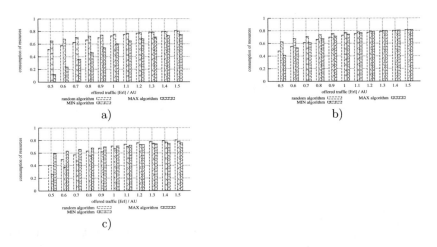

Fig. 6. The average use of resources in System II

As it is observable in Figures 5 and 6, the loss coefficients, regardless of the algorithm applied, are similar. This results from the fact that at the moment of a rejection of a call by one of the subsystems, other subsystems are then analyzed for as long as this call is ultimately admitted for service or all available subsystems reject the call. The differences between the algorithms applied in the study manifest though in the average use of resources of individual subsystems. The random algorithm allocates calls in a uniform fashion with the differences in the size of the subsystems taken into consideration and has the advantage of limiting their overload. The MIN algorithm, due to the traffic distribution mechanism, leads to a very high usage of resources in the smallest subsystem even when the traffic load is slight. A change in the values of the availability parameters for particular call classes within each of the subsystems significantly influences their use for the case of the random algorithm. Variable availability within subsystems can also lead to the total lack of access following a decrease in availability.

Figure 7 case a shows the average number of checked subsystems necessary for a call to be admitted in System III, with the application of the MIN and MAX algorithms, whereas Fig. 7 case b presents the number of checks of subsystems by all calls that arrive in System II.

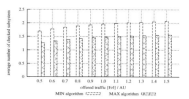
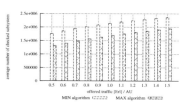

a) The average number of checked subsystems prior to admission of a call (System II)

b) The sum of checks of subsystems (System II)

Fig. 7. Analysis of subsystem checked in systems with differente admission control algorithm

The results obtained in the study clearly show that, with maintaining similar parameters for loss coefficients for individual call classes, the application of the MAX algorithm leads to a limitation in the number of identified subsystems and the number of checks necessary for a call to be admitted. This particular algorithm seems then to be a better solution when large disproportions in the sizes of subsystems occur.

5   Summary

This article presents the results of simulation studies of complex non-full-availability systems. The purpose-made simulator developed by the authors is very easy to use, while the results obtained following its application will provide an opportunity to verify other analytical models in the future and to test algorithms for load equalization in such systems. The experience gained in the process of developing the assumptions and actual implementation of the simulator will also help construct an appropriate simulator of cloud systems.

Acknowledgements. The presented work has been funded by the Polish Ministry of Science and Higher Education within the status activity task "Structure, analysis and design of modern switching system and communication networks" (08/82/SBAD/8229) in 2019.

References

1. Brockmeyer, E., Halstrom, H., Jensen, A.: The life and works of A. K. Erlang. Acta Polytechnika Scand. **6**, 287 (1960)
2. Erlang, A.: Solution of some problems in the theory of probabilities of significance in automatic telephone exchanges. Elektrotechnikeren **13**(5), 189–197 (1917)
3. Głąbowski, M., Hanczewski, S., Stasiak, M.: Modelling of cellular networks with traffic overflow. Math. Probl. Eng. **2**, 1–15 (2015)
4. Hanczewski, S., Stasiak, M., Weissenberg, J.: A queueing model of a multi-service system with state-dependent distribution of resources for each class of calls. IEICE Transact. Commun. **E97–B**(8), 1592–1605 (2014)
5. Hanczewski, S., Stasiak, M., Zwierzykowski, P.: Modelling of the access part of a multi-service mobile network with service priorities. EURASIP J. Wireless Commun. Netw. **1**, 1–14 (2015)
6. Hanczewski, S., Weissenberg, M.: Simulation of queuing systems with non-full-availability server. In: 11th International Symposium on Communication Systems, Networks & Digital Signal Processing (CSNDSP), Budapest, pp. 1–6 (2018)
7. Kaur, R., Luthra, P.: Load balancing in cloud computing. In: Proceedings of Recent Trends in Information Telecommunication and Computing (2012)
8. Kelly, F.: Notes on Effective Bandwidth. University of Cambridge, Cambridge (1996)
9. Kosta, S., Aucinas, A., Hui, P., Mortier, R., Zhang, X.: ThinkAir: dynamic resource allocation and parallel execution in the cloud for mobile code offloading. In: Proceedings IEEE INFOCOM, pp. 945–953 (2012)
10. Mell, P., Grance, T.: The NIST definition of cloud computing. In: National Institute of Standards and Technology (2009)
11. Mijumbi, R., Serrat, J., Gorricho, J., Bouten, N., De Turck, F., Boutaba, R.: Network function virtualization: state-of-the-art and research challenges. IEEE Commun. Surv. Tutorials **18**(1), 236–262 (2016)
12. Nuaimi, K., Mohamed, N., Alnuaimi, M., Al-Jaroodi, J.: A survey of load balancing in cloud computing: challenges and algorithms. In: IEEE 2nd Symposium on Network Cloud Computing and Applications, vol. 12, pp. 137–142 (2012)
13. Tyszer, J.: Object-Oriented Computer Simulation Of Discrete-Event Systems. Kluwer Academic Publishers Group, Norwell (1999)

Simulation Studies of Multicast Connections in Ad-Hoc Networks

Maciej Piechowiak$^{(\boxtimes)}$

Kazimierz Wielki University, Bydgoszcz, Poland
`mpiech@ukw.edu.pl`

Abstract. Ad-hoc networks are an extension of the traditional network infrastructure (cellular networks, wireless LAN, etc) and fit into the concept of the Internet of Things. Routing design in such a networks is a challenge because of limited node resources and nodes mobility. Thus efficient data transmission techniques like multicasting are under scrutiny. The article analyzes and explores the performance of multicast heuristic algorithms without constraints and quality of multicast trees in ad-hoc networks and proves the thesis that well-known multicast heuristic algorithms designed for packet networks have a good performance in ad-hoc networks with decentralized structure.

Keywords: Ad-hoc networks · Topology generator · MANET · Multi-hop networks

1 Introduction

Ad-hoc networks consist of nodes collections placed in dispersed geographical locations with wireless communication between them. The most distinct feature that differs them from other networks is lack of cable infrastructure – the structure is quite decentralized. Nodes in ad-hoc network can work as clients or as routers, or both simultaneously. Last years show increased use of ad-hoc networks and its possible applications. For example, in measurement systems nodes can represent an autonomous sensors or data acquisition devices. Ad-hoc networks can be also used to collect and process data for a wide range of applications such as tensor systems, air pollution monitoring, and the like. Nodes in these networks generate traffic to be forwarded to some other nodes (unicast) or a group of nodes (multicast). In this way, such a networks are part of the paradigm of Internet of Things that in its assumptions integrates and enables several technologies and communication solutions into a borderless network of vehicles, computers, portable devices and all kinds of sensors [1].

In order to save resources of wireless nodes, effective mass data transmission should base on efficient routing protocols. Multicast transmission allows to provide the same data to defined recipients without the need to set up a communication session. The implementation of multicasting requires solutions of many combinatorial problems accompanying the creating of optimal transmission trees

© Springer Nature Switzerland AG 2020
M. Choraś and R. S. Choraś (Eds.): IP&C 2019, AISC 1062, pp. 311–317, 2020.
https://doi.org/10.1007/978-3-030-31254-1_37

[12]. Multicast communication optimization involves the construction of effective routing algorithms whose task is to build a tree with the minimum cost between the transmitting node and the group of receiving nodes. This method of communication prevents excessive multiplication of packets in the network – the data sent only reaches those nodes (routers) that are involved in transmission to the members of multicast group. In the optimization process two approaches are considered: MST (*Minimum Steiner Tree*), and SPT (*Shortest Path Tree*) – tree with the shortest paths between the source node and each of the destination nodes. The relevant literature provides a wide range of heuristics solving this problem in polynomial time and dedicated mostly for packet networks [11,13–18]. In case of MANET multicast protocols, two basics architectures are used: tree-based protocols, where MAODV (*Multicast Ad-hoc On-demand Distance Vector routing*) [7] is the most discussed tree-based protocol and mesh based protocol: ODMRP (*On-Demand Multicast Routing Protocol*) [8]. Multicast routing with QoS was examined in [5,6].

The problem of multicast transmission in the Internet of Things can be simplified when the group signalling is moved to the seventh layer of the OSI model as proposed in the LoRaWAN networks [9]. The energy classes of end nodes are reconfigured, and downlink messages are delivered to the groups of nodes in dedicated time slots. However, this issue goes beyond the scope of the article.

The article discusses the effectiveness of the most commonly used constrained multicast heuristic algorithms as well as their comparative usefulness in ad-hoc networks. The article structure is as follow: Sect. 2 formulates an optimization problem and defines multicast routing algorithms. Section 3 presents the simulation study concept and methodology of research. Section 4 includes the results of the simulation of the implemented algorithms along with their interpretation.

2   Problem Formulation

Let us assume that a network is represented by an undirected, connected graph $N = (V, E)$, where V is a set of nodes, and E is a set of links. The existence of the link $e = (u, v)$ between the node u and v entails the existence of the link $e' = (v, u)$ for any $u, v \in V$ (corresponding to two-way links in communication networks). The cost parameter $c(e)$ is associated with each link $e \in E$. The cost of a connection represents the usage of the link resources – $c(e)$ is then a function of the traffic volume in a given link and the capacity of the buffer needed for the traffic. In simulation tests it is generated randomly from 10 to 1000.

The adopted model of the costs of links between the devices takes into consideration energy used by the antenna system of a device. The proposed implementation assumes that network devices have isotropic radiators. For simplicity, this model bases on the pathloss power law model for radio propagation. With the power law model for radio propagation, and the assumption that transmission power and receiver sensitivity for all nodes is same, the coverage area of any node is a circle with radius r. A node can have direct communication with all nodes that fall inside its coverage area [22].

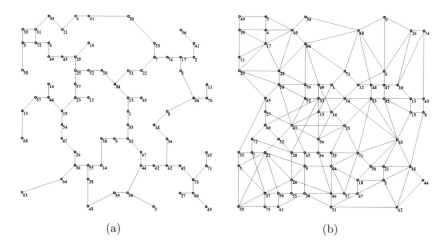

(a)   (b)

Fig. 1. Visualization of ad-hoc networks with 75 nodes obtained using the proposed generator for $r = 100$ units (a) and Waxman model with $\alpha = 0.15$ and $\beta = 0.05$ (b)

For undirected networks with symmetric links, heuristic algorithms have been developed. These algorithms allow to solve the problem of the minimal Steiner tree in polynomial time [23].

The simplest way to implement the routing algorithm for multicast connections is to implement one of the classic algorithms defining the minimum spanning tree: a Kruskal algorithm [24] or Prim algorithm [19]. The spanning tree defined in this way is built for every node in the network, causing excessive increase of network traffic. The PPH algorithm [25] modifies the minimum spanning tree, built by the Prim algorithm by pruning. This mechanism consists in iteratively removing nodes with the degree equal to 1, which do not take part in group transmission.

On the other hand, the algorithms determining the minimum spanning tree are considering where only receiving nodes are taking into account at the stage of creating the tree. In the first stage of the KMB algorithm's [2] operation, a subgraph is constructed in the graph representing the analyzed network. The edges of this subgraph represent all the shortest paths between the sending node and the receiving nodes determined using the Dijkstra algorithm. Then, in the obtained subgraph, the minimum spanning tree is determined using the Prim's algorithm. In the next step, the edges of this tree are replaced with corresponding shortest paths of the considered graph based on the Dijkstra's algorithm.

The PPH and KMB algorithms were selected by the authors as representative algorithms for multicast trees efficiency evaluation in ad-hoc networks.

3   Simulation Study

Using computer simulations, it is possible to turn concepts into more realistic scenarios. It allows to verify ad-hoc models and concepts without the need to implement them in hardware, yet providing a detailed insight. Therefore, author conducted their custom-made ad-hoc generator prepared in C++ as CGI application and PHP [30] for network topology visualization.

The simulation area is a rectangle of 1,000 by 1,000 units where nodes are deployed on a mesh with the granularity of one unit. The maximum radio range of a sensor node is set to 200 units. The proposed generator simplifies network topology model – it provides ad-hoc topologies without nodes mobility (Fig. 1).

The simulation process also uses network topologies represented by random graphs generated by the application of the Waxman method [10]. In order to guarantee the consistency of the graph and create short edges between nodes, boundary values of the Waxman method parameters have been set up ($\alpha = 0.15, \beta = 0.05$). The aim of the authors was to investigate whether the results of multicast algorithms in ad-hoc networks are comparable with results obtained in random graphs with such short edges such as ad-hoc networks. Therefore, they used network topologies generated by Waxman node with an average node degree of $D_{av} = 2(k = 100)$ and $D_{av} = 4(k = 200)$.

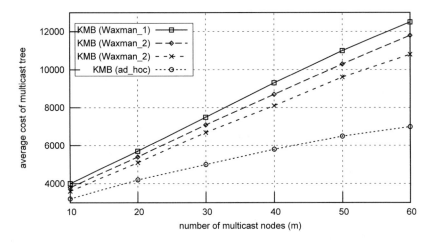

Fig. 2. The influence of the number of the multicast nodes on the average cost of a group transmission tree (m) constructed by KMB algorithm (network topology parameters: $n = 100$, $D_{av} = 4$)

4   Simulation Results

Due to a wide range of solutions presented in the literature of the subject, the following representative algorithms were chosen: KMB [13] and PPH [25]. Their popularity in applications were decisive in their selection. Such a set of algorithms includes solutions potentially most and least effective in terms of costs of constructed trees.

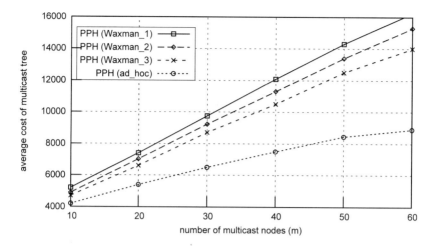

Fig. 3. The influence of the number of the multicast nodes on the average cost of a group transmission tree (m) constructed by PPH algorithm (network topology parameters: $n = 100$, $D_{av} = 4$)

In the first phase of the experiment (Fig. 2) the dependency between the average cost of the constructed trees with an application of KMB algorithm and the number of multicast nodes m was examined. The influence of the β parameter in Waxman model on the costs of the obtained trees is not so significant as in the case when the ad-hoc network is applied.

The KMB algorithm constructs multicast trees with the total cost of 60% lower in ad-hoc networks, on average, as opposed to the same algorithm implemented in Waxman networks ($\beta = 0.95$, Waxman_1), and 40% in Waxman networks ($\beta = 0.05$, Waxman_3) respectively.

In the second phase of the experiment (Fig. 3) the dependency between the average cost of the constructed trees with an application of PPH algorithm and the number of multicast nodes m was examined. The PPH algorithm constructs multicast trees with the total cost of 61% lower in ad-hoc networks, on average, as opposed to the same algorithm implemented in Waxman networks ($\beta = 0.95$) and 41% in Waxman networks ($\beta = 0.05$) respectively.

For each point of the plot shown in Figs. 2 and 3 the average cost of the tree was obtained for 5000 networks generated in 5 series for each routing algorithm. The results of the simulations are shown in the charts in the form of marks with 95% confidence intervals that were calculated after the *t-Student* distribution. 95% confidence intervals of the simulation are almost included within the marks plotted in the figures.

5   Conclusions and Future Work

The literature propose many routing protocols designed for ad-hoc networks. Unicast protocols are dominating set of whole routing solutions while multicast routing algorithms and protocols are in minority and they are still an open topic. First authors' approach evaluates multicast routing algorithms (designed especially for packet networks with Internet-like topologies) implemented in mesh networks with regular structure (grid topology). The results of algorithms obtained in the mesh networks were compared with the results obtained in random graphs. Conducted studies confirmed the effectiveness of examined heuristic algorithms in ad-hoc topologies. Thus the paper proves the thesis statement set in Introduction.

The simulation research methodology proposed earlier [16,17] permit to model networks with wide range of nodes and many network topology parameters. This will constitute the next stage in the authors' research work aiming to define a methodology for testing multicast heuristic algorithms in mesh networks and compare their effectiveness with dedicated algorithms and protocols.

References

1. Reina, D.G., Toral, S.L., Barrero, F., Bessis, N., Asimakopoulou, E.: The role of Ad Hoc networks in the internet of things: a case scenario for smart environments. In: Internet of Things and Inter-cooperative Computational Technologies for Collective Intelligence, pp 89–113, Springer (2013)
2. Kou, L., Markowsky, G., Berman, L.: A fast algorithm for steiner trees. Acta Informatica **15**, 141–145 (1981)
3. Perkins, C., Belding-Royer, E., Das, S.: Ad hoc On-Demand Distance Vector (AODV) routing. In: Network Working Group, RFC: 3561, July 2003
4. Johnson, D., Hu, Y., Maltz, D.: The dynamic source routing protocol (DSR) for mobile Ad Hoc networks for IPv4. In: Network Working Group, RFC: 4728, February 2008
5. Bur, K., Ersoy, C.: Ad Hoc quality of service multicast routing. Comput. Commun. Elsevier Sci. **29**(1), 136–148 (2005)
6. Bur, K., Ersoy, C.: Performance evaluation of a mesh-evolving quality-of-service-aware multicast routing protocol for mobile Ad Hoc networks. Perform. Eval. Elsevier Sci. **66**(12), 701–721 (2009)
7. Royer, E.M., Perkins, C.: Multicast Ad hoc On-Demand Distance Vector (MAODV) routing. In: Network Working Group, RFC: draft, July 2000
8. Lee, S., Su, W., Gerla, M.: On-Demand Multicast Routing Protocol (ODMRP) for Ad Hoc networks. In: Network Working Group, RFC: draft, July 2000
9. FUOTA Working Group, LoRa Alliance Technical Committee. LoRaWAN Remote Multicast Setup Specification v1.0.0. Lora Alliance, 2018
10. Waxmann, B.: Routing of multipoint connections. IEEE J. Sel. Area in Commun. **6**, 1617–1622 (1988)
11. Crawford, J.S., Waters, A.G.: Heuristics for ATM multicast routing. In: Proceedings of 6th IFIP Workshop on Performance Modeling and Evaluation of ATM Networks, strony, pp. 5/1–5/18, July 1998
12. Hakimi, S.L.: Steiner's problem in graphs and its implications. Networks **1**, 113–133 (1971)
13. Kompella, V.P., Pasquale, J.C., Polyzos, G.C.: Multicasting for multimedia applications. In: INFOCOM, strony, pp. 2078–2085 (1992)
14. Mokbel, M.F., El-Haweet, W.A., El-Derini, M.N.: A delay constrained shortest path algorithm for multicast routing in multimedia applications. In: Proceedings of IEEE Middle East Workshop on Networking. IEEE Computer Society (1999)

15. Piechowiak, M., Stasiak, M., Zwierzykowski, P.: The application of K-shortest path algorithm in multicast routing. Theor. Appl. Inf. **21**(2), 69–82 (2009)
16. Piechowiak, M., Stasiak, M., Zwierzykowski, P.: Analysis of the influence of group members arrangement on the multicast tree cost. In: Proceedings of The Fifth Advanced International Conference on Telecommunications AICT (2009)
17. Zwierzykowski, P., Piechowiak, M.: Performance of fast multicast algorithms in real networks. In: Proceedings of EUROCON 2007 The International Conference on: Computer as a Tool, Warsaw, Poland, September 2007, pp. 956–961 (2007)
18. Zwierzykowski, P., Piechowiak, M.: Efficiency analysis of multicast routing algorithms in large networks. In: Proceedings of The Third International Conference on Networking and Services ICNS 2007, Athens, Greece, June 2007, pp. 101–106 (2007)
19. Prim, R.: Shortest connection networks and some generalizations. Bell Syst. Tech. J. **36**, 1389–1401 (1957)
20. Rajaraman, R.: Topology control and routing in Ad Hoc networks: a survey. ACM SIGACT News **30**, 60–73 (2002)
21. Santi, P.: Topology control in wireless Ad Hoc and sensor networks. ACM Comput. Surv. **37**, 164–194 (2005)
22. Grover, P., Gupta, N., Kumar, R.: Calculation of inference in Ad-Hoc network. J. Theor. Appl. Inf. Technol. **16**(2), 105–109 (2010)
23. Piechowiak, M., Zwierzykowski, P.: How to simulate and evaluate multicast routing algorithms. In: Pathan, A.K., Monowar, M.M., Khan, S. (eds.) Simulation Technologies in Networking and Communications: Selecting the Best Tool for the Test. CRC Press (2015)
24. Kruskal, R.: Mininum spanning tree. In: Proceedings of the American Mathematical Society, pp. 48–50 (1956)
25. Piechowiak, M., Zwierzykowski, P., Bartczak, T.: An application of the switched tree mechanism in the multicast routing algorithms. In: 1-st Interdisciplinary Technical Conference of Young Scientists InterTech 2008, pp. 282—286 (2008)
26. Zegura, E.W., Calvert, K.L., Bhattacharjee, S.: How to model an internetwork. In: IEEE INFOCOM 1996 (1996)
27. Piechowiak, M., Zwierzykowski, P.: The influence of network topology on the efficiency of multicast heuristic algorithms. In: Proceedings of The 5-th International Symposium – Communication Systems, Networks and Digital Signal Processing CSNDSP 2006, pp. 115–119 (2006)
28. Piechowiak, M., Zwierzykowski, P.: Performance of fast multicast algorithms in real networks. In: Proceedings of IEEE EUROCON 2007 Proceedings of International Conference on: Computer as a tool, pp. 956–961 (2007)
29. Generator Services Project. http://sftweb.cern.ch/generators/
30. PHP: Hypertext Preprocessor. http://sftweb.cern.ch/generators/
31. Scalable Vector Graphics (SVG). http://www.w3.org/Graphics/SVG/
32. Faloutsos, M., Faloutsos, P., Faloutsos, C.: On Power-Law Relationships of the Internet Topology, pp. 111–122. ACM Computer Communication Review, Cambridge (1999)
33. Watts, D.J., Strogatz, S.H.: Collective dynamics of 'small-world' networks. Nature **12**(393), 440–442 (1997)

V2X Communications for Platooning: Impact of Sensor Inaccuracy

Michał Sybis[(✉)] [iD], Paweł Sroka[iD], Adrian Kliks[iD], and Paweł Kryszkiewicz[iD]

Poznan University of Technology, pl. M. Skłodowskiej-Curie 5, 60-965 Poznań, Poland
{michal.sybis,pawel.sroka,adrian.kliks,pawel.kryszkiewicz}@put.poznan.pl

Abstract. Platooning is one of the prospective applications, where autonomous driving and Vehicle-to-Vehicle and Vehicle-to-Infrastructure communications play a vital role. Various test proved that reduction of inter-car distance withing platoon of trucks results in significant fuel savings. Keeping of such short distance as well as awareness of the platooning cars about the vicinity assumes the presence of highly reliable sensors. In this paper, we briefly discuss the impact of sensor inaccuracy on the overall behaviour of the train-of-trucks driving in the autonomous fashion, supported by the presence of dedicated wireless system utilizing context information stored in databases and maps.

Keywords: V2X and 5G · Sensor accuracy ·
Context information database

1 On V2X Communications for Autonomous Driving

In the context of future wireless cellular systems, the fifth generation (5G) wireless systems is envisioned to provide the so-called ultra-reliable communications (URC) with minimum latency that can be used to ensure various safety or life-saving services [5]. This idea considers, among other issues, the delivery of high-reliability services, such as vehicle-to-vehicle (V2V) communications for autonomous and self-driving/driverless steering of platoons on high-speed roads. The autonomous driving is foreseen as an enabler for vehicle platooning, which is a coordinated movement of a group of autonomous vehicles forming a convoy led by a platoon leader. In such a case, the group of short-distanced vehicles, typically trucks, mutually exchange numerous control and steering information enabling the autonomous movement of the whole platoon with no human intervention. The increased interest in platooning is driven by the expected potential revenues, e.g. fuel savings of 7 to 15% for trucks travelling behind the platoon leader [1]. Additionally, the fuel savings translate to a substantial reduction of CO_2 emission, and, according to the study performed in the Energy ITS project [11], when market penetration of truck platooning increases from 0% to 40% of

The work has been realized within the project no. 2018/29/B/ST7/01241 funded by the National Science Centre in Poland.

© Springer Nature Switzerland AG 2020
M. Choraś and R. S. Choraś (Eds.): IP&C 2019, AISC 1062, pp. 318–325, 2020.
https://doi.org/10.1007/978-3-030-31254-1_38

trucks, CO_2 emission along a highway can be reduced by 2.1% if the gap between trucks is 10 m, and by 4.8% if the gap is reduced to 4 m.

Intuitively, automated driving has to be supported by on-board sensors and inter-vehicular communications to provide faster responses than human drivers, hence improving the road safety. However, it specifies the extreme requirements on the reliability of the wireless communications as well as the quality of applied on-board sensors. In the first case, the information exchange between the platoon members and between the platoon and other devices is done utilizing short-range wireless communications schemes, such as dedicated short-range communications (DSRC), cooperative intelligent transport system (C-ITS) or cellular networks (cellular V2X, CV2X). In case of the on-board sensors, they are necessary to facilitate the operation of an autonomous vehicle considered in platooning and their accuracy mainly depends on the class of the mounted device and can result in various levels of deviation of information.

The generic scenario considered in this paper is visually presented in Fig. 1, where one can observe the road train driving autonomously on a high-speed road in various environments (urban and rural). The communication within the platoon has to be stable and reliable, as well as the functioning of on-board sensors has to be stable enough to reliably monitor the trucks parameters and inter-car distances in real time. In order to meet this goal, the platooning trucks exchange information utilizing available communications infrastructure, mainly Road Side Units, but also small-cells and macro-cells. The fulfillment of highly stringent requirements on the communications reliability and low-latency entails the need for efficient management of prospective interference between the platoons (see Fig. 1) and also between the platoons and other (external) sources of interference.

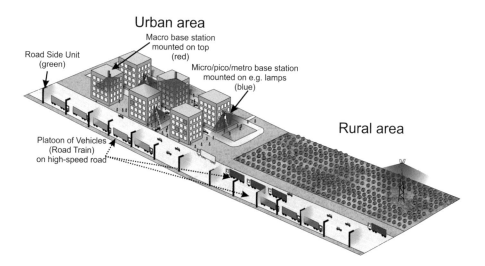

Fig. 1. Vehicle platooning and surrounding wireless-communications environment

In this paper we focus on the impact of the sensor inaccuracy on the overall efficiency of V2X communications. In the second chapter we discuss the practical implementation issues of the adaptive cruise control mechanisms. In chapter three, we analyze the quality of various sensors and discuss briefly the its impact on the system reliability and performance. In chapter four we outline the main challenges related to application of platooning using wireless communications. Sect. 5 concludes the work.

2   Implementation Issues

From the implementation point of view, platooning is very challenging. It requires the coordination of drive control algorithms, actuation, sensing, and wireless communications used to exchange information between platoon vehicles. The key factor limiting the performance of platooning algorithms, and hence impacting the inter-vehicle distance, is the delay of the control system and the accuracy of information describing the surrounding environment, obtained via sensors or wireless communications. The introduction of the cooperative adaptive cruise control (CACC), which relies both on sensor measurements and wireless communications with other vehicles in the platoon, helped to further increase the reliability of the autonomous controller; however, it requires frequent, highly reliable, and low-latency V2V transmission [3,7].

The idea of autonomous driving of vehicles forming a platoon using CACC has attracted a lot of interest in recent years. A number of empirical studies have been performed to evaluate the performance of platooning supported by IEEE 802.11p-based wireless communications. The SARTRE project [1] conducted an experiment with a platoon of two trucks and three cars driving autonomously, with the result showing that the platoon could drive at speeds of up to 90 km/h with a 5–7 m inter-vehicle distance. On the other hand, in the framework of the Energy ITS project in Japan, a platoon of three fully-automated trucks driving at 80 km/h with a 10 m inter-vehicle gap was tested on an expressway [10,11].

One of the main problems of CACC when employing current state-of-the-art wireless communication protocols is the reliability of information exchange between platoon vehicles. A study on the use of CACC with communications based on the IEEE 802.11p standard revealed that even a moderate increase in road traffic on a motorway can lead to wireless channel congestion, and consequently, prevent the automated controller from reliable and stable operation [8,9]. A solution to this problem, suggested in [9], may be the use of a dual-band transceiver that can operate simultaneously in two different frequency bands. However, one should note that even when two sub-bands of the ITS frequency range are used, channel congestion is still possible due to the high density of communicating vehicles and the large amount of exchanged data. A solution to this problem may be to use the spectrum white spaces to improve the reliability of wireless communications, and, consequently, performance of CACC-based vehicle platooning, by finding new, less utilized frequency bands.

3 Role of Sensor for Wireless Communications in Platoons

On-board sensors are an essential part of a CACC-controlled vehicle system. They measure the basic physical quantities necessary for the proper operation of the control algorithm. Sensors used in a vehicle can be divided into two types. The first one is used to monitor the vehicle's surrounding environment, while sensors of the second type are responsible for the evaluation of the vehicle itself. The first class of sensors includes Long/Medium Range Radar (LRR/MRR) sensors and Forward Collision Warning (FCW) sensors. They determine the distance to the preceding vehicle and its instantaneous speed. The second group are the sensors that determine the instantaneous speed and acceleration of the vehicle in which they are mounted. Sensor inaccuracies can be determined in different ways. Magnitudes describing sensor errors can be as follows: absolute accuracy of the sensor, sensor error distribution, error values corresponding to the given percentile of error distribution or output noise.

In order to add inaccuracies of the analyzed sensors to the considered system, it is assumed that sensor inaccuracies can be modeled using the normal distribution truncated to $\pm 3\sigma$. For the assumed sensor accuracy distribution model, standard deviations that correspond to the given sensor error have been determined. For the purpose of this paper we investigated the off-the-shelf average quality sensors that are mounted or could be mounted as an on-board sensors.

Investigated sensors can be divided into following groups: distance sensors, relative velocity sensors, ego velocity and acceleration sensors. Depending on their tasks, these sensors can have different ranges and different measurement accuracy. The purpose of distance sensors is to determine distances to adjacent objects that can be both moving vehicles and obstacles. Relative speed sensors evaluate the speed of surrounding vehicles with respect to the ego vehicle. In addition to the sensors that monitor the position of adjacent vehicles, sensors that determine the speed and acceleration of the vehicle on which they are mounted are also required (for the ego vehicle velocity measurement, it is assumed the electro-mechanical sensors, whose accuracy depends on the quality of vehicle tires). Summary of the investigated sensors is presented in Table 1.

Table 1. Sensors' parameters

Sensor	Mean error	Standard deviation
Distance	0	0.2794
Relative speed	0	0.1439
Ego speed	1%	0
Acceleration	0	0.0410

The performance of the platoon with considered sensors was analyzed and verified in system-level simulations. It is evaluated in terms of average inter-vehicle spacing. Inter-vehicle spacing corresponds to the measured bumper-to-bumper distance, averaged over time, over all pairs of consecutive vehicles in the platoon, and over all simulation runs for a chosen target CACC distance. It is expected that the average inter-vehicle spacing closely approximates the target CACC distance (the minimum target distance that provides collision free platoon performance).

In the investigation, we consider a single homogeneous platoon traveling along one lane of a highway. The platoon consists of a predefined number of cars. The first car is the so-called platoon leader. The platoon is preceded by a jamming vehicle which affects the fluent platoon movement by periodic breaking and acceleration. With the help of control mechanisms, cars in the platoon try to respond to the jammer's behavior by adjusting their speeds and accelerations to maximize the platoon performance (i.e., attempt to minimize the platoon length) and to avoid crashes. The leader's acceleration is controlled by the adaptive cruise control (ACC) algorithm, whereas accelerations of the remaining cars in the platoon are controlled on the basis of the CACC algorithm. In the ACC algorithm the speed and distance of the preceding car are measured by built-in (or on-board) sensors, while for the CACC, in addition to sensor measurements, messages from other vehicles that are transmitted by radio are used. In this work the communication between cars is performed according to the IEEE 802.11p standard.

The jammer vehicle moves according to a predefined pattern which is cyclically repeated during the whole simulation runtime. Each cycle lasts 30 s and consists of three phases: braking with a linear speed decrease from 130 km/h down to 30 km/h (deceleration factor equal to 0.3 g), acceleration until it again reaches the speed of 130 km/h (acceleration factor equal to 1.5 m/s^2), and, finally, constant speed movement for about 2.04 s.

Figure 2 presents the observed inter-vehicle distance with average class sensors applied with respect to the reference system (assuming ideal sensing). One can notice that distance sensors imperfection has little impact on the increase of spacing between vehicles. On the other hand, velocity sensing inaccuracy results in significant increase of distance, with the spacing almost doubled in case of inaccurate ego velocity sensor. For acceleration sensors, the imperfections have minimal impact on the performance. Hence, one can conclude that sensors' imperfections have significant impact on CACC performance.

4   Wireless Communication System - Challenges

As it was mentioned, wireless communications is usually considered to be realized either through the application of cellular networks (such as 5G) or IEEE 802.11p compliant systems. However, in case of high number of users (cars, trucks, platoons) the utilization of a wireless band may be high resulting in lowering the communication reliability.

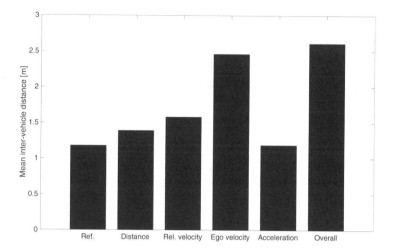

Fig. 2. Mean platoon inter-vehicle distance for acceleration sensors for the CACC algorithm

4.1 Dynamic Spectrum Allocation

One of the solutions to the problem of unreliable communications is to perform transmission in one of the frequency bands known as the white spaces and dynamic spectrum access (DSA). The analysis performed all over the world revealed that a significant part of the licensed spectrum (between 50 and 80%, depending on location) is underutilized and could be used more effectively when dynamic spectrum sharing with other systems is employed. One can assume that when a licensed frequency band is currently not occupied with the transmission of the primary user (the licensed one), it could be used by an unlicensed user (known as the secondary user) [12], which in the considered scenario might be the platoon of vehicles. Moreover, as more attention has been drawn recently to higher frequency bands (above 6 GHz), more spectrum can be made available for highly reliable short-distance transmission, such as in V2V communications [2].

4.2 Context Information Databases

The access to timely and accurate information about the surrounding vicinity may be of highest importance in V2X communications. When cars are aware of the current situation on road ahead of them, they can modify the route or adjust the driving parameters to either e.g. minimize the fuel consumption or to avoid crashes. However, in the context of dynamic spectrum selection of V2X communications, the presence of dedicated context-information databases is necessary, where various data structures will be stored and available for download (depending on the application). In order to efficiently manage the utilization of secondary band, the information about the presence of primary users as well as of other secondary user (other platoons) is necessary. One may imagine the situation, when the platoon leader communicates with the database, who in turns provide the guidelines which frequency band shall be used in order to avoid interference with other (approaching) platoon. However, the information about the spectrum usage is just the straight-forward realization of the database. It is envisaged that the context-information databases can store such information as number of trucks in each convoy, its current length, speed, but also typical speed on given road at given weather conditions and season. In terms of structure of the database system, the hierarchical approach seems to be possible, where long-term information about the environment are stored in the remote databases, and short-term information are stored in the local ones [6].

4.3 Impact of Sensor Inaccuracy on Usage of Information Available in Databases

As shown in Sect. 3, sensors inaccuracy directly impacts CACC performance. Erroneous sensing information, combined with limited reliability of wireless communications, may make effective platooning impossible due to information inaccuracy. Therefore, novel advanced approaches need to be sought, such as the use of additional frequency bands for V2V transmission. One of the solutions considered in the literature is to use mmWave bands for V2V communications to exchange e.g. raw sensing data between vehicles to improve sensing accuracy (cooperative sensing) [2,4]. In such a case the required data rates for wireless communications might increase up to 100 s of Mbps assuming the same levels of latency and reliability. The use of mmWave bands offers the possibility of significant bandwidth increase, however, it also poses serious challenges, such as the reduced transmission range, which is crucial in case of platooning. Therefore, the support of lower frequency bands is also envisaged [2] to support the high data rate communications. Hence, a mechanism of dynamic spectrum allocation is required to manage the selection of frequency bands that provide the best transmission conditions. In such a case also the use of licensed spectrum can be also considered. However, such process of dynamic selection of spectrum needs to be aided with context information from databases, to take into account multiple parameters and aspects impacting the perceived performance in a given frequency band.

5   Conclusions

In this paper aspects of autonomous vehicles platooning using CACC have been considered, with the main focus on sensing inaccuracy and limited reliability of wireless communications. Impact of on-board sensors inaccuracy on performance of platooning has been discussed, with different types of measured parameters considered. Moreover, taking into account the limited accuracy of sensors, challenges for the future wireless V2V communications have been formulated, including constraints on latency, reliability and perceived cooperation. A suggested potential solution to increase the reliability of CACC operation is to use dynamic allocation of multiple frequency bands, including mmWave, to improve the data rates and decrease the traffic load of current ITS bands.

References

1. Chan, E.: Overview of the sartre platooning project: Technology leadership brief, October 2012. https://doi.org/10.4271/2012-01-9019
2. Choi, J., et al.: Millimeter-wave vehicular communication to support massive automotive sensing. IEEE Commun. Mag. **54**(12), 160–167 (2016). https://doi.org/10.1109/MCOM.2016.1600071CM
3. Jia, D., et al.: A survey on platoon-based vehicular cyber-physical systems. IEEE Commun. Surv. Tutorials **18**(1), 263–284 (2016). https://doi.org/10.1109/COMST.2015.2410831
4. Lien, S., et al.: Latency-optimal mmwave radio access for v2x supporting next generation driving use cases. IEEE Access **7**, 6782–6795 (2019). https://doi.org/10.1109/ACCESS.2018.2888868
5. Osseiran, A., et al.: Scenarios for 5G mobile and wireless communications: the vision of the metis project. IEEE Commun. Mag. **52**(5), 26–35 (2014). https://doi.org/10.1109/MCOM.2014.6815890
6. Perez-Romero, J., et al.: On the use of radio environment maps for interference management in heterogeneous networks. IEEE Commun. Mag. **53**(8), 184–191 (2015). https://doi.org/10.1109/MCOM.2015.7180526
7. Ploeg, J., et al.: Design and experimental evaluation of cooperative adaptive cruise control. In: 2011 14th International IEEE Conference on Intelligent Transportation Systems (ITSC), pp. 260–265, October 2011. https://doi.org/10.1109/ITSC.2011.6082981
8. Sroka, P., et al.: Szeregowanie transmisji wiadomosci typu BSM w celu poprawy dzialania kooperacyjnego adaptacyjnego tempomatu. Przeglad Telekomunikacyjny, Wiadomosci Telekomunikacyjne **2017**(6), 350–355 (2017)
9. Sybis, M., et al.: Communication aspects of a modified cooperative adaptive cruise control algorithm. In: IEEE Transactions on Intelligent Transportation Systems, pp. 1–11 (2019). https://doi.org/10.1109/TITS.2018.2886883
10. Tsugawa, S., Jeschke, S., Shladover, S.E.: A review of truck platooning projects for energy savings. IEEE Trans. Intell. Veh. **1**(1), 68–77 (2016). https://doi.org/10.1109/TIV.2016.2577499
11. Tsugawa, S., Kato, S., Aoki, K.: An automated truck platoon for energy saving. In: 2011 IEEE/RSJ International Conference on Intelligent Robots and Systems, pp. 4109–4114, September 2011. https://doi.org/10.1109/IROS.2011.6094549
12. Yucek, T., Arslan, H.: A survey of spectrum sensing algorithms for cognitive radio applications. IEEE Commun. Surv. Tutorials **11**(1), 116–130 (2009). https://doi.org/10.1109/SURV.2009.090109

Author Index

Printed in the United States
By Bookmasters